D0948933

lonely planet

YOUR TRIP STARTS HERE

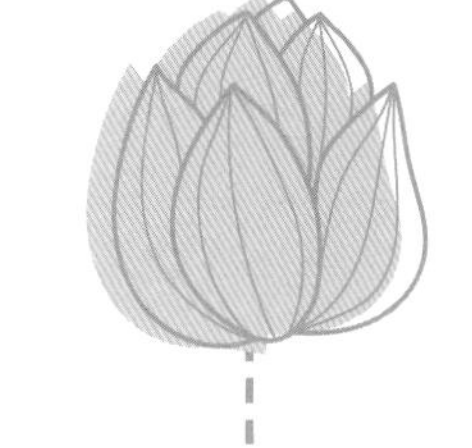

Contents

Americas

Europe

© ALAIN COURTILLAT
© JESS KRAFT/ 500PX
© GEIR OLSEN / BIRKEN
© BLUEJAYPHOTO / GETTY IMAGES

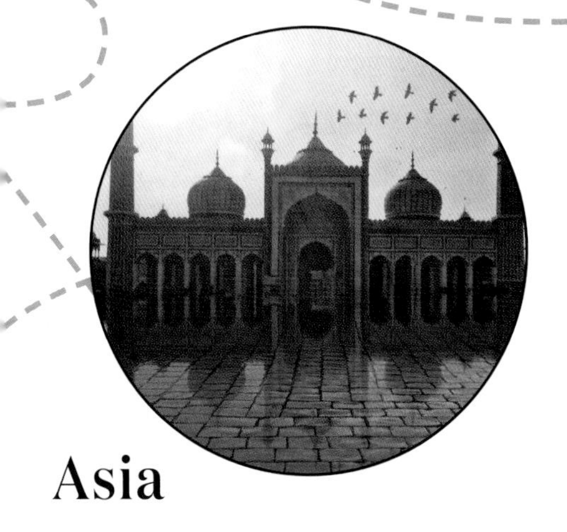

Asia

Oceania

Africa & the Middle East

Right: Canada's
Icefields Parkway route

Introduction

Clockwise, from above: ancient Stonehenge in England has mystical significance; the POLIN Museum of the History of Polish Jews in Warsaw; a statue of Buddhist monk Kōbō Daishi on the island of Shikoku in Japan

Some nine centuries ago, Buddhist pilgrims first trod a path around the Japanese island of Shikoku, hoping to attain nirvana by passing through the stages of awakening, austerity, discipline and then enlightenment. Over the years, the pilgrimage evolved to take in 88 temples spaced along a 1200km (745-mile) route, following in the sacred footsteps of monk Kōbō Daishi and allowing more than enough time for the pilgrim to reflect on their life and decide on what about it to change.

Travel has always been a catalyst for personal development. Anthony Bourdain wrote of how travel leaves its mark on us, on our body or our heart. And no form of travel is more likely to do that than taking a long journey. This can be an adventurous trek, rafting expedition or bike ride, a road trip to explore the history or culture of another place, or simply a traditional pilgrimage. We have all kinds in this book but what they have in common is the potential to transform travellers in numerous ways, whether that is building mental flexibility, confidence and resilience – our brains thrive on dealing with unpredictable new experiences – or encountering other perspectives from people we wouldn't have otherwise met. Travel also allows us to try out other versions of ourselves and maybe retain some of those new qualities once back home.

The journeys of self-discovery in this book generally fall into one of three different categories. At the most straightforward level, adventures by boat, bicycle, horse or on foot, offer varying degrees of challenge. Place yourself at the mercy of the Yukon, Colorado, Zambezi or Franklin rivers as you float downstream on rafts or in a canoe. Pedal the length of New Zealand or surf your way around Sri Lanka. Or trek by horse or on foot through Kyrgyzstan, South Korea, the French Cévennes, or the whole of Jordan.

Next, there are pilgrimages, which may have a spiritual dimension, if you wish to embrace that. Complete a circuit of Mt Kailash in Tibet, walk the Celtic Way from Glastonbury to Stonehenge or follow St Patrick's Causeway in Ireland, the Camino de Santiago in France and Spain, or the Via di Francesco in Italy. Finally, some of the journeys in this book, perhaps the most confronting, explore history and culture by, for example, tracing the Harriet Tubman Underground Railroad Byway in the US; walking with the Goolarabooloo people in Australia; or discovering Warsaw's Jewish history.

Each entry in this book suggests how the journey might change you: emotional or spiritual growth (the heart icon); the insights and understanding it might bring (the book icon); and the physical skills and experience gained (the hand icon). A comprehensive account, by a writer who has first-hand experience of the trip, describes each stage of the journey, its highs and lows. And a detailed map, coupled with practical information, helps you picture yourself embarking on the adventure and begin to plan it. The themes on the following pages provide a starting point.

Longer trips are an opportunity to step outside of our daily routines. But, to get the best out of meaningful travel, two more factors help: slowing down and leaving your comfort zone. The first allows you to linger and take stock of new experiences. The second is sure to be transformative.

© JOAOCCDJ / SHUTTERSTOCK

© DAVID MADISON / GETTY IMAGES

© PIOTRBB / SHUTTERSTOCK

Themes

Culture

Road Trips

Hikes

Off the Beaten Path

Ocean
Tingstadjordet
Russia
Europe
Warsaw
Frankfurt
Le-Puy-en-Velay
Florence
Rome
Central Asia
Rotfront
Gangotri
Darchen
Ovacık
rmo
Asia
Jeju-si
Ryōzen-ji
Manali
Dana
Lhasa
Delhi
Thimphu
Desert
Middle East
Muscat
Syabrubesi
North Pacific Ocean
Salalah
Bangkok
Africa
Arabian Sea
Gunung Mulu
National Park
Colombo
Singapore
Nairobi
Darwin
Minyirr
(Gantheaume Point)
Chirundu
Alice Springs
Indian Ocean
Australia
Cape
Reinga
Adelaide
Wellington
Collingwood River
New Zealand
Ross Island
Antarctica

Themes

These are the corrected page references for the suggestions in the Themes section. Overleaf is a chart that shows the starting point of every journey of self-discovery.

Culture

Road Trips

Hikes

Off the Beaten

Adventure & Sport

Pilgrimages

History

Wildlife & Nature

Greenland

Whitehorse

Banavie

Ballintubber Abbey

Jasper

Edmonton

Ferns

Mynydd Llanllwni

Rocky Mountains

North America

Charlottetown

Glastonbury

Camden

Caen

Irún

Marse

Cambridge

North Pacific Ocean

Lees Ferry

Nashville

Memphis

North Atlantic Ocean

New Orleans

Hawai'i

Morelia

Central America

Caribbean Sea

Cartagena

Georgetown

Isla Santa Cruz

Amazon Rainforest

Andes

South America

Uyuni

Rio de Janeiro

Pacific Islands

South Atlantic Ocean

Santiago

Buenos Aires

South Pacific Ocean

Southern

Adventure & Sport

Pilgrimages

History

Wildlife & Nature

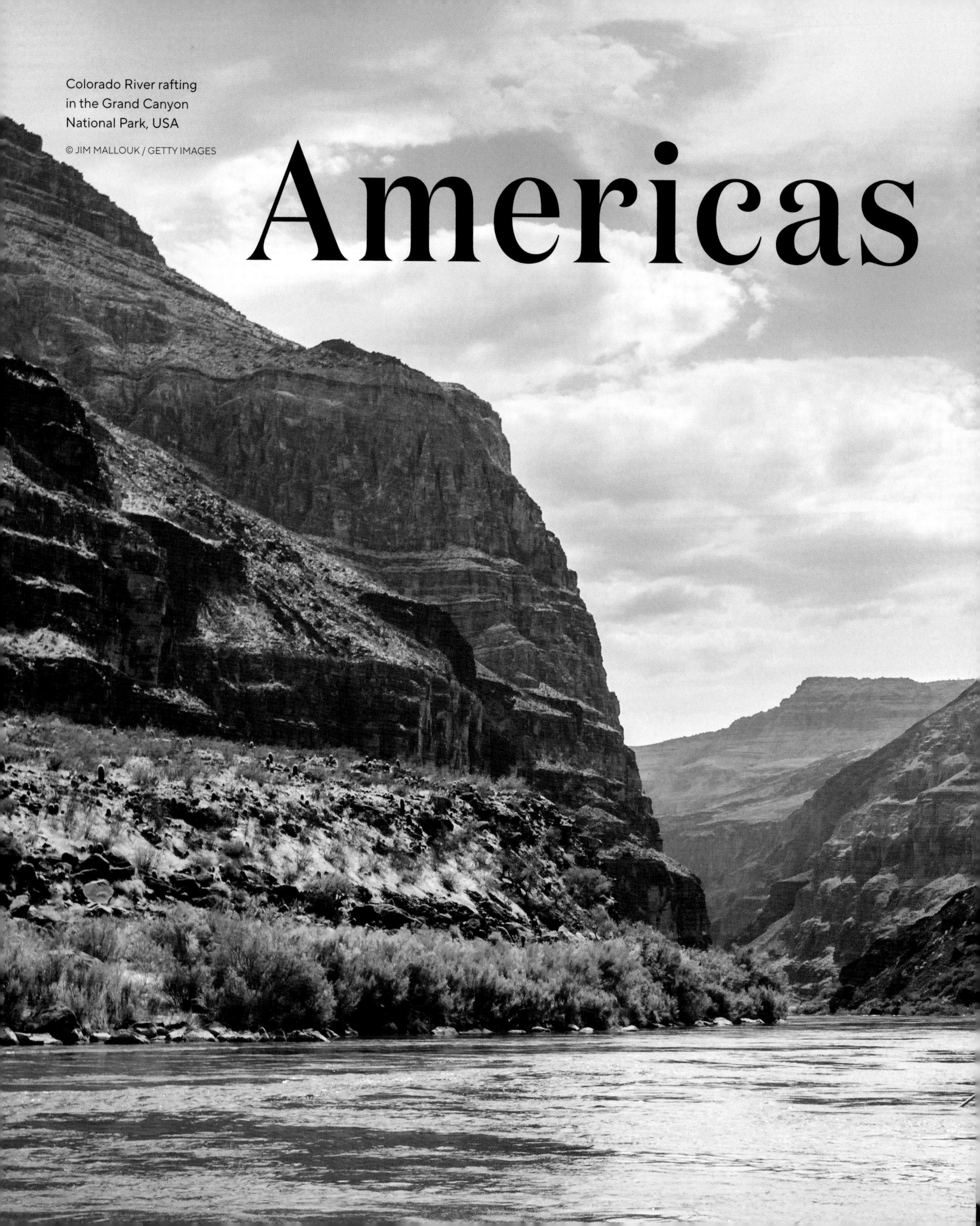

Americas

Colorado River rafting in the Grand Canyon National Park, USA

Remembering the dead on a Mexican road trip

Join in Día de Muertos (Day of the Dead) festivities and witness an incredible butterfly migration on this journey around Michoacán state.

During Día de Muertos, Mexicans celebrate the return of spirits of the dead to the living world. You may find comfort and solace in this age-old way of honouring and communing with passed loved ones.

Día de Muertos provides rich insight into ancient Mexican spiritual beliefs and customs; a visit to Pátzcuaro offers the chance to learn about the Purépecha people and appreciate local folk crafts.

As well as seeing the massed monarch butterflies, hiking through the steep and rocky terrain of high-altitude El Rosario reserve provides an opportunity to test your muscles.

The parade approaches the candlelit cemetery to the raucous sound of a mariachi band of guitars, accordions, trumpets, trombones and clarinets. Walking beside them are families carrying sections of their *ofrendas* – altars to departed loved ones, studded with masses of bright-orange marigolds, fruits and other vividly colourful blooms. Women twirl to the music, wicker baskets of food perched atop their heads; men in straw cowboy hats toss back glasses of tequila and *pulque*, a milky-looking tipple made from fermented agave.

This cacophonous, joyful celebration is typical of those you'll encounter in Tzintzuntzan every year on 1 November, Noche de Muertos (Night of the Dead), and the following day, Día de Muertos (Day of the Dead). Although it's a nationwide celebration, Día de Muertos is celebrated with special fervour in the Pátzcuaro, and in nearby villages such as Tzintzuntzan.

© JIA LIU / GETTY IMAGES

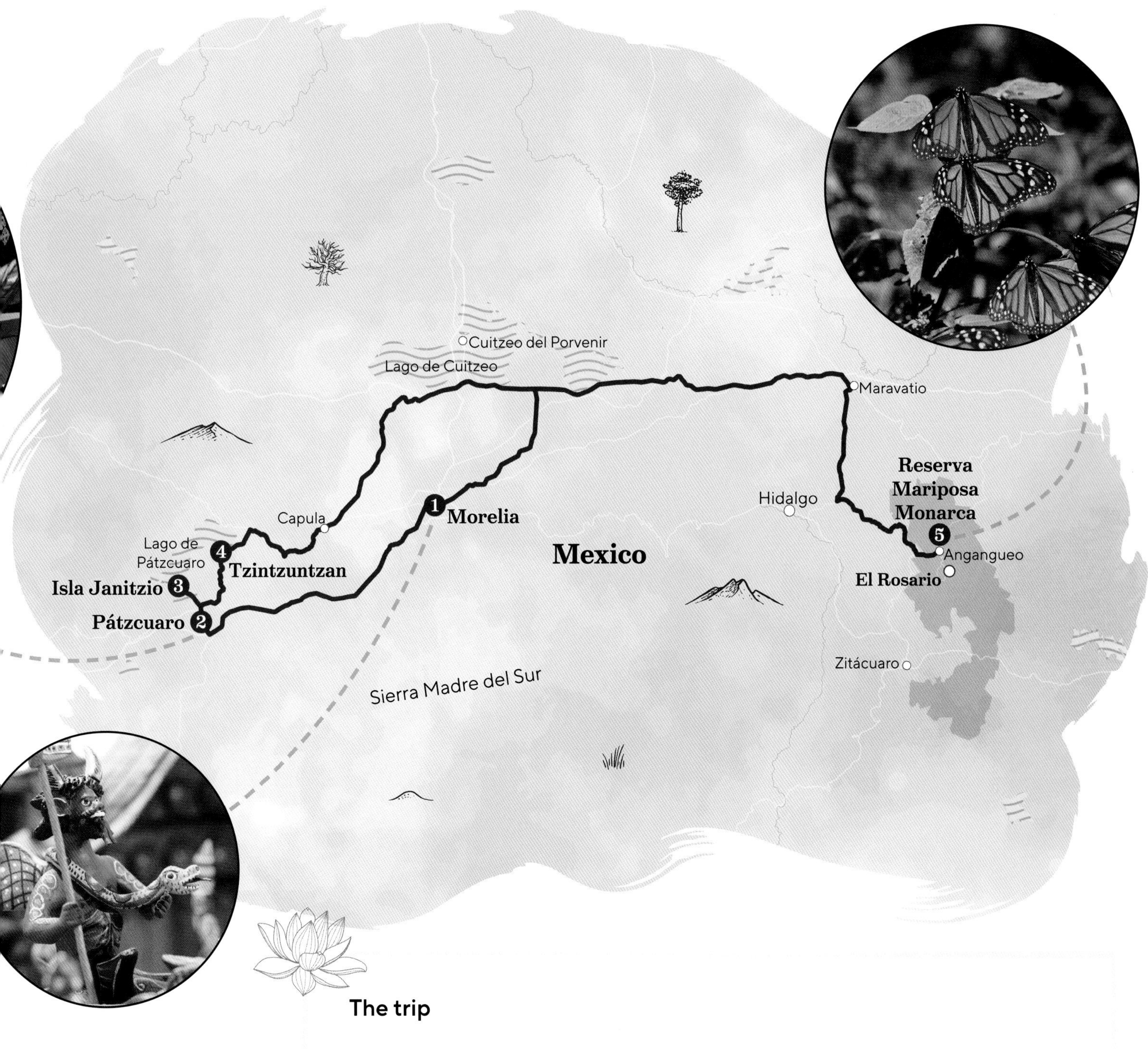

The trip

Clockwise, from left: Isla Janitzio, Lago de Pátzcuaro; Día de Muertos parade, Pátzcuaro; monarch butterflies at El Rosario reserve; devil figurine, Morelia

Distance: 436km (271 miles)
Mode of transport: Car, taxi or bus
Difficulty: Easy

1 Morelia The colonial-era centre of the Michoacán state capital is a Unesco World Heritage site

2 Pátzcuaro This wonderful *pueblo mágico* (magical town) has a long history of producing fine folk crafts

3 Isla Janitzio Picturesque island dominated by a giant statue of Mexican Independence leader José María Morelos y Pavón

4 Tzintzuntzan This ancient Purépecha capital has a sprawling cemetery that comes alive for Día de Muertos

5 Reserva Mariposa Monarca Vast forest reserve that encompasses the butterfly-watching sanctuary of El Rosario

Simon Richmond, Mexico traveller

It's fascinating how Día de Muertos has seeped into contemporary culture through films such as *Coco* and *Spectre*. But nothing could prepare me for experiencing the festival in Mexico itself – and particularly so in Pátzcuaro, where its rituals and traditions are so strong. Observing families gathering in the local cemeteries, sharing quality time at the beautifully decorated graves of their loved ones, is a humbling experience and a deeply touching expression of enduring love.

A trip to Pátzcuaro can also be combined with the Unesco-listed Reserva Mariposa Monarca on the eastern edge of Michoacán state. Every November, millions of orange-winged monarch butterflies migrate from as far north as Canada to small patches of this 562 sq km (217 sq mile) reserve to overwinter and reproduce – it's one of nature's greatest spectacles.

Satisfying the spirits

The roots of Día de Muertos can be traced back thousands of years to the spiritual beliefs of Mexico's Indigenous peoples; Catholic conquistadors assimilated the holiday, moving it from August to coincide with All Saints' Day and All Souls' Day at the start of November. Mexicans celebrate the dead the way they celebrate pretty much everything else – with food, drink, music and much colourful decoration. An *ofrenda* decorated with photos and masses of *cempasúchil* (marigolds) for the deceased is set up in a house or, as some families prefer, in the graveyard. In front of it are placed candles, *calaveras* (skulls made from sugar), *pan de muerto* ('bread of the dead': a loaf made with egg yolks, mezcal and dried fruits), other favourite foods and a container of water, as the spirits arrive hungry and thirsty after their journey.

The route of the dead

Start your journey in Morelia, Michoacán's state capital. Here, the stone-paved, pedestrianised Calzada de Fray Antonio is lined with *ofrendas*, and there are special Day of the Dead displays at the El Museo del Dulce, a candy shop and museum near the city's historic centre.

Move on to Pátzcuaro, the nucleus of an area rich in Indigenous culture. Over five centuries, the towns and villages circling Lago de Pátzcuaro have become

© AURIBE / SHUTTERSTOCK

> “Mexicans celebrate the dead the way they celebrate pretty much everything else – with food, drink, music and colourful decoration”

mini craft empires known as *pueblos hospitales*, with master artisans producing work sought after across Mexico, from copper vases to psychedelic sculptures and guitars. Pátzcuaro's Día de Muertos celebrations officially start on 31 October and continue until 2 November, but it's worth coming a week earlier as there are many events around the lake at this time. You can also observe people building their *ofrendas* and decorating the graveyards, their homes and the streets.

Although it's bombarded by tourists (and particularly so during Día de Muertos), boarding a *lancha* (ferry) to sail out to Isla de Janitzio in the middle of Lago de Pátzcuaro is a not-to-be-missed side-trip. The main attraction of this wonderfully picturesque car-free island is the 40m-tall (131ft) statue of Independence hero José María Morelos y Pavón.

A mysterious migration

After Día de Muertos, venture east via Morelia to witness one of nature's most incredible sights: millions of monarch butterflies overwintering in Reserva Mariposa Monarca. It's wondrous to imagine how these fragile creatures have flown some 4000km (2485 miles) to reach the reserve, and even more so when you factor in that each year, the journey is made by a new generation that have never before been to Mexico. This migration south is completed by a super-generation that can fly and live eight times longer than regular monarch butterflies. The following spring it's their offspring that journey back to the summer breeding grounds in North America.

Of the four areas within the reserve open to the public, the simplest to visit – and also most popular – is El Rosario, 8km (5 miles) from the town of Angangueo. After about an hour of trekking up a steep track into the forest, you'll begin to see masses of monarchs coating the fir trees, conserving their energy in the cold morning air. As the sun appears, its warmth wakes the resting butterflies. The sight of so many delicate orange-and-black wings fluttering in the air is magical, and it's easy to see why some Mexicans believe the monarchs are the returning souls of the dead.

© EVE OREA / SHUTTERSTOCK

From top: Día de Muertos at Tzintzuntzan cemetery; *calavera* candy skulls, *pan de muerto* and marigolds laid out on an *ofrenda*

Practicalities

Region: Michoacán, Mexico
Start/Finish: Morelia

Getting there and back: Morelia's airport, 26km (16 miles) northeast of the city, has direct flights to Mexican hubs and the US. Pátzcuaro is a 1hr drive from Morelia.

When to go: Visit in the last week of October to observe the build-up to Día de Muertos on 2 November. The monarch butterfly migration also starts early November and lasts through to February.

What to take: The end of October and November can be wet and windy in Mexico's western central highlands; pack suitable all-weather clothing and footwear.

Where to stay: Día de Muertos festivities in Pátzcuaro attract huge crowds of visitors, so it's wise to book accommodation and tours well ahead.

Tours: Operators include Zócalo Folk Art & Tours (zocalofolkart.com), based in Pátzcuaro and offering a 10-day itinerary in the region; and Mich Mex Guides (michmexguides.com.mx) in Morelia, which offers an overnight tour from that city that includes the lakeside villages of Santa Fe de la Laguna and Tzintzuntzan, and the less commercial Isla Pacanda – it also has tours to the monarch butterfly reserves. Mexico City-based Aztec Explorers (facebook.com/AztecExplorers) organises separate tours to Pátzcuaro for Día de Muertos and to Reserva Mariposa Monarca.

Essential things to know: The US government has a 'Do Not Travel' advisory for Michoacán because parts of the state are dangerous due to the activities of crime cartels. However, Morelia itself is exempted and the areas around Pátzcuaro are fine so long as you stick to Hwy 15D and other major roads.

Seeing South America by bus

Savour South America on a journey that takes you across the entire continent without ever stepping foot on a plane.

South America will likely test your street smarts at one time or another, which only helps to hone your decision-making skills and build confidence.

The opportunity to learn Spanish or improve your language skills on this journey may serve you well on future trips – 20 nations count Spanish as an official language.

There's nothing like completing a punishing multiday hike – whether high in the Andes or deep in the jungles of Colombia – to remind you what your body and mind are capable of.

'¿A que hora sale el autobús para Bariloche?' ('What time does the bus leave for Bariloche?') you ask, as you book your next onward transport. You may only be a few weeks into your journey along the route known affectionately as South America's Gringo Trail, but you're already beginning to feel like a local – a little, at least. Until the next international border crossing confronts you with a new culture, cuisine, dialect

The trip

Distance: 15,000km (9320 miles)
Mode of transport: Bus
Difficulty: Moderate

1 Iguazu Falls, Brazil and Argentina Marvel at the continent's most famous waterfall

2 Buenos Aires, Argentina Feast, drink and dance into the wee hours

3 Atacama Desert, Chile and Bolivia Journey across the world's driest nonpolar desert

4 Rurrenabaque, Bolivia From this forest-fringed town, immerse in the otherworldly ecosystem of the mighty Amazon Rainforest

5 Sacred Valley, Peru Explore the ruins dotting the heartland of the Inca Empire

Clockwise, from left: Iguazu Falls from Argentina; volcano hiking near San Pedro de Atacama, Chile; *chiva* bus, Colombia; *gaúcho* herder near Mendoza, Argentina

Essie Morrison, overland traveller

One of the biggest takeaways from my South America trip was the benefit of approaching travel (and life) with an open mind. This paved the way for me to meet a diverse group of people from all over the world who added so much depth to my South America experience – shout-out to the Colombian family who saw my credit card get eaten by an ATM, then took me on their family holiday!

and currency to get your head around. But that's all part of the fun of journeying across the continent the traditional way: harnessing the region's excellent bus networks.

Copa, Copacabana...

Beginning your adventure in Brazil is a thrilling opportunity to rip off your South America travel-anxiety band-aid and dive right in – especially during Carnival, when the spectacularly located city of Rio de Janeiro morphs into one giant fiesta. But there's a festive atmosphere at Rio's beaches year-round: order a caipirinha from one of the beach kiosks and join the party before jumping on the first of many buses to continue your journey.

Choosing which destination to book your bus ticket to will likely by the first of many difficult decisions on your journey. There is no 'right' way to complete this trip, but you'll need at least three months if you don't want to rush (much). Embrace the freedom to forge your own path, which for those with limited time might mean making your way inland towards Iguazu Falls.

After soaking up panoramic views of the falls from the Brazilian side, challenge your comfort zone with an excursion into off-the-beaten-track Paraguay. Or cross into Argentina to marvel at the immense power of Iguazu from a fresh angle before bussing south to Buenos Aires. Sumptuous steak dinners, stately buildings, tango shows and myriad other delights tempt an extended sojourn in Argentina's lively capital, from where travellers who skipped

"There is no 'right' way to complete this trip, but you'll need at least three months. Embrace the freedom to forge your own path"

Uruguay can easily pop over by boat to the charming historic town of Colonia del Sacramento. Then get a taste of Patagonia in the Argentinian adventure hub of Bariloche or follow a love of Malbec to the wine capital of Mendoza, from where one of the world's most scenic cross-border bus journeys carries you into Chile.

Shoot for the moon

The energy of a country moving on from its tumultuous past is palpable in Santiago, Chile's youthful capital, where the Museo de la Memoria y los Derechos Humanos (Museum of Memory and Human Rights) offers a sobering insight into the nation's story before you travel on to the street-art capital of Valparaíso. Then settle in for the long journey north to San Pedro de Atacama, a surreal, high-altitude moonscape ripe for adventures – among them the multiday 4WD trip across the rugged altiplano to the crunchy-white salt flats of Bolivia's Salar de Uyuni.

Your pesos stretch further in Bolivia, which might inspire you to take your time travelling from the border town of Uyuni to the dizzying capital of La Paz, 3640m (11,942ft) above sea level. Test your nerve mountain-biking down the infamous 'Death Road', shop for souvenirs in the atmospheric Mercado de las Brujas (Witches' Market), and make new mates while fishing for piranhas on a detour to Rurrenabaque, one of the continent's most affordable gateways for exploring the Amazon Basin.

Feel your spirit stir on a boat trip to Isla del Sol, birthplace of the sun in Incan mythology, on the Bolivian side of Lake Titicaca. Then head into Peru and the pleasant colonial-era capital of Arequipa, where the nearby Cañón del Colca may prompt an Inca Trail training hike – or simply a moment of reflection as Andean condors circle overhead.

Which Inca Trail?

The former capital of the Inca Empire, the Andean city of Cusco lies at the heart of the Gringo

©PHILIP LEE HARVEY / LONELY PLANET

Clockwise, from above: Rio's Copacabana Beach, Brazil; Valparaíso, Chile; tango dancers in Buenos Aires, Argentina

© RAFA BAHIENSE / 500PX

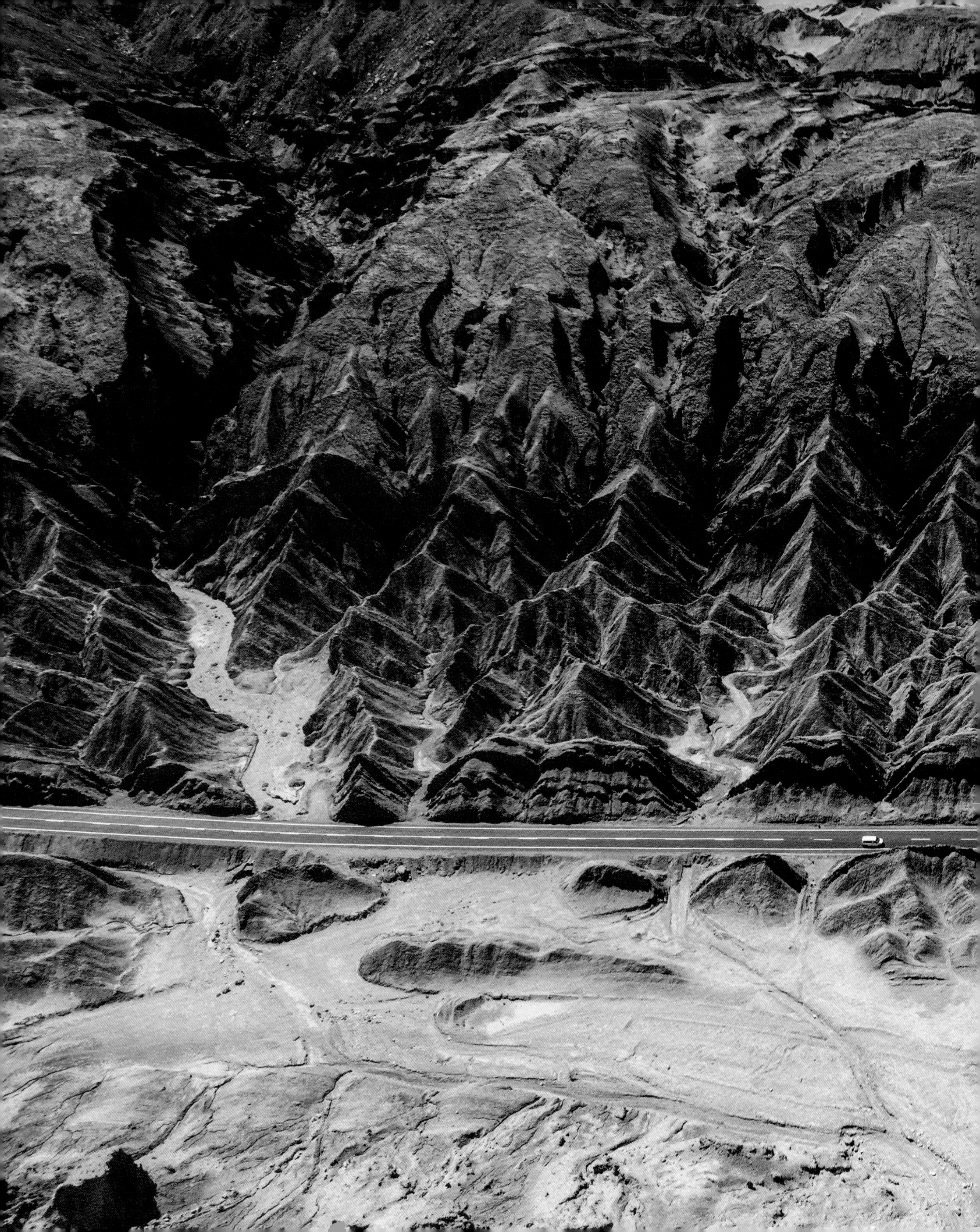

On the road through
Chile's Atacama Desert

The culinary journey

Among the most pleasurable ways to experience South American cultures on this journey is through sampling its smorgasbord of culinary delights. From eat-until-you-burst Brazilian barbecue sessions to Patagonian lamb *asado*, hearty meat dishes tend to dominate the southern end of the continent, complemented by some of the world's finest red wines, from Malbec to Carménère. South American staples, including maize and potatoes, get a good workout in Peru and Bolivia. Peru may be the birthplace of ceviche, but Chile's Santiago has embraced Japanese-influenced Nikkei cuisine, and now rivals the likes of Lima, São Paulo and Buenos Aires for the mantle of the continent's culinary capital. Stuffed with vegetables, meat and cheese, *empanadas* are a convenient on-the-go snack, while the ever-present *menú del día* (set menu) tradition offers another cheap and satisfying way to eat along the way.

Trail. Between its archaeological sites, colonial architecture, Quechua culture and colourful markets alone, there's plenty to see and do within the city limits, but this vibrant tourism centre is the jumping-off point for exploring the Inca ruins of the Sacred Valley – not least the most famous of them all, Machu Picchu. But before you book a classic Inca Trail hike, consider other, less crowded options like the high-altitude Salkantay Trek. Or be among the first to experience recently opened hiking trails on the Great Inca Road, a network of Inca trade routes weaving through the Andes northwest of Cusco.

Then it's onward to the culinary hub of Lima, giving your tastebuds a treat at some of the world's best restaurants before embarking on the 19-hour bus ride (you'll be used to these long journeys by now) to the northern coastal resort town of Máncora; the social atmosphere here offers solo travellers a great opportunity to buddy up with fellow overlanders heading on to Ecuador.

Time to get tropical

The first logical stop in Ecuador is the relaxed city of Cuenca, known for its handicraft traditions; you might instead gravitate to the jungle town of Baños, another affordable entry point to the Amazon and a popular base for the challenging hike to the 5897m (19,347ft) snowcapped summit of Cotopaxi. But this active volcano is a similar distance from the capital, Quito, from where you could treat yourself to a Galápagos Islands side-trip before making your way north to Colombia.

© JESS KRAFT / JESS CRAFT

> "Salsa into Cali, tour the cultural riches of Bogotá and party in Medellín, or take it slow in the likes of Villa de Leyva"

If you've been savvy enough to travel this far without falling victim to petty crime or being forced to tweak your itinerary due to political unrest (protests shut down Machu Picchu in early 2023), this is not the time to drop your guard – the Colombian refrain 'no dar papaya' translates as 'don't give papaya', meaning avoid giving others a chance to take advantage of you. Suitably prepared, get ready to fall in love with the final country on your journey.

Salsa into Cali, tour the cultural riches of Bogotá and party in Medellín, or take it slow in the likes of Villa de Leyva, where lazy mornings are best spent sipping coffee as *caballeros* (horsemen) clip-clop along 16th-century streets. Further north, the adventure hub of San Gil offers a chance to tackle Class IV rapids on a whitewater rafting tour – if you dare – while the gruelling jungle hike to Ciudad Perdida (Columbia's 'Lost City') can be more moving than Machu Picchu. Arriving at its moss-encrusted mountainside terraces, which date back to the 9th century, it's easy to imagine you're the first person on the scene since the Indigenous Tairona people were forced to flee this magical place in the 1500s.

An intoxicating mix of musical traditions – from cumbia to porro, merengue to mapalé – slice through the tropical heat in coastal Cartagena as you stroll the cobbled streets of its Unesco-listed centre. And as you soak up the colours, cultures and history of the final stop on your trip, take the time to appreciate just how far you've come – in every way – to savour this special moment.

Left: Cartagena, Colombia
Right: from Peru's Cusco, hike the Inca Trail to Machu Picchu or discover the less-travelled Salkantay Trek

Practicalities

Region: South America
Start: Rio de Janeiro, Brazil
Finish: Cartagena, Colombia

Getting there and back: Flying into Rio de Janeiro, São Paulo or Buenos Aires to begin this trip will make logistical sense for most travellers. When you reach Cartagena, consider sailing to Panama and travelling on overland through Central America.

When to go: South America is a year-round destination – your dates will depend on where you want to go and what you want to do. Aim for February to March for Carnival in Brazil, September to April for hiking in Patagonia, late July to early September for skiing in Chile and Argentina – and if you want to trek the classic Inca Trail, plan around the annual February trail closure.

What to take: Switch your suitcase for a backpack; your body will thank you for keeping its weight under 15kg/33lb (hint: opt for e-books). A comprehensive first-aid kit, including all medications you could possibly need, is essential, and you might be surprised by how useful a roll of duct tape can be. Swerve single-use plastic by investing in a water purifying device (or pack tablets).

Where to stay: South America has an excellent hostel network, many with private rooms. Book at least a few days ahead, more during holiday and festival periods. Homestays offer wonderful opportunities to practice Spanish and experience local lifestyles. You won't regret splurging on a *full cama* (fully reclining) seat on overnight bus journeys.

Essential things to know: Brazil tourist visas typically need to be secured in advance. Most other South American countries offer a 30-day visa-on-arrival to most nationalities, though Australians now need to prearrange a tourist visa to enter Chile.

Sail on a windjammer in coastal Maine

In coastal Maine, a fleet of historic windjammer ships keep local traditions afloat, welcoming 'flatlanduhs' onboard for a hands-on sail around Penobscot Bay.

When anchors need heaving or sails need hoisting, neither stagnation nor isolation is an option; explore where you might need to pull up anchor and ask for help to move forward.

Windjammers have long supported a rich way of life in Maine. Sail trips continue this today, and can inspire visitors to bolster local traditions wherever they travel to.

Windjammer crews work astoundingly hard – hoisting, baking, swabbing, jibbing – and passengers are encouraged to take on ability-appropriate tasks. Push yourself, at your level, and see what you're able to do.

Many times when we travel, we're looking for a glimpse into a world that's new to us: a different landscape, perspective, era or way of life. Or maybe we're looking to find a different side of ourselves. Once in a great while, we find a journey that allows us to do all of the above.

From the 1870s through to the 1940s, life along the rocky shores of the hardscrabble Maine Coast was tough – for lighthouse keepers, loggers and lobstermen. For mariners, especially, the work was backbreaking and the journeys gruelling. Today, Mainers are intensely proud of the lifestyle that their hardworking ancestors built – the state isn't unofficially nicknamed Vacationland for nothing. And us landlubbers are welcome not only to experience this coastal vibe for ourselves, but to help support the traditions created by this way of life – through taking a trip on one of Maine's elegant windjammer ships.

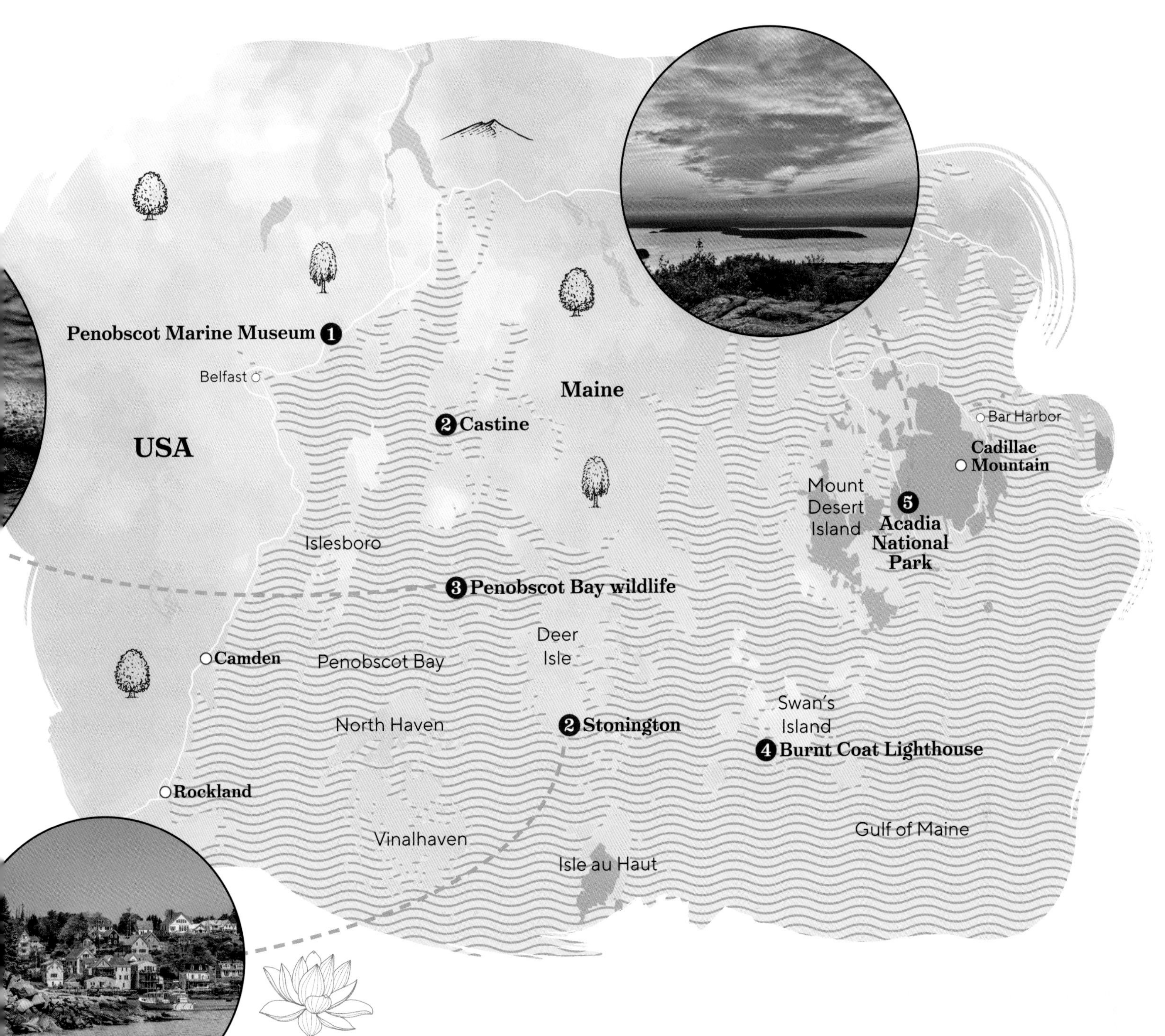

The trip

Distance: Sails cover 10km to 20km (6-16 miles)
Mode of transport: Sailing boat
Difficulty: Easy

1 Penobscot Marine Museum Start your journey early: join an online talk or class before you even set sail

2 New England villages Drop the 'Gram and send a postcard from the meanderable 17th-century villages of Stonington or Castine

3 Penobscot Bay wildlife A highlight of a windjammer itinerary is the lack of an itinerary – just enjoy the bay's scenery and wildlife, from owls to porpoises and seals

4 Burnt Coat Lighthouse One of a dozen historic lighthouses guiding boats around the bay

5 Acadia National Park Add an extra day to watch North America's first sunrise from craggy Cadillac Mountain

Clockwise, from left: windjammer cruising; scout for harbour seals in Penobscot Bay; Cadillac Mountain sunrise, Acadia National Park; a bayside stopoff at Stonington

Emma Matus, J&E Riggin passenger/ first mate

I'm a history nerd – maritime history especially – but I'd had zero experience on a boat. I know more about sailing now, obviously; but more importantly, I feel so much more comfortable in my body. [Sailing a windjammer] pushes you pretty far sometimes. At times I'd think: 'No way could I do that.' A month later, I was doing it without hesitation. It's a great feeling to find out you're more capable than you previously thought.

These tall-masted schooners, with their weathered wooden hulls and polished brass fittings, have three to five masts with plenty of rigging, and anchors heavy enough to rival an elephant. Each day, mainsails must be raised, anchors heaved. Crew members are experts at assigning tasks for each guest – even very young children can help by raising the flag.

A day in the life

These days, the windjammers all serve as passenger vessels, with private cabins and shared common areas. You can charter an entire ship, but most guests come as individuals or couples, with friends or family. (Several windjammers have single cabins – no single supplement applied.) Each person or group gets their own cosy stateroom, and only a few cabins and ships have 'heads' (toilets) for individual cabins. Usually, there are two or three shared heads for 20–40 guests. It's more like camping than cruising; more broad vistas than Broadway revues. Many of the sails these days have themes, from music, knitting and photography to autumn foliage or yoga.

Your day might be filled with many things – perhaps a hike up to a lighthouse, or a sea shanty sing-along. Need a shower? Why waste precious fresh water when you have the entire Penobscot Bay as your personal, natural bathtub? Sure, you can visit plenty of living history museums, but on a windjammer, you are the living history. Passengers aren't technically required to pitch in on sailing, but that's why you're here, right?

Roll with the elements

The route is set by the wind and tide instead of appointments and schedules. The wind points your windjammer's compass ENE at 12 knots? Well, then you're spending the day hiking on the rocky shores of Swan's Island and its stark Burnt Coat Lighthouse on Hockamock Head, where you can read logs written by hardy lighthouse keepers 150 years ago.

© E.J.JOHNSON PHOTOGRAPHY / SHUTTERSTOCK

"Windjammer passengers aren't technically required to pitch in on sailing – but that's why you're here, right?"

These days, just under a dozen windjammers ply the waters off the Maine coast. They leave from either Camden or Rockland and sail for 2–8 (usually 3–6) days around Penobscot Bay, stopping off in small towns, at lighthouses and at rocky shores on uninhabited islands. What every single windjammer sail has in common is the final-night lobster bake. Spend the day hiking and beachcombing on an uninhabited isle in the bay as your meal – lobsters wrapped up in the beach's seaweed you gathered yourself, potatoes and melting butter – steams in a beachfront pit.

Sustainable travel sustains traditions

These stately ships were built in an era before fibreglass and plastic. The 1927 *J&E Riggin* dredged for oysters; the 1871 *Lewis R French* carried cargo. All have small engines or a supporting yawl boat for help navigating into port or when the wind dies, but generally rely almost entirely on wind power – and many also compost, use reusable dishware and serve food made with local or organic ingredients.

Not only do the award-winning windjammers offer some of the most ecologically sustainable travel in the United States, they're also some of the most culturally sustainable. You're not just supporting the environment: windjammer tourism brings in both money and a boost to a way of life that has benefited the Maine coast for over 150 years.

Remember to enjoy!

After you've hoisted or jibbed, hiked and taken a dip, you're welcome to sit back and enjoy. Bring books, journals, watercolours, instruments. On a windjammer, you don't need to take work home to impress the boss. Sailed through the day, dropped anchor for the night? You're done! Play rummy with your shipmates, just sit back and enjoy the brisk sea air. You've earned it.

© DANIEL GRILL / TETRA IMAGES / GETTY IMAGES

From top: setting sail from Camden; Maine lobsters

Practicalities

Region: Maine, USA
Start/Finish: Sails start in Camden or Rockland, and finish in the same location.

Getting there and back: The closest major international airport to Camden/Rockland is Boston (3hr 30min to 4hr), but Portland, Maine, is closer (1hr 30min). Both airports are connected to Camden and Rockland by coach services or shuttle buses.

When to go: Windjammers sail from late May through to October, with themed itineraries available. If you can withstand the chilly temperatures, fall foliage trips are particularly memorable. For living history, the Windjammer Days festival in the last week of June might float your boat.

What to take: Set your comfort expectations to glamorous camping not luxury cruising: quarters are either historic or historically accurate. Bring books, art equipment or an instrument, but also seasickness medication, earplugs, lots of warm layers and hiking boots.

Where to stay: Guests all get private cabins, but most will share a 'head' (toilet). Quarters rarely have full queen-sized beds, so get cosy in a double (or two singles).

Where to eat and drink: Maine windjammers are renowned for their cuisine, and many captain-chefs have written their own cookbooks. You'll eat well, but options are understandably limited. If you have an allergy, you might want to pack some things for yourself.

Tours: The fleet is represented through the Maine Windjammer Association (sailmainecoast.com), which currently has nine windjammers; and Maine Windjammer Cruises (mainewindjammercruises.com), which has three.

Essential things to know: Maine. Is. Cold. Sailing in Maine is colder. Plus, it's not overkill to book popular weeks over a year in advance.

Overland in Guyana, from Atlantic coast to remote rainforest

A journey into Guyana's interior is a powerfully immersive experience, led by Amerindian guides who uncover the wild wonders of the interior.

Overlanding offers few home comforts, but many opportunities to build resilience. Eco-lodges are welcoming but often simple, and journeys between them at the mercy of the elements. The fight against mosquitoes is futile, road travel is rough and distances long. Let it go. Ditch digital devices for sleeping in hammocks, gazing at the stars, rising with the sun. This place is all about the here and now.

Guyana's vast interior is home to one of the last pristine rainforests on Earth. Mindfulness is made easy here: the feral buzz, vivid hues and pungent humidity of rainforest, swamplands and savannah are all-encompassing, and you can learn infinite lessons from the natural world.

Georgetown is a mind-boggling place to be. This 'Garden City of the Caribbean' is a storm-blown port that's actually on the Atlantic, sitting on the fringes of both South America and the Caribbean, but in truth resembling neither. Street names are Dutch, heritage buildings British, and while English is the official language, myriad tongues are spoken, often interchangeably – from Creole and Guyanese Hindustani to numerous

© GAIL JOHNSON / SHUTTERSTOCK

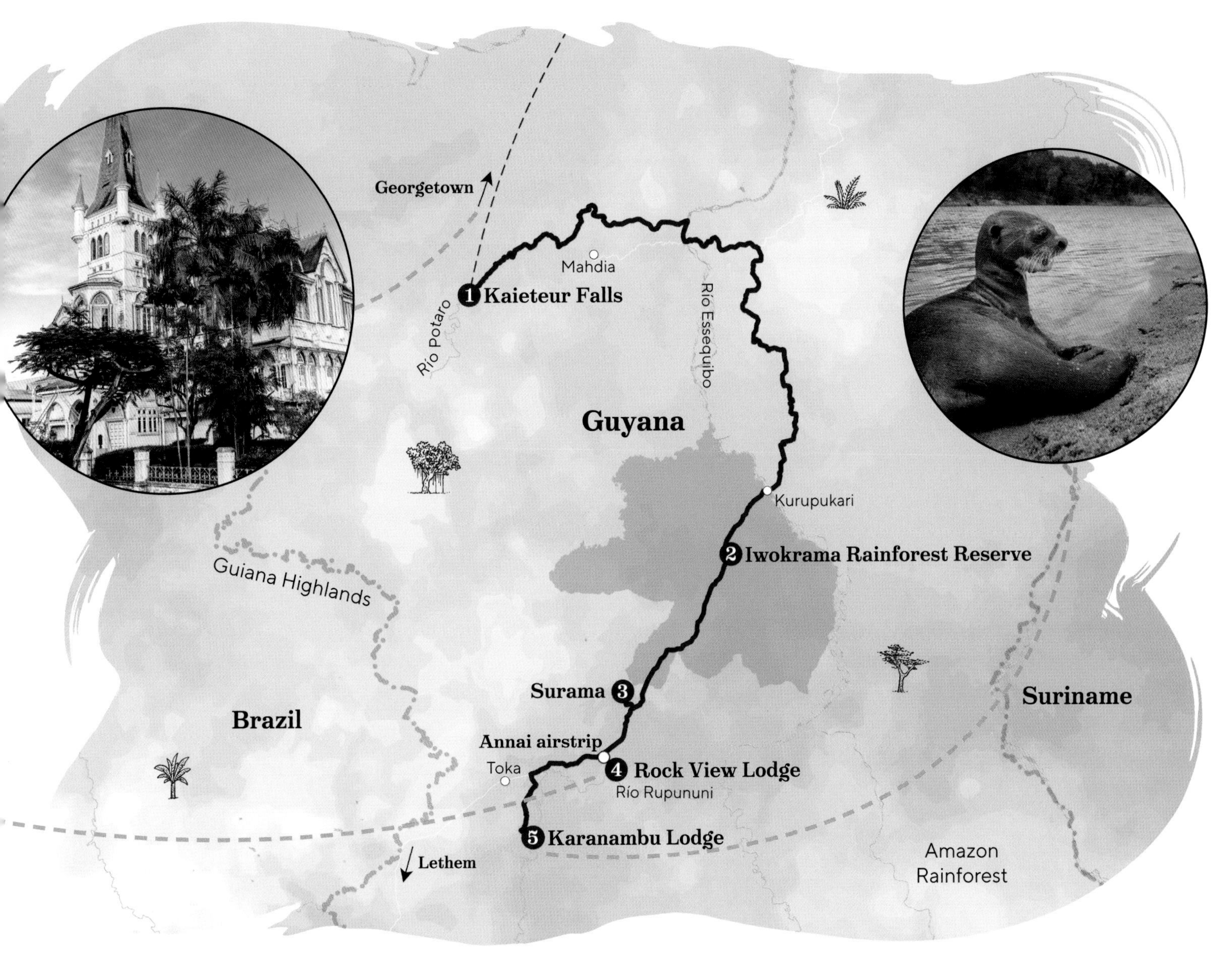

The trip

Distance: 550km (342 miles)
Mode of transport: On foot, 4WD, canoe and river ferry
Difficulty: Challenging

1 Kaieteur Falls The world's largest single-drop waterfall is a stunning start to the journey

2 Iwokrama Rainforest Reserve The green heart of Guyana, where trails and canopy walkways reveal remarkable endemic flora and fauna

3 Surama Stay in an Amerindian village in the north Rupununi, where jungle and grasslands meet savannah

4 Rock View Lodge On the southern fringes of the Rupununi, this is a perfect base to spot the magnificent cock-of-the-rock bird

5 Karanambu Lodge Surrounded by seasonally flooded forest, Karanambu is a wildlife-watching hub: spot black caiman, otters and tapirs

Clockwise, from left: Surama, in the north Rupununi; spot cock-of-the-rock birds from Rock View Lodge; St George's Cathedral, Georgetown; giant river otter, Rupununi

Sarah Barrell, Guyana overlander

Even for South America travel veterans, Guyana's interior is unique. The wildlife-rich jungles and savannahs offer the ultimate lesson in humility. Human life is scant, highly adapted to coexisting with nature. To survive here you must listen, bow to a greater ecological power that can teach us vital lessons. Travel here is a privilege, a jaw-dropping, sweaty, insect-bitten adventure that changed the way I saw my place on the planet.

Amerindian languages. With its patchy collage of colonial architecture – faded demerara shutters and flaking filigree porticos are the vestiges of plantation-owning Europeans who claimed Guyana as their own from the 17th to 19th centuries – Guyana's capital feels like a frontier town for its vast, largely uninhabited interior.

'I've never seen that part of my country,' says a craftsman selling traditional Berbice planters' chairs on Regent St. 'Never left the coast.' It's a sentiment echoed by many living along Guyana's coastal plain, a narrow strip that represents just 10% of the country's total landmass, but which is home to some 90% of the population. From the capital, the hinterland fans out around rivers webbed towards Venezuela, Brazil and the Amazon – Georgetown is a jumping-off point for some serious overland adventure.

Into the rainforest

You'll hear Kaieteur Falls long before you see it. Almost five times the size of Niagara, this single-drop cascade is Guyana's best-known wonder, but has next to no tourist trappings – including guard rails. The thrill of witnessing one of the most powerful waterfalls on Earth thundering 226m (741ft) off a tepui (table-top) mountain is what draws many here from Georgetown, via an hour-long flight that sets down on a scratch of airstrip in the rainforest. Many overlanders get here via a more arduous journey by road and then push on, in search of adventure, into the rainforest wilderness that inspired Arthur Conan Doyle's *The Lost World*.

©MANUELA SCHEWE-BEHNISCH / EYEEM / GETTY IMAGES

"Slow down, give in to the jungle's tightening hold and notice, at every moment, the forest's shifting, intricate wonder"

Making inroads into this largely impenetrable interior often doesn't include roads at all. Bone-shaking red-dirt tracks tunnel through the forest to meet abrupt jungly ends, from where you'll make sweaty, insect-bitten progress on foot and by canoe. A six-hour overland journey south of Kaieteur, Iwokrama Rainforest Reserve reveals a supersized natural world. Here, densely woven rainforest hosts giant variants of anteater, river otter and freshwater turtle, along with false vampire bat (South America's largest bat), harpy eagle (the Americas' largest eagle) and jaguar (the Western Hemisphere's largest cat). Up on Iwokrama's walkway, rope-strung through the canopy, howler monkeys crash through the trees at dawn accompanied by the wolf-whistle wake-up call of the screaming piha bird.

But as with any jungle safari, patience is all. It takes a guide's eye to spot most natural treasures, and tenacious, trouser-penetrating mosquitoes are constant companions. Sudden downpours see hiking trails become rivers, soaking clothes but offering brief relief from the knife-cut humidity. Try to battle this place at your peril. Instead, slow down, give in to the jungle's tightening hold and notice, at every moment, the forest's shifting, intricate wonder: the flash of red-bellied toucans glimpsed through the trees; the vegetal, incense-like scent rising from leaf-mulched ground after the rains.

South to the savannah

Further south still, Surama Eco-Lodge offers a warm welcome to local life and simple lodgings in traditional Makushi Amerindian huts, with clay-brick walls and thatched roofs. Immensely knowledgeable local guides lead explorations into the surrounding expanse of jungle-fringed savannah, pointing out trees that provide balm for snake bites and vines from which it's safe to drink. They also accompany dawn paddles along the Burro-Burro River, aboard dugout canoes hewn from purpleheart trees. Bats swoop low overhead, the watchful eyes of caiman

Below: Kaieteur Falls **Left:** look for harpy eagles on jungle hikes from Iwokrama Rainforest Reserve

glint through the gloom and guides' torchlights cast gothic shadows around the immense buttress roots of mora trees, known as 'lords of the forests'.

From Surama, flat-bottomed balahoo boats transport 4WD vehicles south, across piranha-rich rivers, to bring travellers, blinking, into the sun-scorched savannah. Once terrain for *vaqueros* (cowboys) and gold-panners, the Rupununi grasslands are a prime wildlife-watching spot. 'Mick Jagger, Evelyn Waugh, the royal family – they've all been here,' says Colin Edwards, owner of Rock View Lodge and its surrounding cattle farm. The British expat came to Guyana for its diamond mines and stayed for its other natural riches. Neighbouring Annai airstrip makes Rock View easy-pickings for those who charter planes to get here, but its sights – ground-stomping mating displays from vivid-orange cock-of-the-rock birds; tapir that leave bear-sized footprints – feel like just rewards for road-worn overlanders.

River-running in the Rupununi

At Karanambu Lodge, deeper south into Guyana's Rupununi heartlands, canoes navigate flooded forests blooming with Victoria amazonica lilies, their flowers the size of hubcaps. 'River dogs' (giant river otters), almost 2m (6ft) long, bob alongside those who swim to shore – the chance to rinse red road-dust from clothing an unexpected bonus.

"In the Rupununi heartlands, canoes navigate flooded forests blooming with lilies, their flowers the size of hubcaps"

But by now, few travellers care about appearances, attuned instead to the marvels of this wild natural world. Here, the work of Brit expat conservationist Diane McTurk, who spent a lifetime rehabilitating orphaned river otters until she died in 2016, is continued by her family. Otters abound, as do sightings of comically bushy giant anteater and – notably around neighbouring Caiman House Field Station – thriving numbers of black caiman.

The Rupununi's wildlife attracted such pioneers as David Attenborough and Gerald Durrell in the 1950s. Back then, Attenborough's *Zoo Quest* BBC series brought the unseen animals of what was then British Guyana to rapt TV viewers back home. But today, those who embrace adventure can encounter these inimitable creatures in their own exceptional habitat.

Below: the Rupununi River meanders through Guyana's interior **Right:** explore wildlife-rich rainforests from Surama Eco-Lodge

Practicalities

Region: Rupununi, Guyana
Start: Georgetown
Finish: Karanambu Lodge

Getting there and back: Georgetown is served by international flights from the UK and Europe, usually via Caribbean islands or the USA. There are several daily flights (1hr) between Georgetown and Lethem, the closest airport to the journey's finish point at Karanambu Lodge.

When to go: Most of Guyana is tropical rainforest, so it's very hot with some rain, year-round. The interior's rainy seasons – May to August and, to a lesser degree, December – tend to be a bit cooler, but more humid and with more insects. Overlanding is easier outside that time, but conversely the wet season affords more opportunities for river travel, one of Guyana's greatest pleasures.

What to take: Tropical safari gear, including dun-coloured, collared cotton shirts with long sleeves, quick-dry trekking trousers, a lightweight hat and breathable waterproof trekking boots. Pack a basic first-aid kit: bite cream, antiseptic spray for cuts and scratches (which can fester in the humidity), antihistamine, high-factor sunscreen, high-strength insect repellent.

Tours: Guyana's interior is a remote wilderness, and overland journeys here are best done via an organised tour, which will cover lodge bookings, transfers, transport and guided wildlife safaris. Recommended specialist operators include Journey Latin America (journeylatinamerica.com), Reef & Rainforest Tours (reefandrainforest.co.uk), Responsible Travel (responsibletravel.com) and Wilderness Explorers (wilderness-explorers.com).

Essential things to know: Overland conditions can alter rapidly with changes in the weather, and a destination specialist will know best how to navigate these, switching to an internal flight or river ferry/canoe when roads become impassable, or opting for river travel when wildlife and weather are optimum.

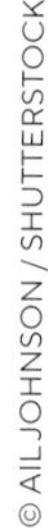

Rafting the Grand Canyon

Wild rapids and ancient rocks are a given on a whitewater trip through the Grand Canyon; friendships born of adventure are a cherished surprise.

Disconnect from tech and connect with fellow river-runners, pondering universal questions in the midst of soul-stirring beauty.

Learn the geologic story of the Colorado Plateau and consider the nature of time beside the Great Unconformity, a mysterious billion-year-old gap in the rock layers. The effects of the ongoing drought are evident at lakes Powell and Mead.

Loading and unloading the rafts becomes easier, as does setting up camp. Completing daily hikes can produce a satisfying sense of accomplishment. When motivation flags on the river, remember this rallying cry: 'You can't do it tomorrow!'

The Colorado River drops a welcome mat at Lees Ferry, where the water is backdropped by the crumbling red beauty of the Vermilion Cliffs. Crossing this mighty river was a challenge for Native Americans, explorers and pioneers, all stymied by the steep, soaring walls of the surrounding Colorado Plateau. But thanks to an upward flex in the ancient rock layers, which exposed more weatherable rocks, erosive forces have carved a

© JIM MALLOUK / SHUTTERSTOCK

The trip

Distance: 446km (277 miles)
Mode of transport: Motorised pontoon raft
Difficulty: Difficult

1 Lees Ferry Launch point for Grand Canyon rafting trips, overlooked by the aptly named Vermilion Cliffs

2 Marble Canyon Layers of the Earth's crust are dramatically revealed along the Colorado River

3 Little Colorado River Take a leap of faith and jump into swift-moving turquoise waters

4 Blacktail Canyon The Vishnu schist rocks here, some of the canyon's oldest, are links to the ancient past

5 Deer Creek Canyon This slot canyon with waterfalls and a 'patio' is a spiritual oasis

Clockwise, from left: paddling through the Grand Canyon at sunrise; resting at the waterside; Deer Creek waterfall; Little Colorado River

Dr Elizabeth Knapp, geology professor and rafter

On Grand Canyon rafting trips you're disconnecting to connect. These days you just don't have enough opportunity to spend a week-plus away from the rest of your life and access to your phone, your emails, your work – just that ability to let go and use your senses in a different way. I find my hearing changes. I smell things differently. It just feels like you have more connection to everything around you, which you just don't get in your daily life any more.

wide valley at Lees Ferry, making it the best crossing point for hundreds of kilometres.

Mormon leader John D Lee established a ferry service here in 1873; today Lees Ferry is the launching point for multiday whitewater trips through the Grand Canyon. From here, at River Mile 0, the Colorado carves through time itself, exposing our planet's geologic story as it drops 1.6km (1 mile) below the Earth's crust. Slot canyons, soaring waterfalls, sun-baked trails, ancient petroglyphs and more than 160 rapids await. Easy escapes do not. On a motorised rafting trip, it will be eight days of remote boating, hiking and camping. Lees Ferry is the last place to change your mind. Once you push off, it's too late. Or maybe, it's just the beginning.

Marble Canyon: the adventure begins

The bulk of the first day is spent on the raft, running the Colorado and its rapids. Configurations differ slightly, but motorised pontoon rafts are typically 8m to 12m (25ft to 40ft) long and divided into sections. Guides steer from the motorwell in the back while passengers – up to 14 per raft – huddle in the front and middle.

Marble Canyon, a steep-walled crack in the Colorado Plateau, plunges south from Lees Ferry. As the river drops, the stacked sediments that comprise the Earth's crust are revealed layer by layer, beginning with a white blanket of Kaibab limestone. Dramatic sites come in rapid succession: the soaring Navajo Bridges, Redwall Cavern, Puebloan granaries, nautiloid fossils and mysterious petroglyphs.

"Lees Ferry is the last place to change your mind. Once you push off, it's too late. Or maybe, it's just the beginning"

A test of courage on the Little Colorado

Stepping into the brilliant blue waters of the Little Colorado River is easy. What's hard is surrendering to its exuberant mini-rapids, which swoosh through a watery gauntlet: a travertine chute, face-pounding waves, butt-bumping rocks and a hard swim to shore across the current. Sourced from a spring, the Little Colorado joins the Colorado at Mile 62 (100km in), between Marble Canyon and Grand Canyon. Its warm waters shimmer a dazzling milky blue in late spring and early summer thanks to an infusion of calcium carbonate. These minerals are eventually deposited as travertine, a chalky limestone that accentuates the blue of the water.

This ethereal place is within the Navajo Nation and is sacred to the Hopi, who believe that life originated here. Efforts to alter its pristine beauty, through dam-building or mass tourism, have so far failed – but rumours of impending development grow louder.

Civil War veteran and geologist John Wesley Powell led the first known river expedition through the canyon in 1869. His tiny flotilla of battered wooden boats regrouped at the Little Colorado confluence before tackling the upcoming – and unknown – Grand Canyon. In his journal, Powell wrote that his crew joked before the launch. But Powell was less carefree, noting, 'to me the cheer is sombre, and the jests are ghastly'. The Colorado's rapids are still pretty ghastly, ranked on a scale of 1 to 10. Below the confluence, Class 8 Hance Rapid drops 9m (30ft) along a wild 1km (0.5-mile) stretch, where giant waves douse those holding tight in front. Downriver, Class 10 Lava Falls drops a stomach-lurching 11m (37ft).

Blacktail Canyon & the Granite Narrows

Geologic wonders come fast and furious in the Inner Gorge of the Grand Canyon, perhaps inspiring thoughts about one's relevance in the universe. One favourite of geologists is Blacktail Canyon, at Mile 120 (193km in). This otherworldly slot canyon provides an up-close look at an enduring geologic mystery: the Great Unconformity.

Below: the Colorado River winds through the Grand Canyon
Left: tackling the rapids

Water woes

The Colorado River is at a historic low due to climate change and a persistent drought, which began in 2000. This unfurling crisis is being witnessed first-hand by boatmen like Art Thevenin, who says: 'The water is a really, really huge thing. There are 40 million Americans [whose] day-to-day life is supplied by water out of the Colorado River...That water source is drying up because of drought and misuse.' So how bad is it? One example is route's-end Lake Mead, the Colorado River reservoir behind the Hoover Dam and a vital power source for California, Nevada, Arizona and Native American reservations. Mead was 317m (1040ft) above sea level in 2022, its lowest level since its creation. If its waters drop below 292m (959ft), it cannot be used to generate electricity. Since the seven states that rely on the river cannot agree on usage reductions, the federal government will be forced to impose cuts. Expect legal battles and headlines.

'I have to admit I cried upon seeing the Great Unconformity,' says Dr Elizabeth Knapp, resident geologist on several Grand Canyon rafting trips. 'That's where the oldest rocks, the 1.7-billion-year-olds – the Vishnu schists and the granites – are in contact with the Tapeats sandstone, which is 525 million years old. It's a break of a billion years of geologic time, and you can stand there and put your hands on it.' Here, the dark and fine-grained Vishnu schist, marked by vertical and diagonal bands, vividly contrasts with the horizontal layers above. Knapp calls it a 'really powerful place to be. Just beautiful. The sounds in there are really amazing.'

The canyon's oldest and deepest rocks, known as the basement layer, are also visible within the Granite Narrows at Mile 135 (217km in). 'That is where the river is only 23m [75ft] wide, and you're surrounded by the cliffs of the Vishnu schist, the really black rock with the pink Zoroaster protruding into it,' says Knapp. 'It's been polished and you're deep in the canyon. To me that was also incredibly awe-inspiring.'

A cold beer on the raft is a time-tested antidote for lingering existential confusion. Beers are typically stashed in nets tossed over the side of the boat, in the river's chilly waters.

Coffee, camping & shared adventure

'Coffee's on!' is a phrase both loved and reviled along the river. Echoing across the pre-dawn chill, with stars still twinkling overhead, this wake-up call promises caffeine, breakfast and another day of adventure.

> "Most first-time river-runners lose their need for constant news and distraction, and gradually embrace life's simpler joys"

Soon, a faint sunrise glow will snuff out the stars and illuminate the canyon with golden light.

And the campsites? Think Gilligan's Island, but with clothes draped on rocks, sarongs hung for privacy and sleeping cots exposed to the stars. These ephemeral camps cling to scrappy sandbars that line the canyon walls, most fringed with boulders, cottonwoods and riparian scrub. Camaraderie, born of shared toil and adventure, develops quickly at camp. Passengers line up to load and unload gear from the rafts. They circle up in chairs on the beach for meals and storytelling, impromptu yoga classes, late-night beers or deep chats by the water. Even the morning line for the Groover – a portable toilet planted just out of camp – becomes a hotspot for good conversation.

'The biggest thing that people have a problem with is unplugging from the regular world,' says Art Thevenin, longtime boatman for Grand Canyon Expeditions. But it eventually happens: most first-time river-runners lose their need for constant news and distraction, and gradually embrace life's simpler joys. Watching this transformation is what brings Thevenin back here, year after year.

Deer Creek Canyon: a spiritual spot

The trail along the Deer Creek tributary, at Mile 136 (219km in) meanders past a 30m-high (100ft) waterfall that plunges into a shimmering pool. From here, the sun-baked path climbs to an overlook with a bird's-eye view of the river and the Kaibab Plateau.

In the Deer Creek Narrows just beyond, the trail clings to the upper wall of a slot canyon that carves through Tapeats sandstone and curves below ancient pictographs. This tranquil stretch of water and canyon is sacred to the Southern Paiute. The trail soon lands at the 'patio', an oasis-like resting point beside the creek's burbling waters and small pools; it's a scenic and spiritual highlight for many. But the trip doesn't end here: you've still got 227km (141 miles) to cover, tackling Havasu Creek, Vulcan's Anvil and Lava Falls before disembarking at Lake Mead – sunburned, exhausted and totally content.

© WILLIAM EUGENE DUMMITT / SHUTTERSTOCK

From top: campsite life along the Colorado River; take a break from the boat for a slot-canyon hike

Practicalities

Region: Arizona & Nevada, USA
Start: Lees Ferry
Finish: Lake Mead

Getting there and back: Outfitters will typically shuttle you to and from Las Vegas, Flagstaff, Arizona or the South Rim of the Grand Canyon.

When to go: April through October. The Little Colorado River shimmers a bright turquoise blue in May and June. Spring and autumn offer the coolest temperatures; June to August is the busy season.

What to take: River sandals made for hiking are your key piece of footwear. A sarong can be hung for privacy at camp, and dampened to keep legs cool and sun-protected while on the raft. A rain suit, comprising a waterproof jacket and pants, is another must-bring. Lees Ferry is 25km (15 miles) downriver from Glen Canyon Dam, and the water powering from the dam is snowmelt from the Rockies – so the river is cold, hovering around 10°C (50°F).

Tours: Grand Canyon National Park (nps.gov/grca) has approved 15 whitewater outfitters for guided trips through the canyon. Trips vary by type of boat (handcrafted wooden dories, oar-powered rafts, paddle-steered rafts and large motorised pontoon rafts), trip length and start- and end-points. Most companies offer full- and half-canyon trips; the latter require a hike in or out of the canyon, typically on the 13km (8-mile) Bright Angel Trail from the South Rim. Full-canyon is recommended if you've got the time.

Essential things to know: Trips are all-inclusive. On motorised rafting trips, buffet meals are prepared and served by boatmen and their assistants, called swampers. Meals are famously good. Most rafters sleep on simple cots directly under the stars.

Reconnecting with nature on the Galápagos Islands

From giant-tortoise encounters to snorkelling with sea lions, experiencing the fauna and flora of this equatorial archipelago will inspire wonder for our world.

The Galápagos may inspire awe and help you see the Earth with fresh, childlike eyes. Every day presents unique ways of experiencing the natural world, whether it's snorkelling with sea turtles or watching flocks of blue-footed boobies plunging into the sea.

There's much to discover on a visit to the archipelago, from learning about the formation of these volcanic islands to apprehending the unusual evolutionary paths taken by the plants and animals that found their way here.

As well as hiking and snorkelling, you can also rent bikes on the three main islands, and head off on kayaking adventures.

Few places conjure as much as awe as the Galápagos. The remote volcanic archipelago, called 'Las Islas Encantadas' (the 'Enchanted Islands') by early explorers, captivated Charles Darwin, who spent five weeks here in 1835. The English naturalist marvelled at the unusual plant and animal life he encountered – species that had developed in isolation from the continent, but which often differed from island to island. The experience sparked his imagination, and featured prominently in his groundbreaking work *On the Origin of Species*, which forever changed our understanding of natural history.

On Isla Santa Cruz, the word Darwin seems to be everywhere: affixed to restaurants, clothing stores, travel agencies – even the main waterfront street in Puerto Ayora. The most important 'Darwin' on the island, however, is the Charles Darwin Research Station. The road east of town ends at this expansive

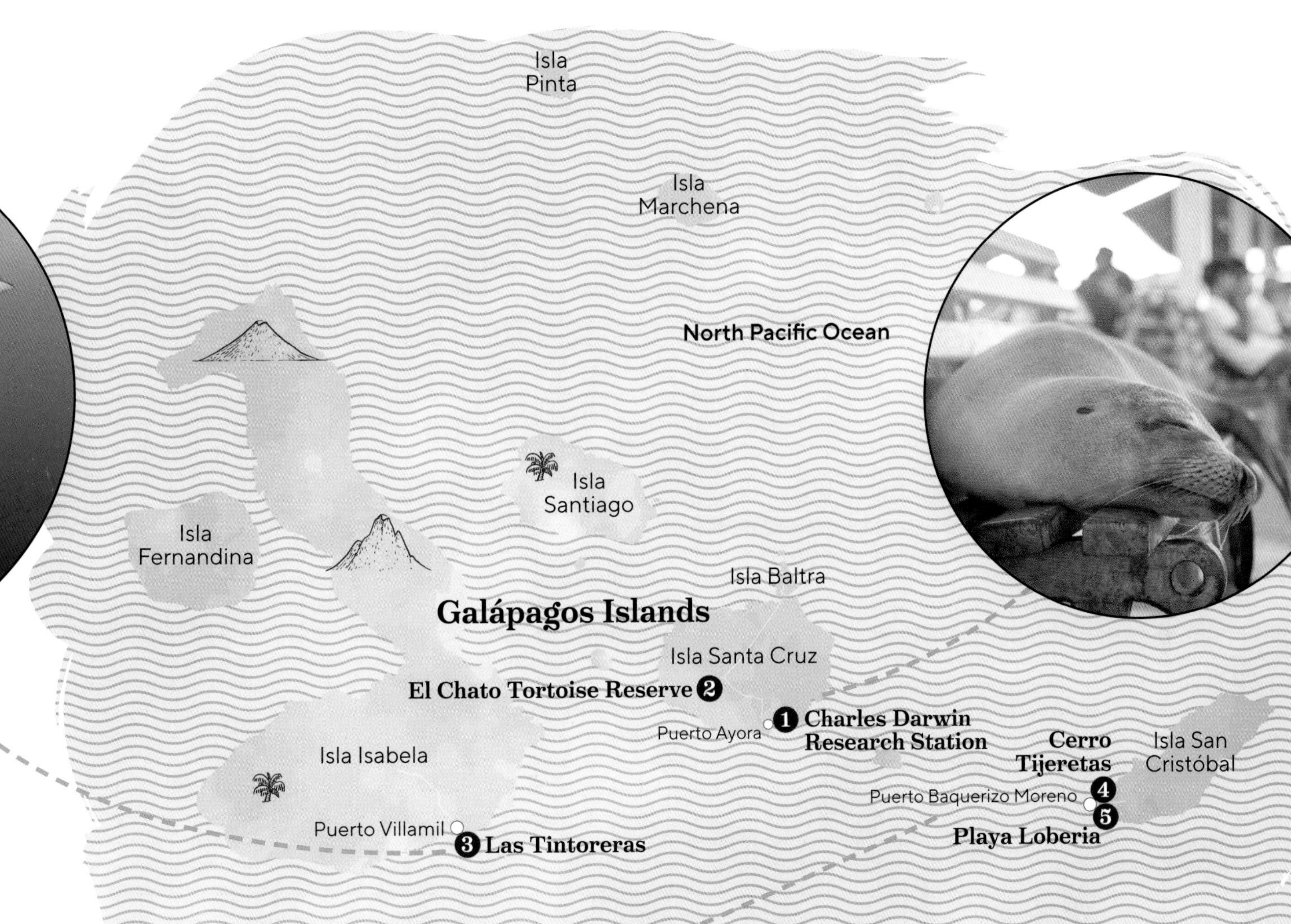

The trip

Distance: 240km (149 miles)
Mode of transport: Boat/on foot
Difficulty: Moderate

1 Charles Darwin Research Station Learn all about giant tortoises and their connection to the Galápagos at this iconic Isla Santa Cruz site

2 El Chato Tortoise Reserve Walk through the Santa Cruz highlands while observing lumbering reptiles and rare birds

3 Las Tintoreras Head to this islet, just offshore of Isla Isabela, to snorkel past volcanic formations amid sea turtles, rays and penguins

4 Cerro Tijeretas Hike up to this San Cristóbal viewpoint to watch diving blue-footed boobies and magnificent frigate birds riding the thermals

5 Playa Loberia Sea lions bask on the sand as marine iguanas emerge from the sea at this San Cristóbal beach

Clockwise, from left: Las Tintoreras, Isla Isabela; eagle rays in the waters around Las Tintoreras; sea lion at Puerto Ayora, Santa Cruz; blue-footed boobies, Cerro Tijeretas

Regis St Louis, Galápagos traveller

On a two-week trip to the Galápagos, each morning seemed like it brought a new adventure, a deeper connection to nature. Returning home, I was determined to keep cultivating that sense of awe, even while out for a walk in my hometown. Wherever you are, wondrous but oft-overlooked things are happening, from huge bird migrations to dramatic seasonal transitions. Being in the Galápagos inspired me to bring that nature-watching zeal home.

facility, where scientists breed giant tortoises and later release them into the wild. Naturalist guides lead tours past enclosures featuring a dozen or so species. The Exhibition Hall delves deeper into the lives of these astonishingly long-lived reptiles.

Behemoths of the highlands

From Puerto Ayora, drive up into the highlands to see the free-roaming residents of El Chato Tortoise Reserve. Against the backdrop of misty scalesia forests, the western Santa Cruz tortoise (*Chelonoidis porteri*) lumbers through the verdant fields. The reserve's winding trails offer ideal vantage points for watching them in their native habitat, and a peaceful setting for contemplation of their extraordinary presence on the island. Like all the other Galápagos reptiles, tortoises originally came from the mainland, and arrived clinging to vegetation washed out to sea during heavy storms. Once on the islands, they followed their own unique evolutionary path and, owing to the lack of predators and competing herbivores, grew to gargantuan size.

Aquatic adventures in Isabela

Back in Puerto Ayora, public boats bob along the dock, waiting to make the journey to other inhabited islands in the archipelago. The 80km (50-mile) ride across the bay to Isla Isabela can be rough, but during the rainy season, when the seas are calmest, you might see bottlenose dolphins riding the bow-waves.

Near the village of Puerto Villamil, the islet of Las Tintoreras teems with life. Zipping through the water like miniature torpedoes are tiny Galápagos penguins – the only penguin species found north of the equator. The crystalline waters are also home to sea turtles and rays that wing slowly past, while white-tipped reef sharks glide through narrow coral-lined fissures in the volcanic rocks. But the stars of the undersea world are sea lions. Inquisitive and playful, these sleek, whiskered creatures sometimes twirl around snorkellers and may come in for an eye-to-eye look before rocketing off. Such close encounters can be a strange feeling, equal parts euphoric and frightening – much like so many other facets of the natural world.

© NWDPH / SHUTTERSTOCK

> "Close encounters with sea lions can be a strange feeling, equal parts euphoric and frightening"

Wonders of sky & sea in San Cristóbal

The last leg of your Galápagos journey means a 160km (100-mile) boat ride from Isabela (via Isla Santa Cruz) to reach San Cristóbal, where the tiny town of Puerto Baquerizo Moreno feels like it's been taken over by sea lions. They bask on docks and dinghies, and waddle in and out of the water off Playa Mann, just steps away from human sunbathers. Around sunset, hike up to the Cerro Tijeretas lookout. Spiky opuntia (prickly pears) and teetering columns of candelabra cacti dot the path, and the grey-white bark of Palo Santo trees scents the air. Even in this harsh environment of once-barren volcanic rock, life found a way to flourish. Like the scuttling lava lizards and the giant tortoises of Santa Cruz, the plants here all arrived from elsewhere, their seeds blown by storm winds, pushed along by sea currents or ingested by birds and deposited in their droppings. Over the eons, sun-baked lava rock was transformed into biologically rich landscape.

From town, it's also an easy 3km (2-mile) bike ride out to Playa Loberia. Orange and blue Sally lightfoot crabs scuttle along the shoreline and, on the rocks, dark shadows emerge from the waves. With a scaly gaze, sharp claws and jet-black skin, the marine iguana looks like a tiny dragon. The world's only true sea-going lizard can spend up to 30 minutes feeding on algae (using its powerful claws to hold tight against the fast-moving tide) before coming up for air. The power of the Galápagos is experiencing these quiet moments far from the din of human civilisation, amid plant and animal life that followed a unique evolutionary path. There's so much to learn from the natural world – if only we take the time to look.

© ANDREW PEACOCK / GETTY IMAGES

From top: snorkelling with sea lions, Las Tintoreras; hike through the realm of giants at El Chato Tortoise Reserve, Santa Cruz

Practicalities

Region: Galápagos Islands, Ecuador
Start: Isla Santa Cruz
Finish: Isla San Cristóbal

Getting there and back: Flights to the islands typically go through Guayaquil to either Isla Baltra (off the island of Santa Cruz) or Isla San Cristóbal.

When to go: January through April is the wet season, which means occasional downpours but generally sunny skies. The seas are typically calmer and warmer, which makes for more pleasant boat travel and snorkelling trips. From June to November, the dry season has slightly cooler temperatures, choppier (and colder) seas and cloudier skies. On the plus side, it's a great time to visit seabird colonies, which are quite active in these months.

What to take: Bring motion-sickness tablets if rocky boats make you queasy. Don't forget to throw in binoculars for wildlife viewing, and good walking shoes (the volcanic rocks can shred thin-soled sneakers).

Where to stay: The islands of Santa Cruz, Isabela and San Cristóbal all have a good selection of lodging options for both modest and high budgets. You'll need to book well ahead to snag unique stays such as the eco-friendly Finch Bay Hotel near Puerto Ayora, Santa Cruz.

Where to eat and drink: You'll find plenty of options in the main towns.

Tours: Many first-time visitors book all-inclusive boat-based tours around the islands. These include food and lodging as well as multiple excursions (hiking, snorkelling, birdwatching) led by naturalist guides. Reputable operators like Happy Gringo (happygringo.com) can arrange passage on dozens of different catamarans and yachts.

Essential things to know: You can save money by basing yourself on the inhabited islands and booking day-trips to nearby wildlife hotspots. However, you won't be able to reach more remote islands this way.

A soulful music pilgrimage through the USA's South

Road-trip through a musical wonderland, where museums, live venues and historic sites offer a joyful and insightful window into the South's culture and history.

From hearing live blues, jazz and country music in the lands that shaped the genres to learning to dance the two-step to Cajun fiddlers, a musical road trip through the US South offers infinite challenges, lessons and insights.

From musical giants through Civil Rights struggles to urban renewal, social history has been shaped across the US South. Travelling in this historic landscape allows for a greater understanding of our world today.

Do you like to dance? To explore? To spend days on a river and nights partying with locals in bars and clubs? This journey is one spent indoors and outdoors – your body won't sit still for long when the music plays!

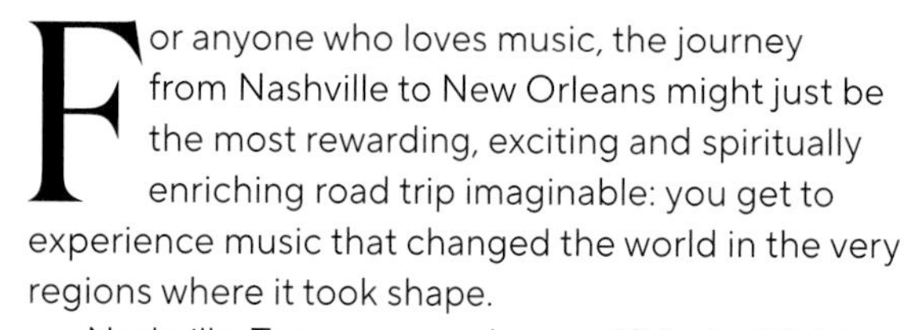

For anyone who loves music, the journey from Nashville to New Orleans might just be the most rewarding, exciting and spiritually enriching road trip imaginable: you get to experience music that changed the world in the very regions where it took shape.

Nashville, Tennessee, nicknamed 'Music City', is both the epicentre of country music and the home of many other genres – indie and Southern rock, rap,

© STAX MUSEUM OF AMERICAN SOUL MUSIC

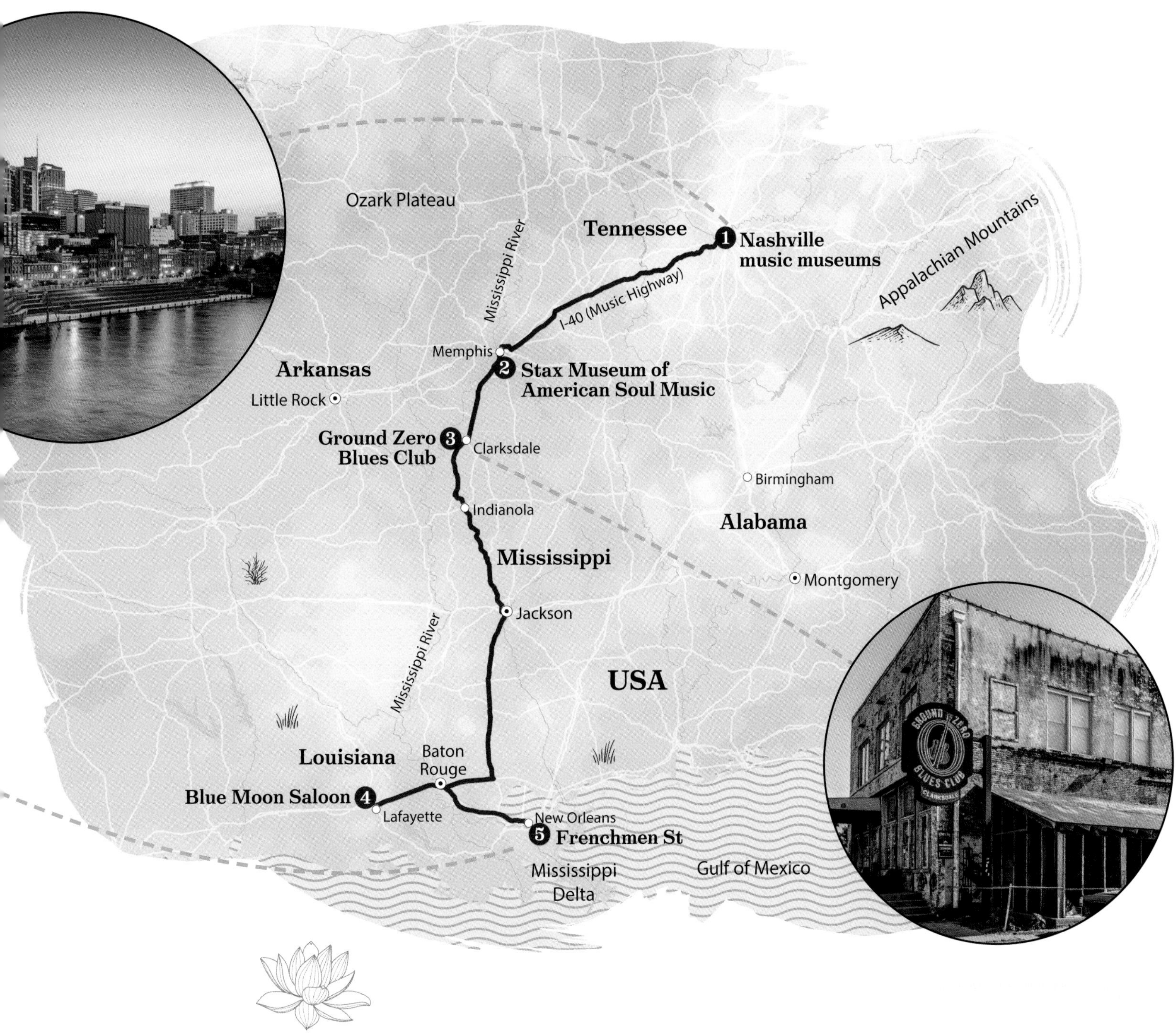

Clockwise, from left: Stax Museum of American Soul Music, Memphis; live music in New Orleans; Nashville skyline; Clarksdale's Ground Zero Blues Club

The trip

Distance: 1308km (813 miles)
Mode of transport: Car
Difficulty: Easy

1 Nashville music museums Get the lowdown on the city's musical heritage at the Country Music Hall Of Fame & Museum and the National Museum of African American Music

2 Stax Museum of American Soul Music, Memphis Built on the site of Stax Records, with displays documenting how soul music took shape

3 Ground Zero Blues Club, Clarksdale Top live blues and R&B in a comfy bar/restaurant

4 Blue Moon Saloon, Lafayette Cajun music headquarters, with great beer and live shows

5 Frenchmen St, New Orleans This lively street hosts a back-to-back array of marvellous live-music venues

Ron Wynn, journalist and broadcaster

I celebrate the Music Highway and the measure of social progress and justice that's been advanced through blues, jazz, soul, rock'n'roll and gospel. The creation and opening of the National Museum of African American Music in Nashville is a direct result of the passion and hard work of many dedicated cultural warriors. On a larger scale, I'm sadly aware many enjoy the music without wanting to appreciate or understand its deeper meanings or its Black heritage.

R&B, blues, electronic, jazz. It's the perfect place to begin a journey through American music.

Nashville is an easy place to engage with Southern culture, with honky-tonk bars, superb concert halls and some outstanding museums: the majestic Country Music Hall of Fame and Museum explores the development of music (country and beyond) and of the Grand Ole Opry radio show; the National Museum of African American Music covers everything from spirituals to rap. There's a great restaurant scene, too: East Nashville is packed with hip hangouts.

The I-40, known as the 'Music Highway', runs from Nashville to Memphis and, as you drive it, signs announce the names of icons who settled or were raised in this fertile region: Tina Turner, Loretta Lynn, Johnny Cash and, of course, the King: Elvis Presley.

City of kings

Elvis was born in Tupelo, Mississippi, and moved to Memphis as a child. Aged 18, he made an impression on Sam Phillips, owner of Sun Studio and Sun Records, a tiny label focused on recording blues. The recordings Phillips and Presley began making in 1954 would ignite a cultural revolution, and today you can tour both Sun Studio and Elvis's sublimely kitschy Graceland mansion. Elvis achieved so much so quickly, yet died aged 42; visiting Sun and Graceland might inspire contemplation on creativity, celebrity and the pressures and loneliness that vast fame and wealth impose. It's hard not to wonder whether your own soul would rise or shrink when confronted with what Elvis encountered.

Right: street musicians in the French Quarter, New Orleans
Below: Sun Studio, Memphis

Speaking of soul, South Memphis hosts the Stax Museum of American Soul Music. Built on the site of Stax Records, the label that introduced Otis Redding to the world, 'Soulsville' emphasises the strength of the spirit, how community and friendship can bring people together to work for change and unity. This story is further developed at the Memphis Rock 'n' Soul Museum over on Beale St, where displays highlight how African American and white music mingled in the Mississippi Delta. A few blocks away is still-functioning Royal Studios, where Al Green cut all his hits. Green now leads Memphis' Full Gospel Tabernacle church – drop by to attend a Sunday morning service and he might be in the pulpit, preaching and singing gospel.

The Mississippi blues

Driving south of Memphis sees you enter Mississippi, a state once notorious for murderous Ku Klux Klan racists. Travelling through these flat lands may prompt often-challenging reflections on the struggle between love and hate, justice and barbarism – and the brutality and injustice that was allowed to reign here. There are no easy answers, but Clarksdale, a former cotton-growing town a little over an hour's drive from Memphis via the legendary Hwy 61, demonstrates how music can bring people together and heal divisions.

Clarksdale is famous for once having been home to celebrated musicians Robert Johnson, Ike Turner and Sam Cooke. Today, this sleepy town is a blues mecca; some say it's home to the mythic crossroads where a devilish bargain was struck in Johnson's famous *Cross Road Blues*. Ground Zero Blues Club puts on top blues and R&B artists; if you want an authentic juke-joint experience, drop into the truly unrefined Red's. Roger Stolle, owner of the Cat Head store – where you can buy Mississippi records, books, folk art and much else – notes that 'for me, Clarksdale is the closest you can get to the heart of the blues. The history is amazing here, the people are like characters in a novel, and the music is truly alive. Because Clarksdale has live blues 365 nights a year, and over a dozen annual festivals, our rustic little downtown has come back.'

Rue St. Pierre
t. Peter

Musical melting pot

American music in all its myriad forms sprang from a collision of cultures as European migrants, African Americans, Native Americans and Mexicans all found themselves sharing the same landmass and hearing one another sing, play and dance. Food and music cross borders more fluidly than any other human creations, thus a true melting pot of sounds and flavours got underway across the South. The creation of blues, gospel, jazz, country, soul, funk, zydeco, bluegrass, rockabilly and rap all share the same mixed heritage, even if they took shape in communities often divided by racial and religious differences. It's a testament to America that the vernacular music of the nation's most marginalised peoples became the true sound of the USA – and is beloved (and copied) across the world. Travelling through the South with ears, mind and heart open offers a road toward wisdom – this is a land where songs serve as teachers.

Clarksdale is home (appropriately) to the Delta Blues Museum – look and learn – and Quapaw Canoe Company, which offers guided canoe trips on the Mississippi River; taking to the water is a great opportunity to unleash your inner Huck Finn.

An hour's drive south takes you to Indianola, another quiet town with a mighty blues history – here is the BB King Museum and Delta Interpretive Center. Honouring native son Riley 'Blues Boy' King, this new museum digs deep into Mississippi history (good, bad and ugly) and allows for meditation on how, via blues, Riley became an ambassador not just for African American culture but human creativity and dignity.

Cajun country, Creole flavours

From Indianola, the highway takes you southwest, through the flat, alluvial plains that once produced (via enslavement, then sharecropping) much of the world's cotton, crossing into Louisiana and arriving in Lafayette. This is Cajun and Creole country: Cajuns are the descendants of French settlers who were expelled by the British from Canada, arriving as refugees in Louisiana (then a French territory); Creole peoples can trace their mixed heritage to 18th-century French and Spanish colonists, African Americans, white Louisianans and Native Americans.

"In Lafayette you are in the USA, but outside mainstream America – it's a good place to think without borders"

Lafayette's live-music scene is stunning: head to the Blue Moon Saloon and the Hideaway on Lee to hear fiddle-led Cajun music, accordion-infused zydeco and R&B-flavoured swamp pop. While everyone in Lafayette speaks English, this Francophone community is the perfect place to delve into Louisiana's unique culture. Many here are proud of their blended heritage – from language and music to food: gumbo and jambalaya are delicious local dishes. Here you are in the USA, but outside mainstream America – it's a good place to think without borders.

Lafayette musician and broadcaster Roger Kash says: 'Lafayette has deep culture with traditions

that have been observed for well over a hundred years – and more great local music per square foot than anywhere else. The level of musicianship is astounding and the youth keep the traditions alive! It's a truly unique place. And don't get me started on the food – the best cooks in the US by far!'

The Big Easy
The drive from Lafayette to New Orleans is stunning, crossing mighty swamps full of cypress trees dripping Spanish moss, on a journey both beautiful and atmospheric. As is New Orleans, the city where jazz was born – local icons include Louis Armstrong, Professor Longhair, Dr John and Irma Thomas. Here, live music echoes from streets and bars day and all night, and there are sensual pleasures aplenty: music, food, drink, festivals, carnivals and partying. The notorious Bourbon St is jammed with raucous bars and best avoided; instead, head to Frenchmen St, where a rich array of venues host local artists (jazz, funk, blues, folk) of a very high calibre. As its Big Easy nickname suggests, New Orleans is a city that encourages visitors to slow down and taste life's sweetness. Engage with the remarkable cultural life all around you and, while doing so, you might just learn a few things about yourself.

Right: brass bands at the Second Line Parade, New Orleans
Left: catch live blues at Red's, Clarksdale

Practicalities

Region: Tennessee/Mississippi/Louisiana, USA
Start: Nashville
Finish: New Orleans

Getting there and back: This is a road-trip and is best experienced by car or motorbike. That said, Greyhound buses link all the major centres, and Amtrak trains run between Memphis and New Orleans.

When to go: The South's musicians play year-round. From May to mid-September the region gets very hot, but air-con keeps venues, restaurants, museums, hotels and so forth accessible. Winter can be cold and wet. Music- and food-themed festivals are held throughout the region across the year.

What to take: Waterproof boots: when it rains in New Orleans, it pours. You'll also need lots of light clothing – outside of winter, the South is sultry.

Where to stay: Accommodation ranges from five-star hotels and boutique options to strip-mall chains, hostels and Airbnbs, all widely available. If your trip coincides with a major festival, book city accommodation in advance, and expect spiked rates.

Where to eat and drink: The roads between New Orleans and Nashville promise a vast array of options for food and drink, with regional styles – Memphis BBQ, Mississippi soul food, Louisiana's Creole dishes – that demand to be sampled. Differing dietary requirements – from vegetarian/vegan to kosher or halal – are now relatively widely available. American beers have improved vastly in recent years; New Orleans is a cocktail mecca.

Tours: Quapaw Canoe Company (island63.com) offers guided Mississippi River trips.

Essential things to know: All big US cities have high crime rates, so stay alert, take taxis to and from unfamiliar neighbourhoods and leave valuables securely at your accommodation. Carry a driver's licence or another form of photo ID – music venues are often strict about 'carding'.

Canoeing the Yukon from source to mouth

Paddle for months through the history and cultures of the north, navigating the wild and majestic Yukon River all the way to the Bering Sea.

There are few places on the planet where you can be further from a road or from human habitation. Learning to travel here equips you for remote journeys anywhere in the world.

Places like Alaska often get called wildernesses, but spend time in the First Nations villages along the Yukon to get a sense of the history of this area and how, for many people, this land is not a wilderness but home.

It's an awful lot of paddle strokes from the Yukon's source to its mouth. Not only will you be a proficient canoeist by the end, you'll have very impressive shoulders.

The Yukon is the longest free-flowing river in North America: there is one small dam near its headwaters, and from there it runs 3190km (1982 miles) to the sea. Canoeing its length is a journey to devote months to – and indeed months is all you have, because between October and May the Yukon is frozen solid. During that time, the locals go by snowmobile or dogsled – and that is a whole other adventure. But for that brief stretch

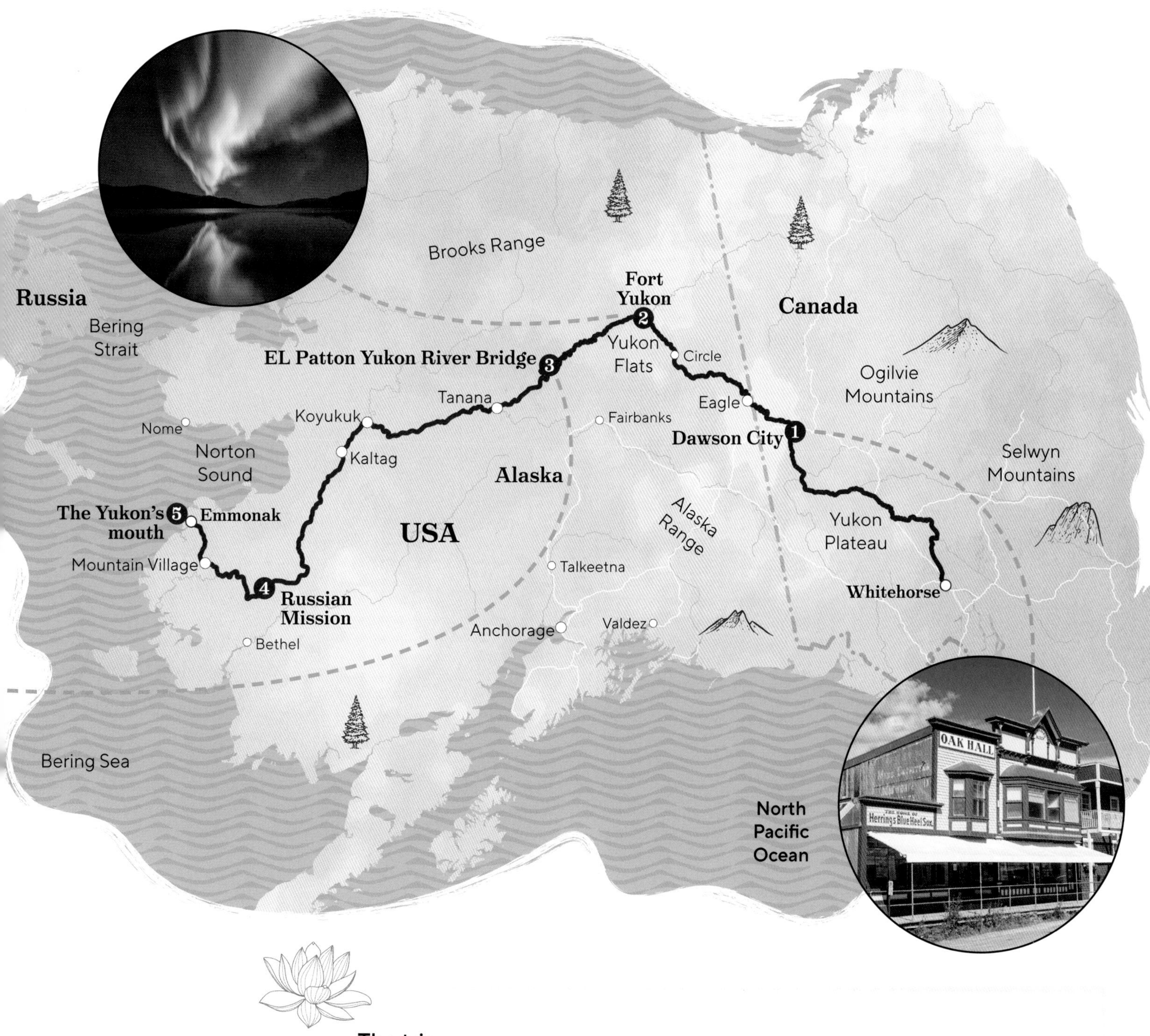

The trip

Distance: 2886km (1793 miles)
Mode of transport: Canoe/sea kayak
Difficulty: Difficult

1 Dawson City Epicentre of the Klondike Gold Rush, and still trading on its nostalgia

2 Fort Yukon Victorian tourists boarded paddle-steamers here for a glimpse of the midnight sun

3 EL Patton Yukon River Bridge The last bridge over the river. Two months to go!

4 Russian Mission Russians were the first Yukon colonisers, and their legacy endures in this former fur-trading post

5 The Yukon's mouth The river is a mighty 11km (7 miles) across where it meets the Bering Sea

Clockwise, from left: the Yukon at Miles Canyon, near Whitehorse; the EL Patton Yukon River Bridge; look out for the aurora on overnight camps near Fort Yukon; Dawson City

Adam Weymouth, Yukon River canoeist

I had scarcely set foot in a canoe before I began, but by the Yukon's end I had had fallen in love with my boat; with how it becomes an extension of your body, so that you can feel the river's moods through its hull, each flex of current and breath of wind. The Yukon made me a canoeist, which is to say that it taught me to appreciate rivers. And for that I am eternally grateful.

when the ice is gone and the sun scarcely sets, the summer days seem to carry on forever.

Setting out

Carry your canoe down to the Yukon's edge at Whitehorse and let the current take it. Drift away from the bank and let the city slip away. Feel your muscles start to work as you slide your paddle into the water. The river is swimming-pool blue, so clear you can see fish, but before long the successive tributaries muddy it to the colour of strong tea, and until it meets the ocean it will not run clear again. The silt, rubbed from distant mountains, whispers at the hull, and if you dip your paddle and hold your ear to the shaft you can hear it clearer still, as though the river is deflating.

Already it feels like a big river, but this is nothing to what it will become. It rushes along, swirling and eddying. Cliffs rise up around you, hazed by ravens calling madly. By the time you cross the border into Alaska (a 10m (33ft) swathe chainsawed out of the forest is the only clue that you've entered the US) it is wider still. Paddling down its middle channel, catching the fastest current, feels like balancing on a tightrope. By the time it makes the ocean, the Yukon River is so wide that you can scarcely see the other bank.

The daily grind

To live like this is work. Each evening you must unload the boat, raise and stake the tent, gather wood to make a fire, cook dinner. You must haul the canoe up the beach, flip it and tie it down to something solid in case the wind picks up in the night – and in the morning you reverse the entire process before you carry on again. But as the weeks pass and your old life falls away, these tasks begin to feel less like chores and more like the necessities of existence: you're just another animal carving out a livelihood here. You find a rhythm with the vagaries of the weather. You paddle on.

Clockwise, from top: brown bears inhabit the remote Yukon watershed; float past forest on the Yukon

Time is travelling in reverse. In 1896 the Klondike Gold Rush brought 100,000 people to Dawson City in search of rumoured riches. In a few years they were gone again, leaving devastation in their wake. Now all down the river are the remnants of that time: broken telegraph wires, rusted dredging equipment, rotting paddle-steamers, ruined cabins. The land is on the march here, clawing its way back.

And yet this is not an empty landscape. Every bluff and eddy has a name. Every few days there is another village, holding to the river's bank. The gold rush has ebbed away but the Alaska Native people are still here: Tlingit, Tr'ondëk Hwëch'in, Athabascan, Inuit. Often far from the nearest road system, the Yukon is their highway, as well as providing vital food in the salmon that migrate up it every summer.

It would be easy to paddle past these villages without stopping. But take some time to meet people, and it doesn't take long to realise that the Yukon is equally fascinating for the cultures of those who call it home. People have made a living in these places for millennia, living semi-nomadic subsistence lifestyles. Since colonisation they have been forced to settle, and in many ways are now as subject to the impacts of industrialisation and global capitalism as city-dwellers. But people here also live lives that are intimately connected to this land – in the food they hunt and how they travel and how they think about their home. To spend time here is to understand how much we are all impacted by our environment. In the cities, insulated from the natural world, this is easy to forget.

To eat & to be eaten

At the Yukon Flats the river unravels, frays like rope as vast wetlands spread out across the landscape. The

Breaking the ice

Each spring, a tripod is erected on the river ice outside of Dawson City, connected by a wire to a clock that is tripped and stops when the ice begins to move and melt. Since 1896, when gold miners with money to waste and time to kill began gambling on the moment as a good way to see in spring, the time of each event has been recorded in a ledger kept in town. The first time that break-up occurred before May was 1940, and then again in 1941. There was not another April break-up until 1989; since then it has occurred nine times. In 2016, the record was smashed by more than five days. All along the Yukon – in a region that's warming up to four times faster than the global average – it is impossible to ignore the changing climate, from the shifting migration patterns of animals to stories of experienced hunters lost through ice that has become unreliable and dangerous.

Yukon's waters ramble across the plains, all urgency forgotten. The Flats have the spacious feel of an estuary, but you're still more than 1000km (620 miles) from the sea. Two lynx gambol on a sandbar in the sunshine; they barely notice you in the canoe. And then after some days the mountains rise up, and the river gathers pace again.

The ecosystem still feels remarkably intact, and not least because your status as a human no longer puts you at the apex of the food chain. Seeing the big mammals is rare, as a grizzly requires a territory of 250 sq km (96 sq miles); that bears can still find a home here is emblematic of a much vaster, unseen wilderness. The Yukon's watershed has a population of one tenth of a person per square kilometre, for a close to a million square kilometres – the equivalent to a pre-agrarian society. If the whole planet were similarly populated, the global population would be that of Istanbul.

A growing awareness

After many weeks outside, paddling on along the river, something happens to the mind. Lie back in the canoe and drift and listen. Look around. You can distinguish different birdsongs now. You can identify the trees by the way they move in a breeze, how the aspen leaves twinkle but the birch shiver. For as long as you can remember, you have spent the whole day paddling, eyes fixed on the landscape. At first it all looked the same. Spruce and alder, aspen and willow, an endless forest. But the more you look, the more you notice difference. Now you can focus in on a far-off fleck of white and spot a bald eagle sitting motionless, scarcely aware how you have done it. This is a wonderful way to travel. Sit back and drift. Sketch or read or write. Make a coffee in the prow. The river knows where it is going.

By September the weather is already turning. The landscape is changing, too. Gone are the forests, replaced by tundra in autumnal red and yellow, stretching off to blue mountains in the distance. Lightning cracks in broody skies, and the geese are migrating south. In a few short weeks the Yukon will be entombed in ice. Dig in the paddle and carry on.

Clockwise, from right: Dawson City; paddling on through the Yukon wilderness

> **"This is a wonderful way to travel. Sketch or read or write. Make a coffee in the prow. The river knows where it is going"**

Practicalities

Region: Yukon, Canada/Alaska, USA
Start: Whitehorse, Canada
Finish: Emmonak, USA

Getting there and back: You can fly direct from Germany to Whitehorse in summer, though an overland journey to the start is an adventure in its own right. Grant Aviation flies out of Emmonak to Bethel, from where Alaska Airlines flights connect to Anchorage.

When to go: The Yukon is typically ice-free between early May and late September. If you don't dawdle too much, it's possible to paddle from one end to the other within that window.

What to take: You'll want a good lifejacket on at all times: the Yukon is very cold and silty, and a single mistake can be fatal. Take a tent in which you can comfortably sit out a few days of bad weather. Gaffer tape can fix most things when you're desperate, but bring a repair kit for the canoe. A Personal Locator Beacon is a crucial emergency precaution.

Where to eat and drink: You'll need to carry several weeks' worth of food between resupply points, which you can supplement at the basic (and very expensive) village stores, and with occasional fishing (buy a rod licence). Cabbages are your friend, and seem to keep eternally fresh.

Tours: The most commonly paddled section is from Whitehorse to Dawson City. Several outfitters (try upnorthadventures.com or kanoepeople.com) will rent you a canoe and gear in Whitehorse and pick you up at the other end.

Essential things to know: You can sell your canoe in Emmonak, but supply and demand means you won't get more than a few bucks: every home has a few of them by now. Alternatively, put it on one of the empty cargo planes that leave town daily, and sell it back in Anchorage. Doing a course in wilderness first-aid before setting out is not a bad idea, neither is learning about bears.

Step into the infinite in the Salar de Uyuni

Bolivia's blinding-white Salar de Uyuni salt flats express the endless – exploring these vast plains draws you into yourself and to the contemplation of infinity.

Amid sparseness, it's natural to appreciate every living thing and realise that the desert isn't so deserted after all. By slowing down and paying attention, you can see the way of life here abides by the harsh rules of survival.

Looking up, starstruck, you can't but marvel at the night sky in the *salar*. Unfettered by city lights, the Southern Cross seems to bestow a five-pointed benediction from ink-black skies.

Exploring the Salar de Uyuni is a bit of a fitness test. Trekking at altitude busts the lungs and strains the hamstrings, but breathe deep and you'll find the rhythm of the altiplano.

As you board the train to Uyuni at Oruro station, you're beginning a journey to what feels like another planet. You're headed for the world's largest salt flats, the Salar de Uyuni, a stark and surreal landscape covering some 12,100 sq km (4670 sq miles) and sitting at an altitude of 3653m (11,984ft) above sea level. From Uyuni, you'll explore the flats on one of the 4WD tours that zip around the *salar*, stopping off at lakes and highlight spots, and taking to the salty surface for exhilarating hikes across this chalkboard expanse.

Nothing is something, after all

The salt flats are an outward and inward challenge, as the brisk night air will surely remind you. There are no trees to block the wind up here, just the jagged, 10m-high (33ft) cardón cacti saluting you with bristly, off-kilter arms. The morning sun glances brilliantly off

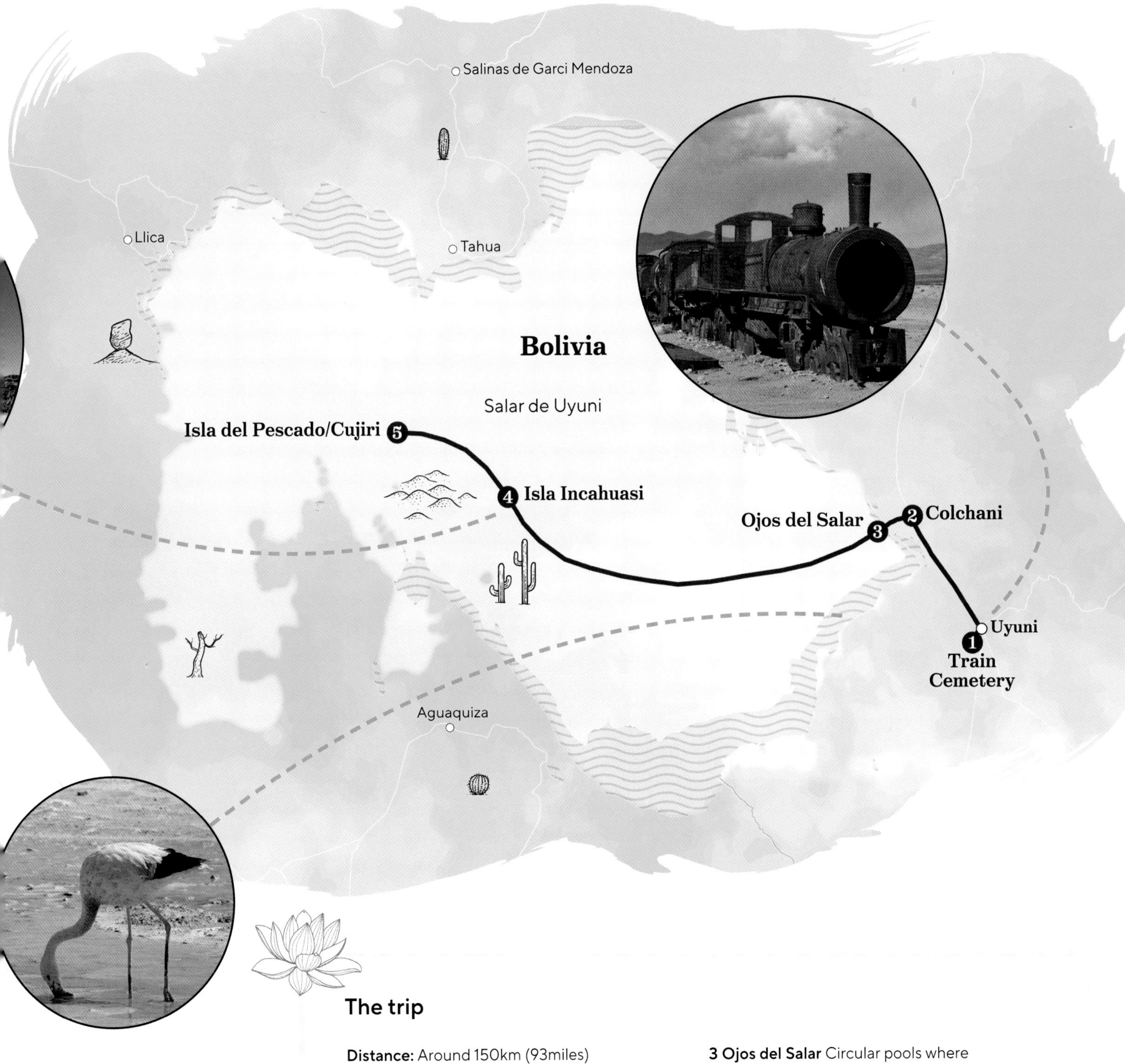

The trip

Distance: Around 150km (93miles)
Mode of transport: 4WD vehicle/on foot
Difficulty: Moderate to difficult

1 Train Cemetery, Uyuni These abandoned iron horses are a monument to forsaken technology

2 Colchani King sodium chloride still rules at this village at the edge of the *salar* – see a local salt processing co-op in action

3 Ojos del Salar Circular pools where subterranean rivers meet the salt-crust surface

4 Isla Incahuasi This volcanic mass arising from the flats promises challenging hikes and colourful birds

5 Isla del Pescado/Cujiri Another awe-inspiring 'island', said to resemble a fish beached in the salt flats when seen from afar

Clockwise, from left: a reflective moment in the Salar de Uyuni; Isla Incahuasi; Train Cemetery, Uyuni; flamingo feeding at a *salar* lake

Patricia Colic, Salar de Uyuni traveller

One of the best things for me about Salar de Uyuni was feeling so connected to the universe. I was in contact with my inner self just looking at the sky at night, and drinking in the vastness of the massive salt flat during the day. Those stars were not only breathtaking but revealing. As it's so clear, it's easy to stargaze and wonder about other worlds; it's as if you are walking among the stars. We even created our own escapade to Laguna Colorada, a bright pink lagoon home to awesome flamingos.

the crusty, too-white surface. In the wet months, a thin sheet of water covers the flats, creating a wondrous mirror-like reflection of the clouds and the blue altiplano sky – forced introspection in high definition (it also makes for unforgettable photos).

Imaginative travellers here may see themselves in *The Last Jedi* (which was filmed here), battling it out on Planet Crait. Or feel themselves flying across the expanse with the Dakar Rally drivers who risk life and limb scrambling around the saline lakes during the South American instalment of the legendary race (there's a stop-off to see a massive salt carving dedicated to Dakar midway through most tours).

A town lost in time

The bus from La Paz might be more comfortable, but the train is the coolest way to get here – though cool is also the watchword, as you arrive in Uyuni at dawn, when temperatures can dip well below zero. In this small town, the landmark clock tower is dwarfed by the landscape: human time writ small against the backdrop of the ages.

Uyuni is not exactly a ghost town, but it is a shell of its former self, having long passed its boom years; these days, tourism is the mainstay of the economy. A can't-miss daylight sight is the Cementerio de Trenes (Train Cemetery), a rusting collection of historic steam locomotives and carriages dating back to the 19th century, when there was a rail-car factory here.

Colchani: people of the salt

Salt is the alpha and omega of these highlands. On the *salar*, it gathers in small piles, formed in irregular, imperfect hexagons and multiplied across the Andean plain. Most tours stop at Colchani, a village of some

© DELPIXEL / SHUTTERSTOCK

> "The salt flats are an outward and inward challenge, as the brisk night air will surely remind you"

600 souls and site of a salt-processing cooperative and a bare-bones Salt Museum. It's not hard to picture life here a century ago, when villagers scraped salt from the land to sell to the now-abandoned tin mines in Potosí. You can feel the pinch of extraction and exploitation, in a zone abandoned once the tin mines passed their mid-century apotheosis, but still a place for salty 21st-century interlopers who mine for lithium to run battery-powered vehicles far away. The unique, nearly unblemished surface of the Salar de Uyuni offers the perfect conditions for scientists to calibrate satellite altimeters, and travellers here may find themselves recalibrated, too – lighter and less burdened by worldly concerns.

Eyes & islands

Tours through the salt flats may also make a stop at the rounded, salt-rimmed pools of underground rivers which bubble up on to the crispy surface. These Ojos de Salar (ojos means 'eyes') are aptly named, resembling staring-blue orbs gazing into the void. Visitors can close their own eyes and take a leap of faith into the pools' mineral-laden waters; locals say they have healing powers.

Journeying on through the *salar*, your mind can start to fly, dizzy with the hallucinogenic combination of altitude and sun and endless blinding white. From seemingly nowhere, Isla Incahuasi rises from the emptiness like a mirage. This rip-rap of jagged volcanic rocks – leftovers of an ancient eruption which arose and was subsequently swallowed up by a lake – is a jolt to the senses. It's an unmissable sight during the dry seasons, offering a challenging hike amid its dense battalions of cacti. Birdwatchers will be thrilled by the avian array: a colourful lineup of black-hooded, greenish-yellow and bright-rumped finches, spinetails, swallows and hummingbirds.

Another similar piece of higher ground, and also on the tour circuit, the Cujiri or Isla de Pescado has a bonus feature: a cave at the summit revealing the startling strata of the surrounding desert, layer by colourful layer.

© SERGIO PESSOLANO / GETTY IMAGES

From top: dawn over the Salar de Uyuni salt flats; cacti on Isla Incahuasi

Practicalities

Region: Potosí Department, Bolivia
Start/finish: Uyuni

Getting there and back: Book ahead for the comfy heated overnight buses from La Paz to Uyuni (todoturismosrl.com; 10hr); you can sleep, have a hot dinner and quick breakfast and save on hotel costs. A flight from La Paz to Uyuni is the fastest but most expensive way in; the scenic option is a bus from La Paz to Oruro (4hr), then an overnight train to Uyuni (8hr; Mondays only), arriving in the brisk desert dawn. There are also trains to Uyuni from Tupiza and Villazon (Thursdays only), and buses from Chile's San Pedro de Atacama.

When to go: In November, three flamingo species arrive to feast on krill in the saline lakes. During the wet months (December to April), parts of the *salar*, including Isla Incahuasi, are inaccessible; stargazing may be more limited, as is access to remote hotels.

What to take: Sunglasses, lip balm, high-factor sunblock, warm clothes, hiking boots that you won't mind discarding after your visit (the harsh terrain take its toll); flip-flops are better on the flats when they're wet. Bring cash (there are few ATMs here and they tend to empty out quickly), as well as bottled water and snacks, both of which are overpriced in the *salar*. Birders will want binoculars.

What to eat and drink: In Uyuni, Tika Bolivia offers veggie options and uniquely Bolivian specialities like peanut soup and llama goulash. Its Tika Palace location offers high-end, high-altitude wining and dining; Tika City is more budget-friendly.

Tours: Red Planet (redplanet-expedition.com) is a reliable Uyuni tour operator.

Essential things to know: Reckless driving does cause accidents in the Salar de Uyuni. Less cautious tour companies add to the risk-and-reward atmosphere; research your operator thoroughly and avoid any with a record of incidents.

The Indigenous roots of the Canadian Rockies

Road-trip from Alberta's capital through Canada's iconic mountains to learn about the diverse heritage and cultures of the region's First Nations and Métis peoples.

This journey across Indigenous territories in present-day Canada is physically easy: it's primarily a driving tour, with walking experiences at each stop. But psychologically, it can be tough. Learning about historical events in which Indigenous people were forced from their homes, separated from their families and taken from their traditional cultures is an often challenging experience.

The Indigenous-led tours on this journey highlight how people across Canada are slowly working towards reconciliation. Approach the experiences with an open mind and don't be afraid to ask respectful questions.

Many Indigenous cultures are oral cultures, passing knowledge from generation to generation through legends, myths and historical narratives. On a road trip through Alberta, in Western Canada, you can learn about the traditional cultures of the Canadian Rockies and surrounding regions through a series of Indigenous-led experiences. As you explore museum exhibits, join guides for forest walks or take a wildlife tour on the Prairies, stop and listen. These are stories that need to be told – and heard.

Starting with stories

Fort Edmonton Park is a recreated pioneer village, mostly concentrated on illustrating the lives of settlers during the 19th century. But its Indigenous Peoples Experience relates a more comprehensive history: that of the people who have called this region home

© JEFF CLOW / 500PX

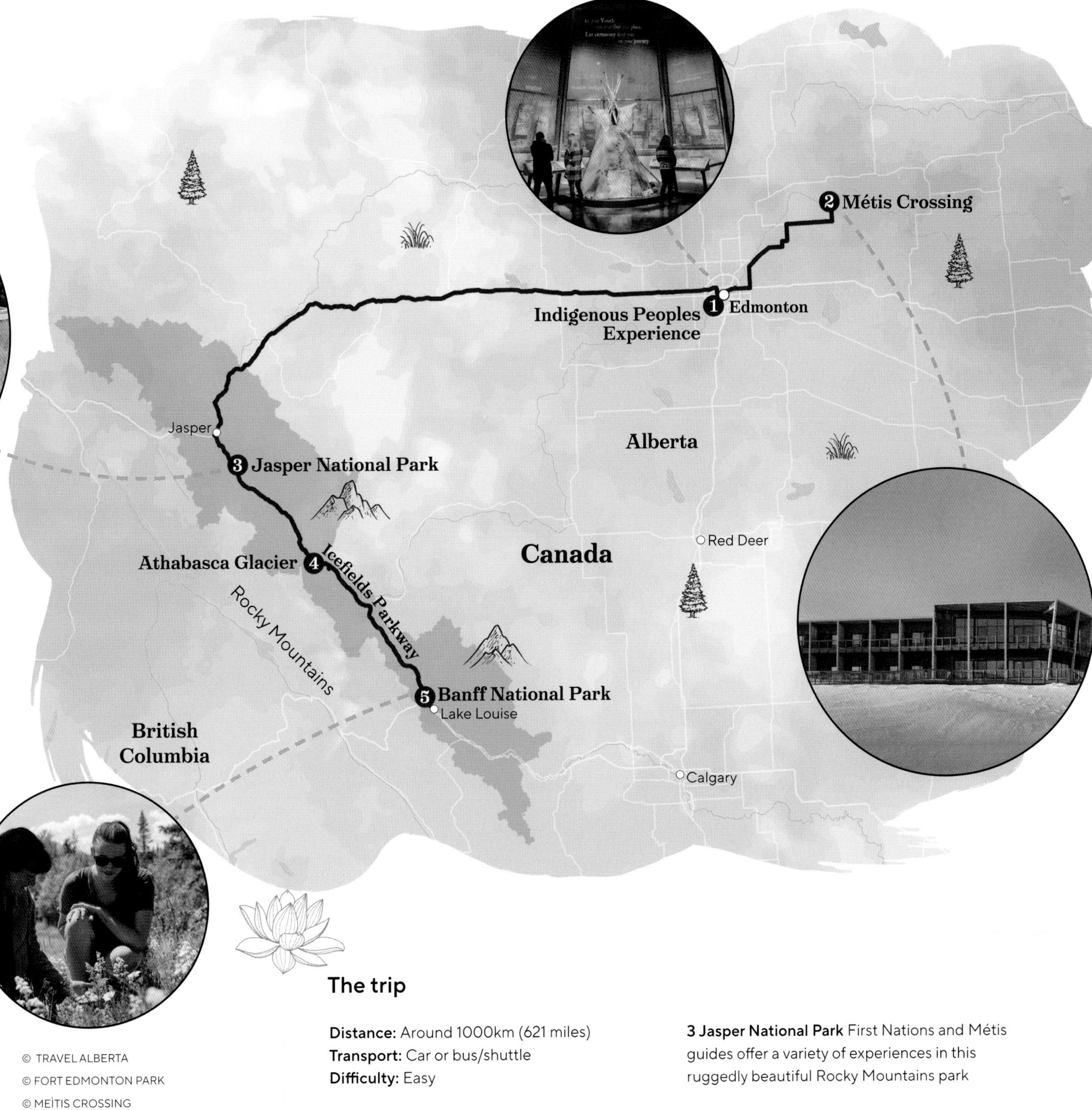

The trip

Distance: Around 1000km (621 miles)
Transport: Car or bus/shuttle
Difficulty: Easy

1 Indigenous Peoples Experience An Edmonton multimedia gallery sharing audio and video stories direct from First Nations and Métis Elders

2 Métis Crossing Learn about the Métis people – and sleep in a stargazing dome – at this innovative cultural centre and lodge

3 Jasper National Park First Nations and Métis guides offer a variety of experiences in this ruggedly beautiful Rocky Mountains park

4 Athabasca Glacier Take an Indigenous-led glacier walk to learn about Athabasca's historical importance and current environmental threats

5 Banff National Park Explore Canada's most-visited national park through the lens of Indigenous guides

Clockwise, from left: Icefields Parkway; Warrior Women plant walk, Jasper; Indigenous Peoples Experience, Edmonton; The Lodge at Métis Crossing; Mahikan Trails hike, Banff

Glacial silt turns the Athabasca River blue on Icefields Parkway

Carolyn B. Heller, Alberta roadtripper

I'm a settler on Indigenous lands. When I moved to Canada, making my home on the traditional territories of the Musqueam, Squamish and Tsleil-Waututh people in present-day Vancouver, I became as much a settler as others who colonised this continent centuries ago. By talking with guides like Matricia Brown, I've learned some of the stories of the Indigenous people who were here first. It's an honour to listen and to learn.

for thousands of years. You can hear audio and video stories that Indigenous Elders have shared, enabling you to hear direct testimonies from the region's Cree, Nakoda, Dene and other First Nations peoples, as well as the Métis, who have mixed Indigenous and European ancestry. Today, of the more than 250,000 Indigenous people who reside in Alberta, roughly 53% represent 45 different First Nations, while most of the others are Métis.

The galleries are organised by the seasons, each representing a different phase of life and its rites of passage. The section exploring European arrival highlights the changes – and hardships – those settlers brought. Inside a Métis cabin, video displays illustrate Métis music, arts and spiritual beliefs. In the circular Meeting Place, video presentations allow visitors to reflect on the challenges that Indigenous people continue to face – including the legacies of the residential school era, when the Canadian government removed Indigenous children from their communities and sent them to often-abusive boarding schools.

Repopulating the bison

Around 110km (68 miles) northeast of Edmonton, you can learn more about Métis communities at the first centre dedicated to their culture. Métis Crossing has a small exhibition area and offers a variety of experiences, from traditional art workshops to snowshoe tours. A highlight is a guided wildlife safari, which allows you to get up close to elk, Percheron horses and three species of bison (including the endangered white bison) that the community is working to repopulate. You can visit Métis Crossing as a day-trip from Edmonton or overnight at the lodge or in a 'Sky Watching Dome', a yurt-like structure that's open to the stars.

© MÉTIS CROSSING

"On a Glacier Ice Walk, guides share how important this mountain region has long been for Indigenous peoples"

Tracing the wood wide web

From Edmonton, travel west towards the Canadian Rockies and into Jasper National Park. Twenty-six Indigenous communities have ties to the land that today makes up the park, which encompasses more than 11,000 sq km (4247 sq miles). When the national park was created in 1907, most of these people were forcibly relocated outside of its boundaries.

Indigenous-led tours can help unravel this complicated history. The Jasper Tour Company offers several adventures in which Métis owner and head guide Joe Urie shares his culture as visitors explore different areas of the park. On a plant walk with First-Nations-owned Warrior Women, Cree knowledge-keeper Matricia Brown explains how the plants that grow here are used for food and medicine, and discusses issues that local Indigenous communities have encountered. She might also talk about how trees communicate, sharing elements such as nitrogen and carbon, through what she calls the 'wood wide web'.

Glaciers & plants

The drive between Jasper and Banff national parks is one of Canada's most spectacular, following the 232km (144-mile) Icefields Parkway as it cuts between the mountains. Part of the Columbia Icefield, the vast Athabasca Glacier spreads across the mountains at the Parkway's midpoint. A unique way to explore the glacier and understand its ecosystem is to head out on to it. On an Indigenous-led Ice Walk, guides take you on a 5km (3-mile) glacier hike, sharing how important this mountain region has long been for Indigenous peoples; they also outline the challenges that the warming climate is posing to the glacier, which has lost roughly half its volume over the past century.

Once you arrive in Banff National Park, continue exploring the region's ecology on a hike with Mahikan Trails, a First Nations-run company that introduces you to the park's Indigenous heritage, plants and animals. As you walk, you might learn how plants such as cedar, juniper and spruce are used as sources of food and medicine.

© WARRIOR WOMEN/ INDIGENOUS TOURISM ALBERTA/ ROAM CREATIVE

Caption: Cree knowledge-keeper Matricia Brown leads Warrior Women trips into Jasper National Park; bison at Métis Crossing

Practicalities

Region: Alberta, Canada
Start: Edmonton
Finish: Banff National Park

Getting there and back: You can do this trip in any direction, but Edmonton is the obvious start point. Its international airport has flights from destinations in Canada, the US and worldwide; ViaRail's Vancouver–Toronto trains stop in Edmonton and Jasper. You'll need a car to travel between Edmonton and Métis Crossing, but most of the other destinations are reachable by public bus or shuttle. Edmonton has an extensive public transport system; Banff's Roam Transit buses can take you to many destinations within the national park.

When to go: While many of these experiences are available year-round, you'll have the best weather between late May and early October. September to October, when the air begins to cool and the summer crowds abate, is a particularly beautiful season in the Canadian Rockies.

What to take: You need a Parks Canada day-ticket or annual Discovery Pass to enter the national parks, available online (parks.canada.ca) or at any park entrance or visitor centre. As mountain weather can be variable, pack layers, no matter when you're travelling.

Where to stay and eat: Edmonton has hotels and restaurants in all price ranges. Lodging options in Banff and Jasper range from camping and hostels to hotels and upscale lodges; dining options are equally varied. The Lodge at Métis Crossing offers overnight accommodation and has a restaurant/cafe.

Tours: Book Indigenous-led tours with Jasper Tour Company (jaspertourcompany.com); Warrior Women (warriorwomen.ca); Athabasca Glacier Ice Walk (icewalks.com); Mahikan Trails (mahikan.ca).

Essential things to know: Indigenous Tourism Alberta (indigenoustourismalberta.ca) and Destination Indigenous (destinationindigenous.ca) provide useful details for exploring Indigenous Canada.

Driving the Harriet Tubman Underground Railroad Byway

Freedom fighter Harriet Tubman risked everything to save others. Find inspiration in her powerful story while following in her footsteps between Maryland and Pennsylvania.

Tubman often cited her faith as instrumental in her missions to free others. From spiritual meeting houses to still-active African American churches going strong since the 19th century, this journey takes you to places ideal for contemplation of living a life with purpose.

Though Harriet Tubman's name is known around the world, many do not fully comprehend the depth of the struggles she and so many others experienced in bondage. This trip along the Underground Railroad Byway underscores both the brutal reality of enslavement and the terrible choices faced by those who fled north, when freedom sometimes meant leaving behind those you loved.

In some ways, the Harriet Tubman Memorial Garden seems like a humble place to begin a journey commemorating one of America's most inspirational figures. Just off the roadway on the outskirts of Cambridge, Maryland, and nearly hidden by oak trees, a small monument pays homage to the woman some called Moses for her role in leading so many people to emancipation. And yet, Tubman probably wouldn't have wanted it any other way. She didn't seek fame or fortune during her lifetime, but rather freedom for those she loved and equality for all people regardless of their skin colour, gender or creed.

Despite standing just 157cm tall (5ft 2in), Tubman became one of the giants of American history. Her great-great-great-nephew, Charles Ross, painted the mural found here. With a hopeful dawn blazing behind her, she faces the viewer, with a quiet

© COURTESY HARRIET TUBMAN BYWAY

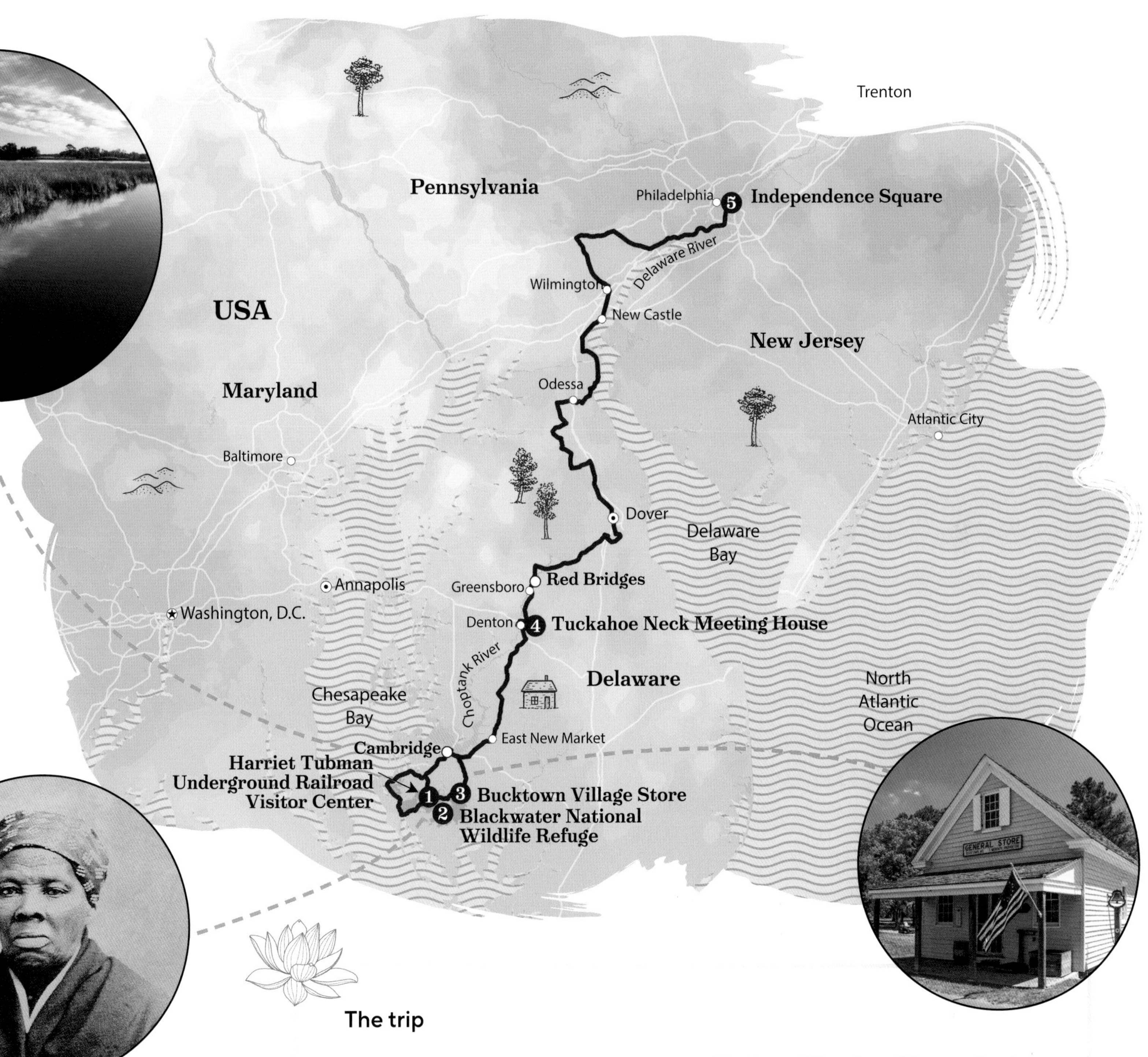

The trip

Distance: 400km (249 miles)
Mode of transport: Car
Difficulty: Easy

1 Harriet Tubman Underground Railroad Visitor Center Learn about Tubman's life, then take time to reflect in the meditation garden

2 Blackwater National Wildlife Refuge Walk amid wildlife-filled wetlands unchanged since Tubman's time

3 Bucktown Village Store Tubman suffered a traumatic injury here, which caused her pain throughout her life

4 Tuckahoe Neck Meeting House Historic gathering spot for abolitionists and supporters of the Underground Railroad

5 Independence Square Reaching Philadelphia brought freedom, but also sorrow for those left behind

Clockwise, from left: Byway Visitor Center; Blackwater Refuge; Bucktown Village Store; Harriet Tubman

Shaniya Jackson, Byway roadtripper

I was going through a rough patch when my sister insisted the two of us make the trip along the Byway. It turned out to be one of the best experiences we've had together. I felt so moved by what Harriet did, her bravery, and how she risked her life saving those she loved. In a big way, this trip helped me be more awake to the people I care about in my life and not to take for granted our time here together on this planet.

unshakeable resolution etched in her expression. It's a proud gaze that has come to represent not only her own story, but that of so many others who faced oppression – and whose lives are commemorated along the Harriet Tubman Underground Railroad Byway, a 400km (249-mile) self-guided driving tour through Maryland, Delaware and Pennsylvania that connects 45 sites and museums related to her life and work.

The horrors of enslavement

A short drive from the garden, the road crosses Cambridge Creek Drawbridge into the sleepy downtown of Cambridge. Though the original building burned in a fire in 1852, a courthouse here still marks the spot where enslaved people were auctioned off to the highest bidder. In 1850, one year after Tubman escaped to freedom, her niece Kessiah was placed before buyers and the bidding began. Through luck and cunning, Kessiah's freed husband managed to whisk her and their two children into hiding, and later take her by boat to Baltimore. There Tubman, who had helped in the planning of their escape, met them and guided the young family on to Philadelphia and later to freedom in Canada.

A few blocks northeast, High St ends at the waterfront. Sailboats bob along the basin, against views of the Rte 50 Bridge stretching across the Choptank River. But spin the clock back to the early 1800s and the idyllic scene becomes a hellscape. Large ocean-going vessels dropped anchor here and unloaded their cargo. Hobbling in shackles off the boats, kidnapped men, women and children – the

> "Tubman's proud gaze has come to represent not only her own story, but that of so many others who faced oppression"

ones who survived the voyage (millions died along the way) – reached the end of their journey, and the beginning of a harrowing new existence in an unfamiliar world.

Exhibitions & an unchanged wilderness

Some 18km (11 miles) south, in Church Creek, stands the showpiece of the Byway. Opened in 2017, the sustainably designed Harriet Tubman Underground Railroad Visitor Center has interactive exhibitions that delve into Tubman's many rescue missions. The museum also sheds light on her involvement in the Civil War as a spy, nurse and battle planner (when she became the first woman in US military history to lead an armed raid), and her fight for women's rights in later years.

Nearby, the 113 sq km (43 sq mile) Blackwater National Wildlife Refuge draws birdwatchers and hikers to its wetlands, open fields and mixed forests. In the 19th century, however, this nature sanctuary was a place of untold suffering for the enslaved people sent to work here – like Minty Ross (as Tubman was then known), aged seven, sent barefoot into the marshes to set traps for muskrats, whose pelts would fetch high prices when sold up north. She toiled in the bitter cold of winter without proper clothing. Despite the hardship, Tubman also gained valuable knowledge that she would put to use in later years. As she worked alongside her father in the forests, she learned important survival skills from him: how to navigate by the stars, read a landscape and travel safely through seemingly inhospitable terrain. Though she never learned to read or write, Tubman deployed her knowledge of the wilderness to save many lives. While walking the paths here, it's worth remembering that we all have gifts to share.

Suffering & hope

East of the Blackwater Refuge, tiny two-lane Greenbriar Rd leads past the farmlands where Tubman lived her early years. A roadside layby overlooks what was once the Brodess Plantation,

© F11PHOTO / SHUTTERSTOCK

© JILL JASUTA / COURTESY HARRIET TUBMAN BYWAY"

Clockwise, from above: Independence Hall, Philadelphia; Blackwater National Wildlife Refuge, Maryland; engraving from William Still's *The Underground Railroad*

Remarkable escape

In October 1857, two astonishing escapes occurred on Maryland's eastern shore, baffling owners of enslaved people and civil authorities. Over a three-week period in 1857, two large groups of enslaved people, some 44 in all, made a run for freedom not far from Cambridge. Comprising five families, the self-liberators had to endure rain and snow in ragged clothing, and some even lacked shoes. They travelled by night and only narrowly avoided capture by bounty hunters near Wilmington. Incredibly, there were a large number of children among them. Daffney and Aaron Cornish escaped with six of their sons and daughters, one of whom was just two weeks old; they had to leave behind two of their boys who had been hired off the farm. Both Black and white underground railroad agents helped in the daring escape, and every one of them ultimately made it to freedom in Canada.

where she may have been born. Though the plantation no longer stands, the unchanged fields provide a quiet spot to contemplate the past and perhaps to reflect on one's own life.

Further up the road, the Bucktown Village Store provides a portal into the 19th century. A pot-bellied stove sits in the middle of the restored shop, its shelves lined with crockery and tins above a countertop with a scale for weighing tobacco and other goods. It was here, in the 1830s, that Tubman, aged 13, was caught in the middle of a dispute between an enslaved man and his overseer. In a rage, the overseer threw an iron weight at his captive, which missed. Instead, he hit Harriet, fracturing her skull and nearly killing her. As a result of the brain injury, Tubman suffered a lifetime of blinding headaches, seizures and insomnia. She also had visions, which she later attributed to God helping to guide her freedom missions. The site provides another example of Tubman's unshakable resolve, and gives visitors the opportunity to examine their own challenges, and perhaps, as she did, to nurture hope in spite of hardships.

Continuing north, the town of Denton represented both promise and horror in the race to freedom. The courthouse (reconstructed after the Civil War) was where runaways and Underground Railroad conductors were held after being captured. Just under 3km (2 miles) northwest, the Tuckahoe Neck Meeting House was a key gathering place used by Caroline County Quakers in the 19th century. Members here played a vital role in supporting the Underground Railroad, and some Quakers operated safe houses for escapees.

Another half-day's walk (or 15 minutes by car), and the journey through Maryland ends near the peaceful stream at Red Bridges. Here, among birdsong and swaying poplars, the Choptank River is shallow enough to cross on foot, which was the preferred route Tubman took as she led her charges into Delaware.

Northward to freedom

From here, it's another 170km (106 miles) up through Delaware and into Pennsylvania, the stepping stone to emancipation for so many making the perilous journey north. In Philadelphia, abolitionists gathered on Independence Square in front of Independence Hall, where years earlier the US Constitution had been debated and signed. Ironically, the document espousing liberty did not apply to enslaved people. Instead help came from unsung heroes like William Still, a leading Black abolitionist who helped finance Tubman's rescue missions to the south.

Still's last residence (now a private home) marks the final stop of the Byway. It was here that he wrote his monumental work *The Underground Railroad*, published in 1872, which commemorates the courageous people who risked everything when fleeing slavery, and those who helped along the way. Among the many lessons that might be inspired by the lives of Still, Tubman and so many others of their era is that there are things each of us can do, both small and large, to help those around us. The first step is deciding to make a difference.

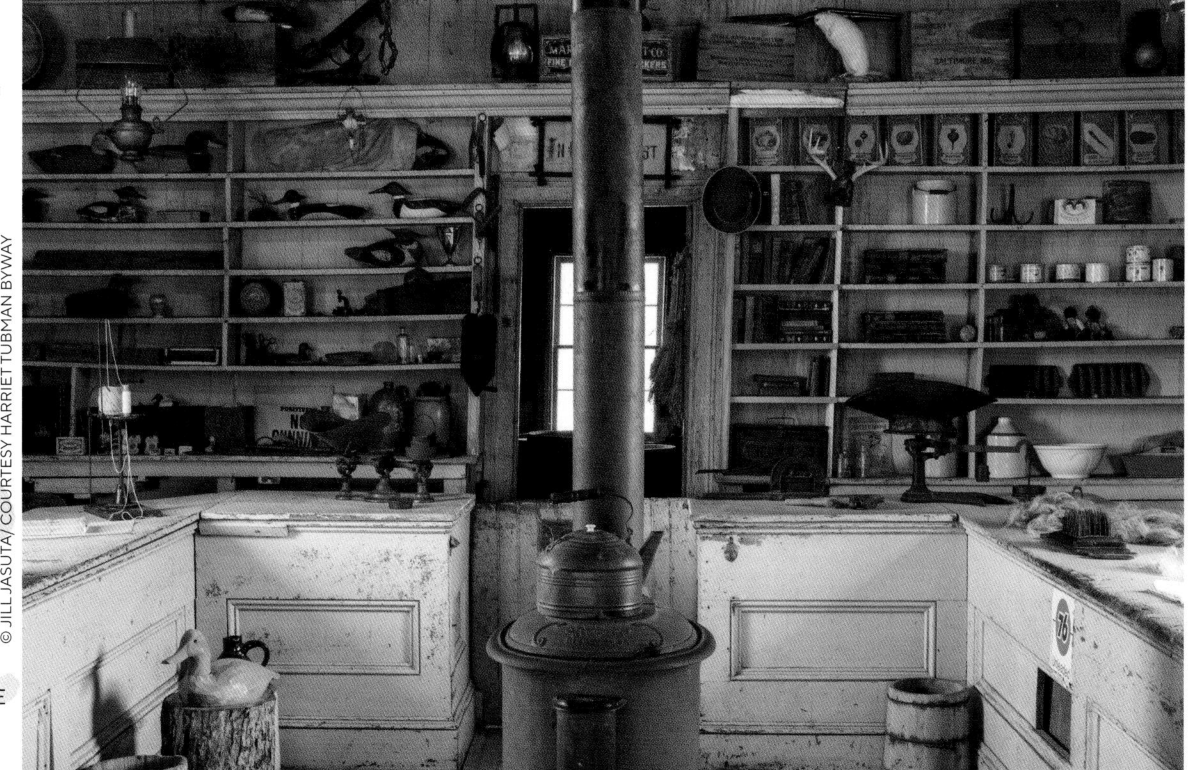

© COURTESY CAROLINE COUNTY TOURISM

From top: Tuckahoe Neck Meeting House; the period interior of Bucktown Village Store

Practicalities

Region: Maryland/Delaware/ Pennsylvania, USA
Start: Cambridge, Maryland
Finish: Philadelphia, Pennsylvania

Getting there and back: Washington, DC, and Baltimore are handy gateways to the region. From either city, it's a 2hr drive to Cambridge via the Chesapeake Bay Bridge. At the finish in Philadelphia it's an easy journey back to Baltimore or DC by car, bus or train.

When to go: The Byway is accessible year-round, though (rare) snowstorms may lead to road closures and frigid temperatures in January or February. Keep in mind the heat, humidity and mosquitoes of summer (June to August) – though travelling in difficult seasons can give a sharp appreciation of the many challenges that ill-equipped freedom seekers faced throughout the year.

What to take: Apart from water and snacks, be sure to pack a good biography of Harriet Tubman to read along the way. Kate Clifford Larson's outstanding *Bound for the Promised Land* is a page-turner and is also available as an audiobook – perfect for the road trip.

Where to stay: You'll find a good range of lodging options along the Byway, with a mix of B&Bs, chain hotels and simple inns scattered across the eastern shore and Delaware. Philadelphia offers even more variety, though prices are higher, especially for city centre lodging. It's wise to book ahead during the summer months.

Where to eat and drink: Bigger towns such as Cambridge, MD, and Wilmington, DE, have a decent selection of restaurants and grocery stores. Pick up picnic supplies before venturing to more remote areas like the Blackwater National Wildlife Refuge.

Essential things to know: There's plenty of background on the Byway's website (harriettubmanbyway.org); before visiting the Maryland section, download the Timelooper Harriet Tubman app (iOS and Android), which has an excellent audioguide.

Cycling the Great Divide Mountain Bike Route

Traversing the spine of North America on a bicycle, from Jasper in Canada to Antelope Wells in New Mexico, will test mind, body and spirit.

Meeting the mental challenge of cycling such a long way and relying on your own resilience may give you more confidence in other areas of your life. And the necessity of meeting and building a rapport with people you meet along the way may help grow social skills. If you're riding in a group, negotiating interpersonal dynamics will be essential – you will each have good and bad days, so supporting and tolerating each other will be key.

The physical skills to be gained on this journey are not about the fitness required (that will come) but more about the organisational experience of planning a day, setting up camp efficiently, eating and drinking enough, and looking after yourself.

Some days, the fatigue will feel crushing and you'll question why you embarked on such an adventure. The wind might push you backwards and mud might suck at your wheels. You might feel cold, hungry, tired or fearful. But, each time, you'll pass through the dark patch and light will dawn again – some friendly assistance, an awe-inspiring view – and hope will be restored as you get a step closer to completing the Great Divide Mountain Bike Route.

The GDMBR was designed by America's Adventure Cycling Association and inaugurated in 2007, when it was the longest offroad cycling route on the continent. The route connects Banff in Alberta, Canada, surrounded by glacial lakes, snow-tipped peaks and pine forests, with Antelope Wells in New Mexico, a land of sculpted sandstone and cactus-spiked desert. It follows mostly dirt or gravel roads along the Great Divide, which separates the eastern

© CALLUM HAYES

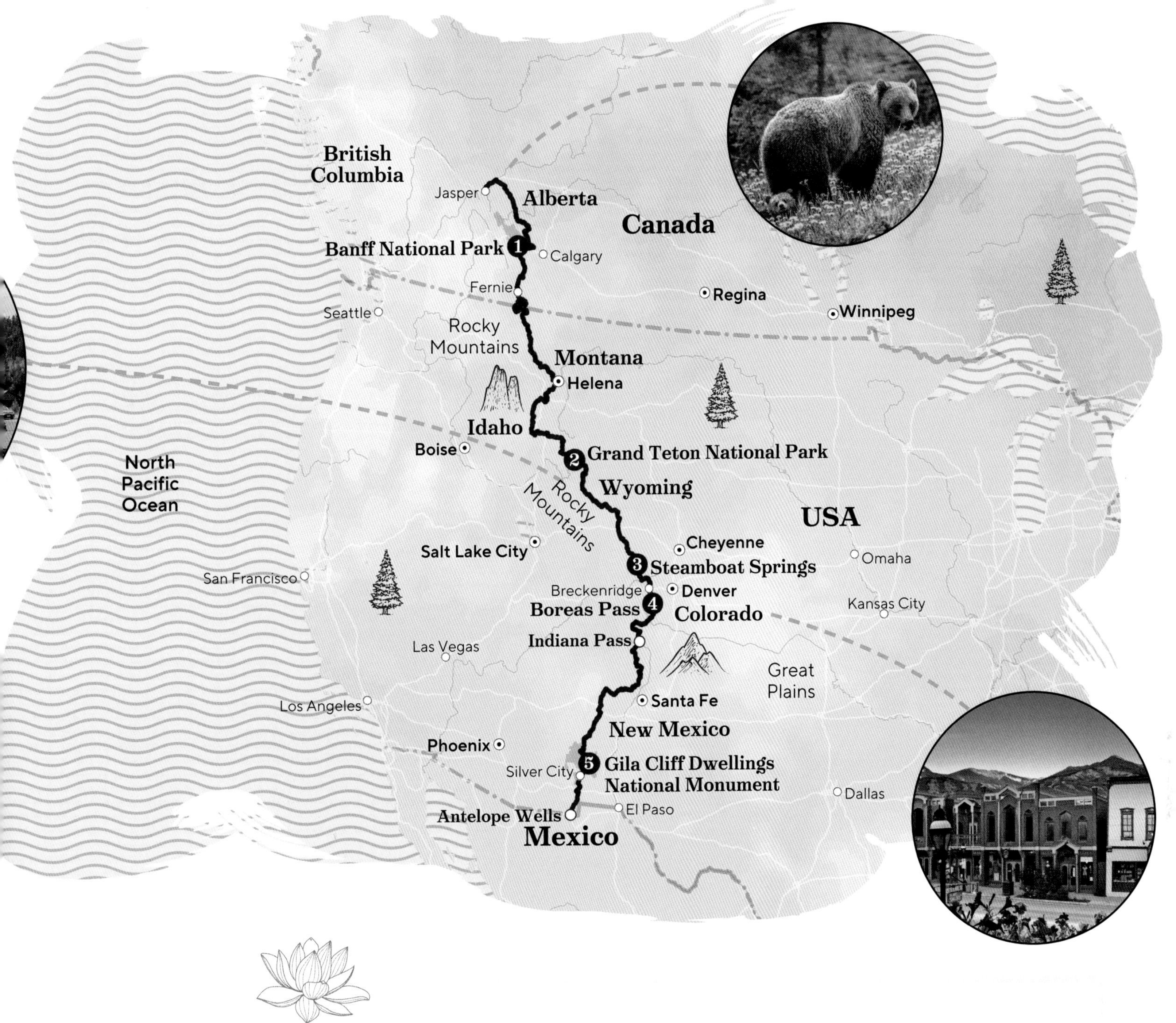

The trip

Distance: 4973km (3090 miles)
Mode of transport: Mountain bike
Difficulty: Hard

1 Banff National Park, Canada Beautiful lakes and big bears abound here, but you'll soon leave the comforts of this tourist-friendly town

2 Grand Teton National Park, Wyoming The giant peaks inspire awe and anxiety, but you'll take passes long used by Native Americans

3 Steamboat Springs, Colorado Discover the mountain-bike culture of this low-key mountain town, which has less of the glitz of other alpine resorts in Colorado

4 Boreas Pass, Colorado This is one of the route's highest and most scenic passes

5 Gila Wilderness, New Mexico Explore Native American culture in this region, where ancient cliff dwellings remain

Clockwise from left: riders and railroads in New Mexico; Lava Mountain Lodge in Wyoming: beware of Jasper's grizzlies; Breckenridge, Colorado

Callum Hayes, bikepacker

The four of us hadn't done any bikepacking before and our preparation went as far as watching some YouTube videos of the route, though we all had mountain-biking backgrounds. The first few days were definitely a learning curve and riding a bike for nine hours a day was tough. The journey showed me how much I can push myself and made me feel a true sense of freedom. You realise how little you really need to create life-long memories. We never expected so much kindness and so many people willing to help us out. We couldn't be more thankful.

and western watersheds along the length of the US. This means that you'll travel through several distinct geographical zones and gain an appreciation of climate, geology and ecology.

Mountain biking is almost the perfect pace at which to journey across a landscape. It's speedy enough that you can cover several hundred miles in a week; and sufficiently sedate that you can smell the flowers, take a break to listen to the birdsong in the trees and meet people along the way. As Callum Hayes, who rode the GDMBR in 2022 with three friends, puts it: 'Whenever we stopped anywhere, passers would chat with us and ask what we were doing, where we'd come from, or if they could help us with anything – conversations like these gave us extra motivation during the day.'

At almost 5000km (3000 miles), the GDMBR is an enormous undertaking. With so much of the route through wilderness, it's not easy to break it into sections either: like the Appalachian Trail, the most satisfying approach is to block out two months and tackle the trip in its entirety as a bikepacker, carrying everything on your bike that you require to survive. This can be, at times, the Adventure Cycling Association suggests, a spiritual experience. But it will demand some cycling experience. Although the GDMBR was Callum Hayes' first bikepacking trip (and about the biggest initiation he could have picked), he had a lot of mountain biking experience. And, as yet, the GDMBR is not set up for e-bikers, meaning that unfortunately it is not suitable for some groups of people. The hazards are real: the wildlife includes black and brown bears, although greater risk will come from exposure to bad weather at altitude.

© CALLUM HAYES

"We never expected so much kindness and so many people willing to help us out in any situation we found ourselves in"

Most people ride north to south, for several reasons, but not least that Canada's Rocky Mountains, despite their epic scenery, offer more amenities than the more arduous and remote sections of desert in New Mexico. If cycling through the national parks of Jasper then Banff can be said to be a soft introduction, they're a good way to iron out any issues. 'Within the first 350km [220 miles], we snapped bottle cages, lost water bottles, and our bags bounced off on every descent,' admits Callum Hayes.

After a farewell to Canada in Fernie, you'll cross the US border and head towards Helena, the state capital of Montana. It's about now that the enormity of the ride ahead sinks in, and it's easy to feel overwhelmed by both the surroundings and the task ahead. Thankfully, most GDMBR riders report receiving a warm and generous welcome from the communities through which they pass. Local people have been known to offer food, a porch to sleep on, or just words of encouragement. Experiencing such hospitality is always a revelation and the best response is simply to pass on the good vibes.

Tough tests await in the Badlands of Wyoming – vast basins of straight roads, strong headwinds and no drinking water. Then the route starts climbing again, high into the Colorado Rockies. You'll quickly learn to listen for summer thunderstorms and descend to safer levels if one is approaching. Mountain towns, from the simple style of Steamboat Springs to the bright lights of Breckenridge, offer welcome respite from the trail before you return to the aspens and alpine meadows.

Once over Indiana Pass, at 3650m (12,000ft), the route descends into New Mexico and becomes more rugged. Expect more mud, rocks and steeper gradients and less water and fewer people. Just before Silver City, stop at the Gila Cliff Dwellings National Monument to explore Ancestral Puebloan history and culture. The people of this ancient culture created roads, communication systems and complex dwellings. It's a reminder that you're neither the first nor the last people to pass through this country.

© TRAVELLER70 / SHUTTERSTOCK

From top: the Gila Cliff Dwellings National Monument is a must-visit stop in New Mexico; riding the gravel roads of Colorado's backcountry

Practicalities

Region: North America
Start: Jasper, Alberta, Canada
Finish: Antelope Wells, New Mexico, USA

Getting there and back: The closest airport is Calgary International, which is about 500km (310 miles) southeast of Jasper. With bikes, the easiest way to get to Jasper is by renting a car one-way if you're a group or booking one of the bus services (bikes must be boxed). You will certainly need a shuttle service from Antelope Wells to either El Paso, Texas or Tuscon, Arizona – two or three firms operate.

When to go: With the extreme terrain, the riding window tends to open in late June and close in early October. Most people ride north to south due to prevailing conditions, but it is possible to start earlier from the south and still finish in Jasper before the snow.

What to take: There are many online accounts of what bikepackers have preferred to carry on the Great Divide. Racers will want to travel fast and light, but those taking their time may not cut back on their baggage as much. As a start, most people will consider navigation, shelter, security (such as emergency beacons), food and water, and a few lightweight luxuries. A means to recharge batteries and a bicycle with a dynamo are useful.

Where to stay: The route passes plenty of camping possibilities in Bureau of Land Management property, national parks and forests, and in the wilderness. Towns at regular intervals also offer the opportunity of a warm shower and a soft bed. Plan ahead rather than take a chance during the summer season.

Where to eat and drink: Carrying sufficient food and water for a few remote stretches will be an important concern. It may be 160km (100 miles) before the next resupply spot, but most days you will be near at least one gas station. Carry a water purification method and the means to carry a day's water. And remember to drink and replace calories hourly.

Essential things to know: Research further at bikepacking.com and adventurecycling.org.

Hike across Canada's smallest province on the Island Walk

This epic coastal route around Prince Edward Island pairs leisurely glimpses of local culture with a multiday physical challenge.

Increasing your regular activities to train for this adventure, and hiking daily on the trail, allow you to build stamina and physical strength.

A long-distance walk isn't only about the physical challenge of putting one foot in front of the other, day after day. When mosquitoes swarm, rainstorms drench you or you realise you should have refilled your water bottle back in the previous town, you may appreciate the importance of staying mentally strong.

Meeting islanders along the route – from fishers and farmers to musicians and artists – offers insight into PEI's seafaring heritage and rural culture.

Not everyone who hikes or cycles the Island Walk considers their journey a pilgrimage. Some choose to take on sections of this 700km (435-mile) trail around Prince Edward Island to enjoy the brisk ocean air, sandy beaches, forests and red-rock cliffs, or to stop for fresh seafood and leisurely chats. Yet for others, circuiting Canada's smallest province is a challenge, an adventure or a personal quest.

The route was the brainchild of Bryson Guptill, a PEI resident and volunteer with the non-profit Island Trails organisation, which develops and maintains local walking paths. After walking sections of the Camino de Santiago pilgrimage route in Spain and France, Guptill wondered why there wasn't a similar long-distance path around his home province. The Island Walk route that he designed is modelled on the Camino, following existing trails, rural roads and larger roadways, and

© ELENA ELISSEEVA/ / 500PX

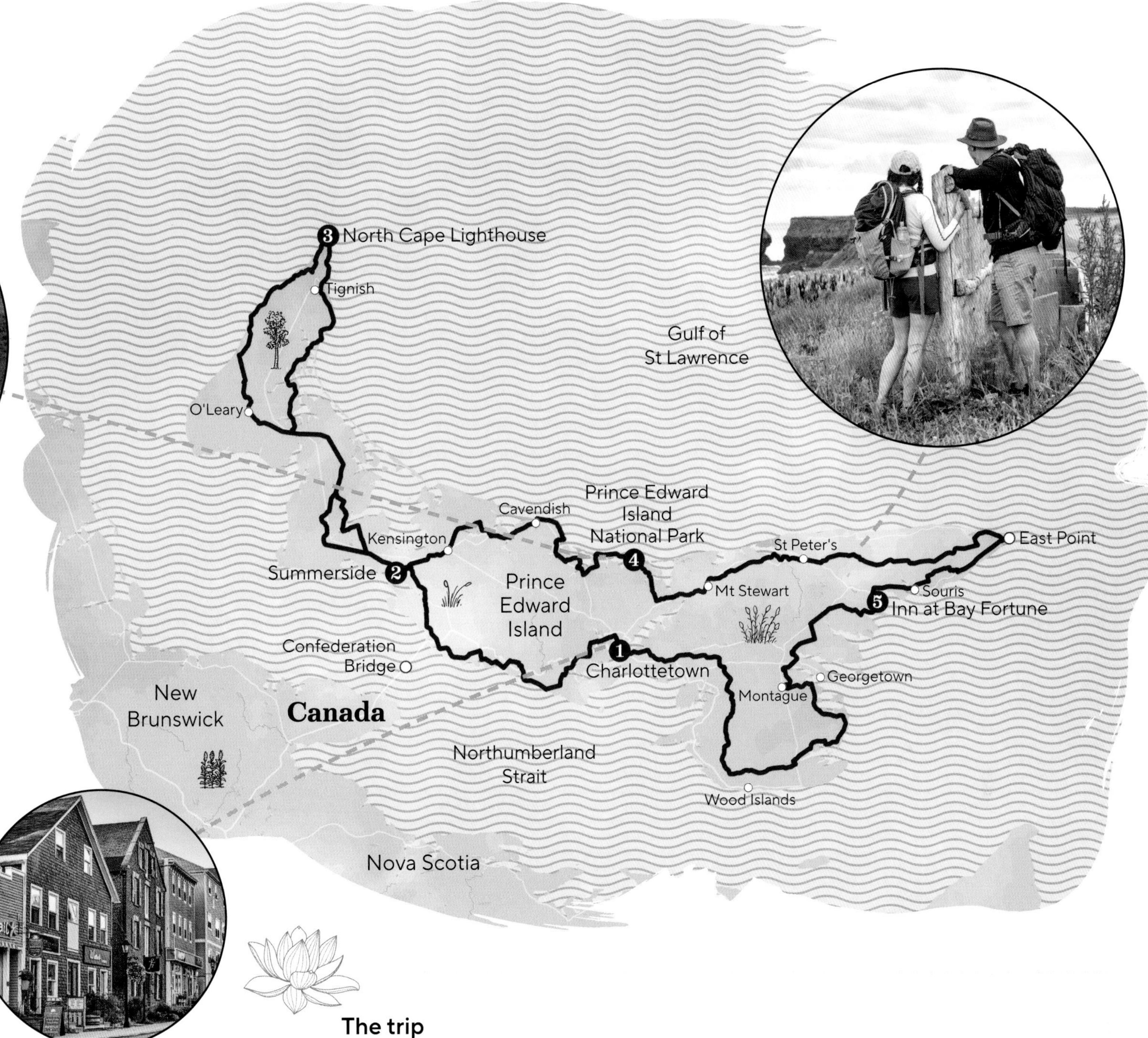

The trip

Distance: 700km (435 miles)
Transport: On foot or bicycle
Difficulty: Moderate

1 Charlottetown The Island Walk starts and ends in Prince Edward Island's waterfront capital

2 Summerside This centrally located community has historic homes, harbourside eateries – and a fox museum

3 North Cape Lighthouse Hike past red cliffs that descend sharply to the ocean to reach this historic lighthouse

4 Prince Edward Island National Park Several Island Walk segments take you through PEI's scenic seafront reserve

5 Inn at Bay Fortune Recharge with a gourmet meal at this upscale eastern PEI inn

Clockwise, from left: sunset over the PEI coast; on the trail through Prince Edward Island National Park; hiking the Island Walk; Charlottetown

Sahari Ruiz, Island Walk hiker

The first few days I was really excited. Then somewhere around day three or four, it hit me – I'm going to be doing this every single day. Wow. Overwhelmed. Some days wear on you. Sometimes my feet hurt a lot. Or other parts of my body were saying, 'I don't want to walk today. I want to rest.' But you have to go through the good days and bad days. Like anything else in life.

divided into 32 sections of 20km to 25km (12–15 miles), each comprising a reasonable day's hike for a moderately fit walker. Cyclists can combine segments or simply take their time for a leisurely pedal.

Hiking the Island Walk

The route begins and ends in provincial capital Charlottetown. Leaving the city, it meanders west along the south shore, with views of the Confederation Bridge, which connects the island to the mainland. It continues into Summerside, PEI's second-largest city, where you can stop to tour historic houses or visit the quirky International Fox Museum, which documents the island's previously thriving fox-rearing industry.

Walking towards PEI's increasingly rural west end, you'll pass O'Leary, where the Canadian Potato Museum highlights an abundant local crop. (The restaurant serves spuds in many forms, plus a unique-to-PEI seaweed pie.) Towards the island's northwest tip, the trail follows a boardwalk through a forest of wind turbines to the North Cape Lighthouse. In the nearby Wind Energy Interpretive Centre, you can learn why you've walked past so many whirling windmills.

The town of Tignish makes a good base on the island's west side, with a historic inn, several restaurants and PEI's largest church, its steeple extending 56m (185ft) high. Beyond Tignish, the walk returns through Summerside before turning north to Cavendish, childhood home of the writer Lucy Maud Montgomery; her early 20th-century *Anne of Green Gables* novels, recounting the adventures of a feisty red-haired island resident, have become Canadian classics. The route then takes you into Prince Edward Island National Park, with its rust-red cliffs and sandy beaches along the north shore.

© DEATONPHOTOS / SHUTTERSTOCK

"I went through all the emotions that you can go through when you're doing hard things... but we had to push through"

One of the most scenic sections follows the water between Mt Stewart and St Peter's via a stretch of the multiuse Confederation Trail. Beyond St Peter's, the trail goes into the woods; walking this section in autumn promises stunning fall colours as the leaves start to change. There are equally striking vistas along the shoreside section that circles East Point.

The Island Walk doesn't have lodgings or food at each segment's end; some sections start or finish in a town, but many are bookended simply by roadside markers. This might mean walking an extra 3km to 5km (2–3 miles) from the section's endpoint to your accommodation. Some people drive over to PEI in an RV; others carry camping gear, overnighting in a mix of campgrounds and B&Bs; some stick to inns or hotels. You can also work with a local outfitter to plan where to stay, and arrange transportation to and from the trail; innkeepers can sometimes provide transfers for a fee.

Trials & triumphs

But however you structure your days on the trail, this long-distance route will likely inspire you in all kinds of ways, offering plenty of time to appreciate both the highs and the lows of walking for days on end. 'I went through all the emotions that you can go through when you're doing hard things,' says Sarahi Ruiz, who hiked the full trail over 34 days with her partner. 'Some days were really hard, but we had to push through.'

For walker Laura MacGregor, travelling to PEI was a chance to consider the next stage of her life. After the death of one of her sons, she spent 31 days walking the trail, describing her hike as 'an intentional pilgrimage, an opportunity to step off the track of life and do some reflection'. Like many who complete this long-distance hike, she experienced it as a 'rare opportunity to experience Canada in a wonderful, meaningful way... slow tourism, national treasures, time alone, and even thankless roads. Because sometimes, you know, life sucks. [But] you don't need to be a hardcore hiker to enjoy a sustained walk. A pilgrimage starts when you step on a trail, and it finishes when you step off a trail. A pilgrimage is anything you make it to be.'

© DARRYL BROOKS / SHUTTERSTOCK

Clockwise, from left: Charlottetown, PEI's waterfront capital; a slice of the Confederation Trail; Green Gables Heritage Place, Cavendish

© V J MATTHEW / SHUTTERSTOCK

Practicalities

Region: Prince Edward Island, Canada
Start/Finish: Charlottetown

Getting there and back: Confederation Bridge connects PEI to New Brunswick and Nova Scotia, and buses travel between the three provinces. From Caribou, NS, ferries make the 1hr 15min crossing to PEI's Wood Islands. Charlottetown's airport is around 6km (4 miles) from the city centre. T3 Transit runs a limited schedule of buses across PEI.

When to go: The best months are September and early October, when the days turn crisp, the leaves change colour, and the mosquitoes are less prevalent. June is another good option, though you may have a greater chance of rain and more bugs. In the peak summer months of July and August, accommodations are busier, there's more traffic on the roads and days can be hot. Avoid the snowy season between November and April.

What to take: Always carry food and water; you won't find places to eat or drink everywhere on the route. Bring rain gear and insect repellent; and as toilets can be few and far between, packing toilet paper (and bags to carry out your waste) is advisable.

Where to stay: Organise lodgings in advance. While PEI has B&Bs, inns, hotels and some camping options, they're not located at the ends of every Island Walk section. The Island Walk website lists accommodations that especially welcome walkers or cyclists.

Tours: Experience PEI (experiencepei.ca) plans Island Walk itineraries and arranges accommodation, transportation and luggage transfers. MacQueen's Bike Shop (macqueens.com) offers similar services for cyclists. Outer Limits Sports (ols.ca) provides walking and cycling packages.

Essential things to know: The Island Walk website (theislandwalk.ca) and active Island Walk Facebook page are the best sources of information for planning your adventure. Tourism PEI (tourismpei.com) has lots of general island info, too.

Memphis to Montgomery: an American Civil Rights journey

Visionary leaders and nonviolent activism powered the Civil Rights Movement, a fight for racial equality centred in the Deep South. Trace the history where it happened.

Historic sites can trigger unexpected emotions, from sadness and anger to empathy and hope. Hard-hitting exhibits recreate the realities of the Jim Crow era of segregation; the impact is visceral, and one you won't get from a textbook.

Civil Rights museums curate a vast story: the fight by Black Americans for justice and equality, ever since the first ship carrying enslaved people landed on these shores. Exhibits provide context while tracing daily oppression and illuminating the power of peaceful resistance. Tour guides often have links to historical events.

A white wreath on the second floor of the Lorraine Motel in downtown Memphis draws the eye towards Room 306, marking the spot where Dr Martin Luther King, Jr was assassinated on April 4, 1968. Dr King was one of many Black Americans who challenged segregationist Jim Crow laws in the 1950s and '60s. Their activism transformed the American South, and a journey between key Civil Rights sites, from

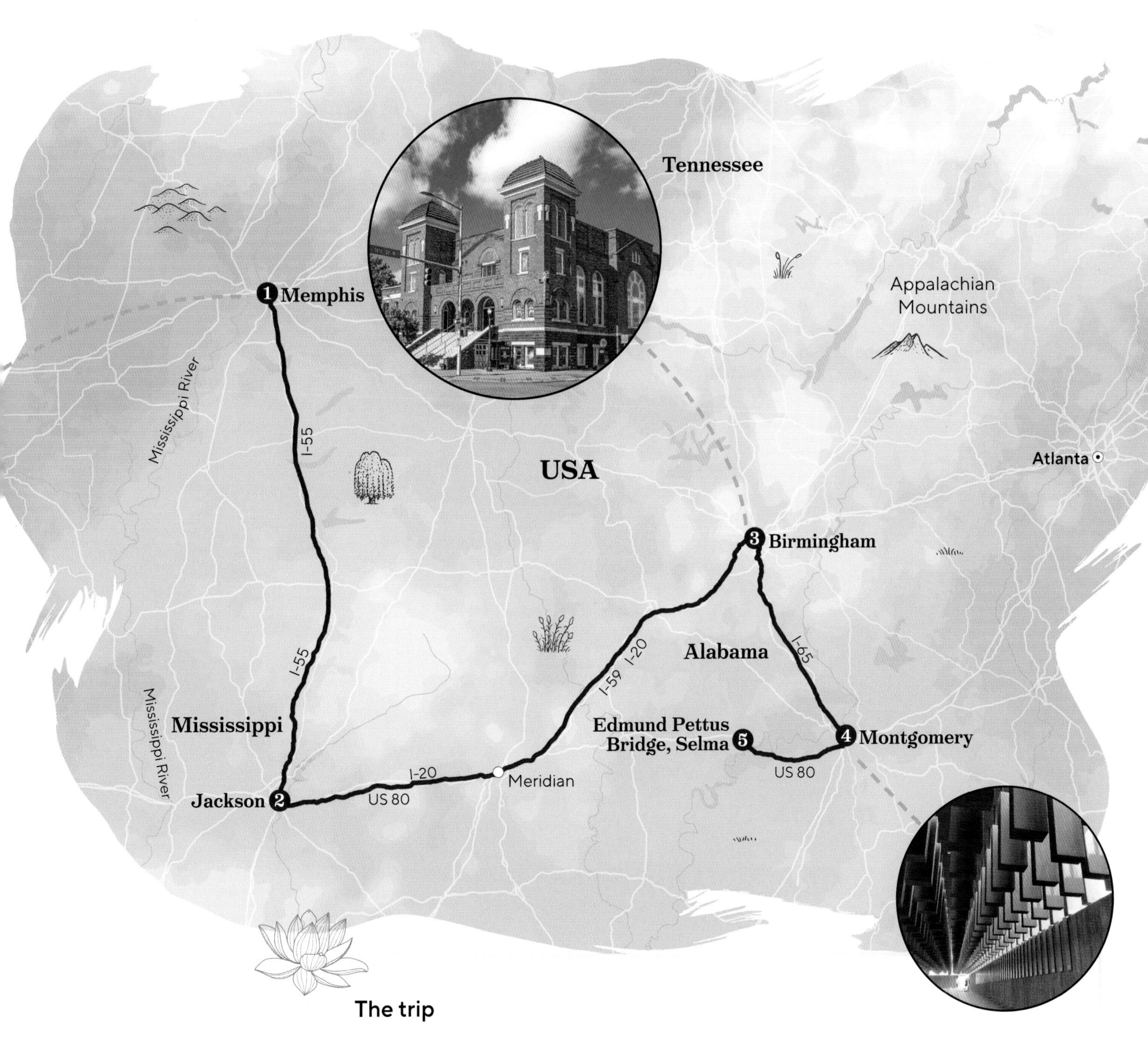

The trip

Distance: 966km (600 miles)
Mode of transport: Car
Difficulty: Easy

1 Memphis, Tennessee Home of the National Civil Rights Museum and its Lorraine Motel, and the Stax Museum of American Soul

2 Jackson, Mississippi Learn about cotton production, enslavement and activists Medgar and Myrlie Evers

3 Birmingham, Alabama Explore a history of violence and segregation at the Birmingham Civil Rights Institute

4 Montgomery, Alabama Museums and memorials spotlight the city's bus boycott and horrific history

5 Edmund Pettus Bridge, Selma Reflect on the struggles that brought about the 1965 Voting Rights Act while crossing this iconic bridge

Clockwise, from left: march from Selma, 1965; Lorraine Motel; 16th St Baptist Church, Birmingham; National Memorial for Peace and Justice, Montgomery

Keena Graham, Medgar and Myrlie Evers Home superintendent

In our country, whenever Black or Brown people get memorialised, they're memorialised because of their death, not because of their life... [We want] to present how day-to-day racism affected people because that was more what everybody dealt with, that was more of a reality than the assassination. So what was systemic day-to-day racism like? Because lots of times in our heads, if somebody's not being beaten every day, then it's not racism.

Memphis to Selma, drops you into this compelling history – and the ongoing story. The Lorraine Motel is the emotional heart of Memphis' National Civil Rights Museum, its iconic mid-century facade evoking an era marked by discriminatory laws and episodes of horrific violence. Steps away, interactive and immersive exhibits – a Montgomery city bus, a segregated lunch counter – viscerally conjures up the realities of the era.

A 10-minute drive from the museum, there's another side of Memphis history to absorb at the Stax Museum of American Soul Music, celebrating the stories and sounds of African American icons from Otis Redding to Isaac Hayes.

Following the Freedom Riders

From Memphis, it's a three-hour drive south to Jackson, a powerful second stop. At the Mississippi Civil Rights Museum here, black-and-white mug shots of the Freedom Riders soar over Gallery 5, showcasing the effectiveness of nonviolent group protest and underscoring the bravery of these individuals, who helped desegregate buses and bus terminals in 1961. Hezekiah Watkins, one of the youngest Freedom Riders, is on staff at the museum. 'He shares his first-hand experience of being arrested and then held on death row at the state prison,' says Pamela DC Junior, museum director. 'His story has a powerful impact on students and adults.'

Mississippi became an epicentre of activism due to its history as a Southern 'slave state', and as a major cotton producer in the mid-1800s; it was later marked by violent resistance to desegregation. As visitors

© F11PHOTO / SHUTTERSTOCK

> "Black-and-white mug shots of the Freedom Riders showcase the effectiveness of nonviolent group protest"

step into Gallery 3, the museum's inspiring hub, a song – *This Little Light of Mine* – grows louder while a luminous light sculpture brightens, symbolising the power of community. 'We want to remind visitors that we all have a light to shine,' says Junior, 'just like the people who were part of the movement.'

Everyday people effecting change

From the museum, it's a short drive through downtown Jackson to the Medgar and Myrlie Evers Home National Monument. Marked by its mint-green exterior, the home of these Civil Rights icons was filled with joy and love – and it's the Evers' everyday lives that Keena Graham, the site's superintendent, wants to share during tours of the home. As well as serving as an important gathering place for movement organisers (Medgar was the first field secretary for the Mississippi NAACP; Myrlie would later become the NAACP national chair), it's the place where the Evers' children gathered blackberries in the yard, and where neighbours dropped by to hang out.

Medgar Evers was assassinated at the house in 1963, gunned down in the carport while his family waited for him just inside. But Graham wants the site to be remembered for more than just this one tragic event. 'This home may look like it's this seemingly ordinary place, but so many great things happened here. It was the nerve centre for the Jackson movement, and people like James Baldwin visited. Sonia Sanchez. Lena Horne. James Meredith used the home to call Thurgood Marshall so that he could plan his strategy on how to get into the University for Mississippi.'

The story of four girls

From Jackson, drive around 230km (143 miles) northeast for a similarly visceral glimpse into history. The Birmingham Civil Rights Heritage Trail begins near 16th St Baptist Church, a former hub for activists and now a stop on a seven-block walk through a city that played a crucial role in the movement. Churches were often one of the few places where Black people

Clockwise, from above: Montgomery's Legacy Museum and National Memorial for Peace and Justice; the Lorraine Motel, part of Memphis' National Civil Rights Museum

Detour to the Delta: the story of Emmett Till

In 1955, two white men abducted 14-year-old Emmett Till, a Black boy from Chicago, from his great-uncle's house in Glendora, Mississippi. Till had allegedly flirted with one of the men's wives at a general store in the small town of Money, and his mutilated body was later found in the Tallahatchie River. At the murder trial in Sumner, an all-white jury acquitted the men, who later confessed to *Look* magazine. Till's mother, Mamie Till-Mobley, insisted on an open casket at her son's funeral so that the world could witness the atrocities done to him. His murder triggered national condemnation and galvanised the modern Civil Rights Movement. In Glendora, the Circle of Truth exhibit at the Emmett Till Historic Intrepid Center tells Till's story. 'It's thought-provoking,' says the centre's executive director, Johnny Thomas. 'We're hoping to create a dialogue and to begin the healing.'

could legally gather in the Deep South, and pastors regularly became Civil Rights leaders.

At 10:22am on September 15, 1963, four young Black girls were killed by a bomb that exploded in the church basement while they attended Sunday School. Twenty others were injured. Across the street, a timeline in the Birmingham Civil Rights Institute museum traces the decades-long effort to bring the perpetrators to trial. Elsewhere, various exhibits – including the bars from the cell where Dr King drafted his *Letter from a Birmingham Jail* – share the city's history of segregation and its importance during the Civil Rights era.

The public are welcome to attend Sunday services at 16th St Baptist Church, which reopened in 1964, and you can also tour the building.

A long legacy in Montgomery

Some 154km (90 miles) south of Birmingham, the movement's story continues in Montgomery, the capital of Alabama where, in 1955, Rosa Parks refused to move to the back of a segregated city bus. Her refusal was the impetus for an ensuing year-long bus boycott, a successful and carefully planned large-scale protest that ultimately saw Montgomery's buses desegregated, and marked a foundational moment in the Civil Rights Movement.

In downtown Montgomery, large glass jars, filled with soil, are stacked high along a wall inside the Legacy Museum. Collected at sites where African Americans were lynched, and labelled with the name of the victim and the location, they bear sombre

© AMY BALFOUR

"We want to remind visitors that we all have a light to shine, just like the people who were part of the movement"

witness to racist terrorism. The museum also examines the trade in enslaved people, poll taxes and racial inequities in policing and the criminal justice system, and displays works by Black artists.

At its companion site, the National Memorial for Peace and Justice, rows of steel sculptures – resembling upright caskets – commemorate the country's 4400 victims of lynching. The 800 sculptures are engraved with the names of those murdered. Many hang above a sloped walkway, evoking rows and rows of hanging victims.

Marching for the right to vote

From Montgomery, an 80km (50-mile) drive through the rolling rural landscape to the west brings you to the small city of Selma. In the early 1960s, an overwhelming majority of Black residents here were not registered to vote as a result of wildly unfair literacy tests, steep poll taxes and threats from local whites. Activist groups and Civil Rights leaders, including Dr King and late US Congressman John Lewis, organised voting-rights activities here.

On March 7, 1965, after activist Jimmie Lee Jackson was beaten, shot and killed, protesters began a 54-mile walk from Selma to Montgomery to commemorate his life and spotlight police brutality. Troopers, some on horseback, attacked the 600 marchers with nightsticks, whips and tear gas as they crossed Selma's Edmund Pettus Bridge. National news outlets captured their brutality, and the event became known as 'Bloody Sunday'. A subsequent 25,000-strong march ended at Alabama's State Capitol. On August 6, 1965, President Johnson signed the Voting Rights Act – one of the key successes of the Civil Rights Movement.

Concluding at the bridge feels appropriate. Though it was the site of a hard-won triumph that instigated lasting change, some say its legacy – preventing the disenfranchisement of Black and minority voters – is under assault from new and proposed voting restrictions in many Southern states. Decades later, the quest for equality, understanding and healing continues.

© MORTON BROFFMA / GETTY IMAGES

Clockwise, from left: Dr King leads Selma activists to Alabama's State Capitol in 1965; Montgomery's Legacy Museum; 16th St Baptist Church, opposite the Birmingham Civil Rights Institute

Practicalities

Region: Tennessee/Mississippi/ Alabama, USA
Start: Memphis, Tennessee
Finish: Selma, Alabama

Getting there and back: Memphis borders the Mississippi River in western Tennessee and straddles the I-40. The city is a daily stop on the City of New Orleans Amtrak rail route between Chicago and New Orleans. The train station is downtown, just two blocks from the National Civil Rights Museum. Memphis International Airport is 16km (10 miles) from downtown. Selma is about 80km (50 miles) west of downtown Montgomery via the US80.

When to go: The museums and sites are open all year, but note that some are closed on Mondays. Summers can be hot and muggy, particularly in Mississippi during July and August. February is Black History month and museums may have special exhibits supporting the theme.

Where to stay: Larger and mid-size cities have numerous hotels, motels, inns and short-term rentals. Smaller towns will have more limited options, and you may want to book a few days in advance.

Where to eat and drink: Memphis and Birmingham have thriving food scenes. Open since 1946, the Four Way in Memphis was a safe space for Civil Rights activists, and a favourite of Dr King. In Montgomery, Pannie-George's Kitchen, adjacent to the Legacy Museum, serves delicious soul food classics.

Essential things to know: A few museums may prohibit backpacks and handbags, while others use metal detectors to scan belongings. Be prepared to leave some personal items in the car. When considering photographs and selfies, situational awareness is key. The Edmund Pettus Bridge is an inspirational spot but other sites are more sombre, memorialising fallen protestors and acts of injustice and cruelty.

© EQUAL JUSTICE INITIATIVE

Europe

Tuscany sunset
near Pienza, Italy

Norway's many-storied skiing escapade

The Birkebeinerrennet ski route invokes an 800-year-old tale of courage and deliverance: it's a rite of passage for Norwegians and a transformative experience for visitors.

Sharing a remote mountainscape with thousands of strangers, all engrossed in the same unfolding story, is an intense experience. The enthusiasm of skiers and onlookers creates an incredible atmosphere.

The Birkebeinerrennet takes you deep into the Norwegian backcountry, an otherworldly realm resonant with folkloric tales of trolls and other creatures that are not intended for children. Shivers are not always caused by the cold.

Norwegians learn to ski as soon as they can walk, and the Birkebeinerrennet route is mentally and physically demanding. To enjoy the experience fully, training is required.

Attracting 16,000 entrants every year, the Birkebeinerrennet – a 54km (34-mile) 'classic' cross-country marathon – is one of the largest skiing races on the planet. But it's so much more than just a race. For Norwegians, it's a cornerstone of their country's cultural calendar. Skiing this rollercoaster route, through the mountains and forests between Rena and Lillehammer, goes beyond a tradition – it's a rite of passage. And it's

© BIRKEN / GEIR OLSEN

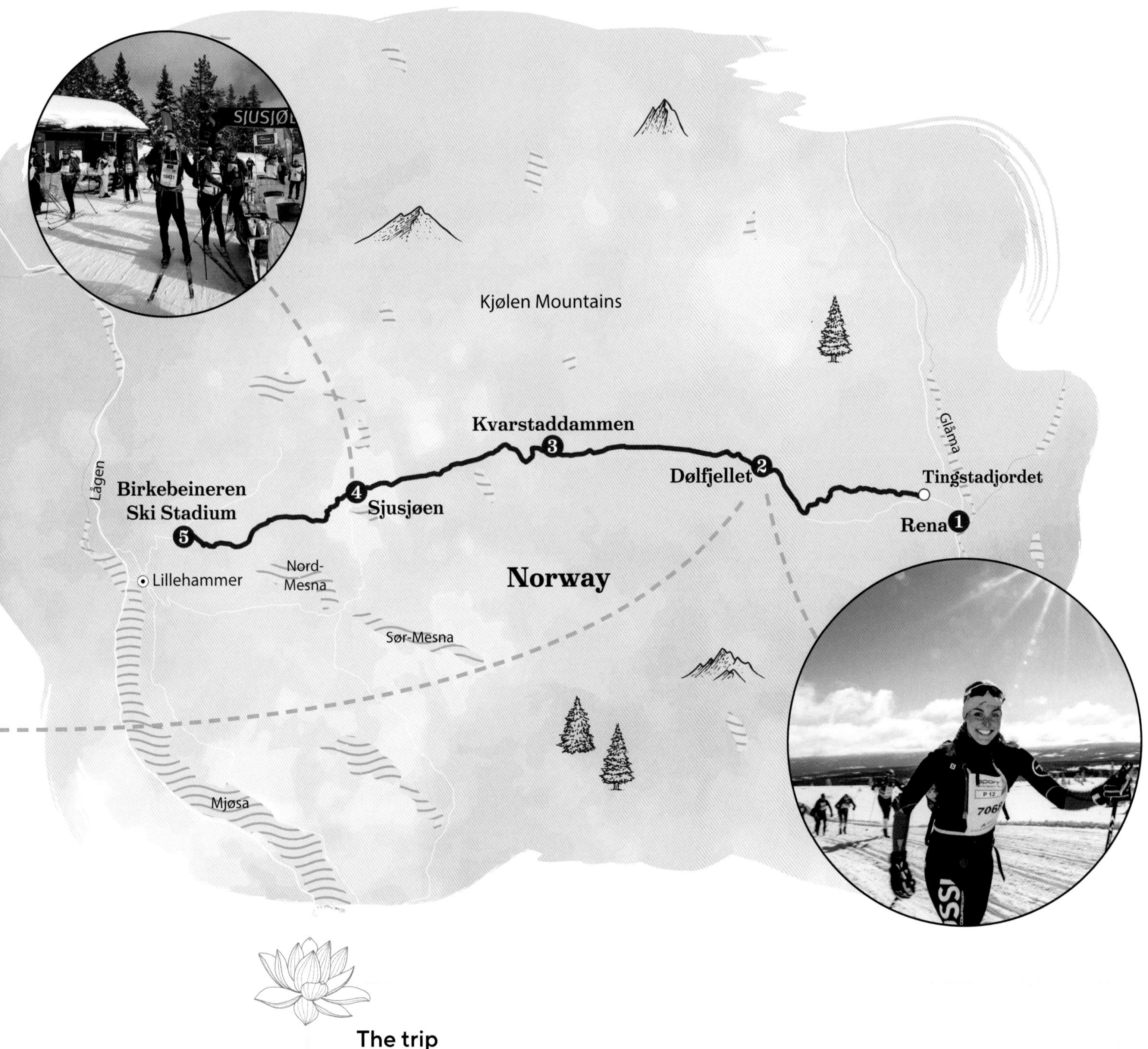

The trip

Distance: 54km (34 miles)
Mode of transport: Cross-country skis
Difficulty: Difficult

1 Rena This Østerdalen Valley village is the beginning of the Birkebeiner Trail; the Birkebeinerrennet race starts at Tingstadjordet, 3km (2 miles) from the centre

2 Dølfjellet The first mountaintop on the route stands 820m (2690ft) above sea level

3 Kvarstaddammen Beyond the woods around Nysætra, this 660m-high (2165ft) peak marks the halfway point of your icy odyssey

4 Sjusjøen The El Dorado of cross-country skiing, where 2500km (1553 miles) of tracks spiral off in every direction

5 Birkebeineren Ski Stadium The endpoint of the trail, just 3km (2 miles) from the centre of Lillehammer

Clockwise, from left: race competitors; support from the sidelines; hydrating at a water station; classic cross-country, Birkebeinerrennet-style

Patrick Kinsella, competitor

Torbjørn Moastuen is a Lillehammer local who has skied the Birkebeinerrennet 20 times. He helped me prepare and train for my own experience at the event. Each time he does the race, Torb carries, as his requisite princely baby weight, a copy of a Norwegian encyclopaedia – the second volume, with 'B' for Birkebeinerrennet in it – and he writes his race time next to the entry. 'The event leaves its mark on the city all winter, and it escalates towards the race,' says Torb. 'Many friends, with whom you train all year, participate, and the atmosphere is very motivating.'

every bit as eccentric as it is extraordinarily popular. But even if you experience the trail outside the excitement of the race itself, tracing this challenging route takes you on a journey of discovery where you'll discover historical figures and folkloric creatures amid a spectacular landscape.

Blast from the past

One reason the Birkebeinerrennet pulls in so many people, perhaps, is that it has a brilliant backstory, and a unique set of rules as a result. The race was inspired by events eight centuries ago, when an infant prince was carried to safety through the snowy mountains by two skiers, who were protecting him from assassins. To honour the actions of these men, the number-one race rule demands that all competitors carry an extra 3.5kg (3.5lb) of surplus baggage – beyond any extra clothing potentially required, and not including food or drink they might consume – to symbolically represent the weight of the infant prince. (Yes, that is a light two-year-old; the rules used to stipulate 5.5kg/12lb, but in 1993 the little prince was put on a diet.)

This strict stipulation applies to all competitors – even the professional skiers taking part in the Birkebeinerrennet as part of the Worldloppet series, which attracts some of the world's best cross-country skiers. And everyone takes this rule really seriously. In Rena on the morning of the race, rows of weighing scales surround the starting line at Tingstadjordet, and skiers nervously queue to check they're carrying enough bulk. There are also spot-checks at the end

"The Birkebeinerrennet is so much more than just a race. For Norwegians, it's a cornerstone of their country's cultural calendar"

to ensure that skiers are still lugging their fair share of pretend baby.

But even without its iconic idiosyncrasies and wonderful historical backdrop, for anyone with a pair of skis and a sense of adventure, the eye-boggling beauty of the terrain and the provocative profile of the route make the Birkebeinerrennet an irresistibly compelling challenge to take on.

Free roaming

In Norway, and across most of Scandinavia, access to the outdoors is regarded as sacrosanct, and *allemannsretten* (the 'right to roam', respectfully, wherever you want to go, on foot, bike or skis) is enshrined in law. So, although you wouldn't experience the extraordinary atmosphere that surrounds the event, it is entirely possible to ski the route of the Birkebeinerrennet outside of the race. And later, in the green season, you can hike, bike or even trail-run it, too. The Birkebeinerrittet mountain-bike version of the race between Rena and Lillehammer takes place each August, and a Birkebeinerløpet cross-country running challenge happens in June.

To tackle the trail independently on skis, you'll have to have good mountain and winter skills, and will need to be proficient in the art of classic cross-country skiing. Unlike skate-style skiing, the classic cross-country version makes use of parallel tracks, which resemble inverted small-gauge railway lines, and from January through to April the whole Birkebeiner Trail is groomed to facilitate this. However, once you climb out of Rena, the terrain is very remote – two-thirds of the course goes across wild mountaintops, with just three road-access points; there are some very exposed, lonely sections.

On race day, it's a very different story. At least as many people come out to watch the Birkebeinerrennet as take part in it, and alongside the trail, as it undulates through the mountains, spectators dig in and make themselves comfortable. Large groups come together to cheer on friends,

Below: crossing the Sjusjøen plateau
Left: since 1930, Birkebeinerrennet racers have weighed in at the start-line to confirm they're carrying the requisite baby-weight

Birken backstory

The storied Birkebeinerrennet commemorates an act of heroism in 1206, when Norway was in the midst of a bitter, century-long civil war. That winter, a two-year-old infant, the son of Håkon III, was smuggled through the mountains and forests between Rena and Lillehammer by two warrior skiers, who were protecting him from assassins. Traversing terrain covered in deep snow, the skiers wore protective leggings made from birch bark, earning them the name Birkebeiner (birch legs). The child grew up to become Håkon IV, who ended the conflict, reigned for 46 years and led Norway into a golden age. In 1930, Håkon Lie, a Lillehammer author, suggested a race of remembrance, and two years later the first event took place. Now, thousands of Norwegians (and a few intrepid incomers) gather annually to retrace the route taken by the Birkebeiners – and the result is a ski event like no other on Earth.

family and complete strangers. Some of the setups are truly elaborate: windbreaks arranged to shield barbecues brimming with burgers; tables, sofas and footrests are sculpted from snow and ice; and seats are lined with sheepskin rugs to prevent bums from getting frozen. Onlookers are typically very well organised, with hours of supplies to keep them going, including various bottles of booze stuck into the snow – social antifreeze to consume as they watch 16,000 skiers slide by. A festival buzz fizzes in the alpine air as they yell 'Heia! Heia! Heia!' (Go! Go! Go!) at passing skiers, who are themselves fuelled by bananas and honey-drizzled *vafler* (waffles), stuffed into their mouths by volunteers at feeding stations.

Great white wilderness

Whether you're taking part in the occasionally riotous race, or just rambling along on your own alpine adventure, the landscape along the route is brutally beautiful. Even deceptively softened by the snow, the terrain is terrific in the true sense of the word – and the traverse is more than tough enough to leave a permanent imprint on anyone who experiences it.

The first 13km (8 miles) are almost entirely uphill, which means digging in and getting your anti-gravity glide going as you pole past Skramstadsetra and reach the 820m-high (2690ft) peak of Dølfjellet. From here it's downhill all the way to Dambua, an exciting section of swooping, whooping and hollering. Another climb follows, as the route rears and steers west, crossing 880m-high (2887ft) Raudfjellet at around 20km (12 miles) from the start.

© SPORTOGRAF

> **"The terrain is terrific, and the traverse is more than tough enough to leave a permanent imprint on anyone who experiences it"**

Entering the woods here, you pass forest-fringed Nysætra en route to the halfway mark at 660m-high (2165ft) Kvarstaddammen.

Troll bait

The trail then winds through trees to the highest point on the course, 910m (2986ft) Midtfjellet, before wending west to the sensational Sjusjøen plateau, a Norwegian nirvana for cross-country skiers and mountain-bikers. In Sjusjøen, the Birkebeinerrennet trail intersects with another backcountry route, the Trolløype (which literally means 'troll track'), and in the distance the dark Jotunheimen range looms large on the western horizon. According to folklore, these moody mountains are home to trolls – and not the cute roly-poly ones from *Frozen*, either, but the tear-you-limb-from-limb kind that populated Norwegian mythology long before Disney hijacked them and added sugar.

Even if the thought of hungry trolls on your tail doesn't encourage you to get your skates on, the terrain leaves skiers little choice but to pick up speed, with the last 8km (5-mile) section of the course plunging downhill. The trail dramatically drops down 400m (1312ft) through a series of narrow tree-lined gullies before delivering you right into the arms of the Birkebeineren Ski Stadium, close to the intimidating-looking Lysgårdsbakkene Ski Jumping Arena, built for the 1994 Winter Olympics.

Independent skiers can continue on to Lillehammer, 3km (2 miles) down the hill, but for racers, the stadium is where the story ends. Here, once officials have checked you still have the correct amount of princely babyweight on board, you are presented with a prestigious finisher's pin. This treasured trophy proves you've eluded the trolls and survived to add your own tale to the ever-expanding narrative of the Birkebeinerrennet.

© SHUTTERSTOCK / HENRIK A. JONSSON

From top: skiing the Birkebeinerrennet trail outside of the event offers a challenging route through spectacular scenery; en route to Birkebeineren Ski Stadium, Lillehammer

Practicalities

Region: Gudbrandsdalen, Norway
Start: Tingstadjordet, Rena
Finish: Birkebeineren Ski Stadium, Lillehammer

Getting there and back: Norwegian airline flies regularly to Oslo from many European cities. From Oslo Gardermoen airport, there are direct trains to Lillehammer (12–18 daily; 2hr). During the Birkebeinerrennet, buses run between Lillehammer and the Rena starting line. Outside of the event, you can get a train from Lillehammer to Rena.

When to go: The Birkebeinerrennet (birkebeiner.no) takes place every March; sign up via the website. The best skiing conditions are between January and April, when the trails are regularly groomed. The Birkebeinerrennet team also run three other races, all detailed on their website. The Rena to Lillehammer SkøyteBirken, on the same course but with skate-style cross-country skis, usually happens the day before the Birkebeinerrennet. In August, the 86km (53-mile) Birkebeinerrittet mountain-bike race runs between Rena and Lillehammer; and in June, the Birkebeinerløpet half marathon maps a course between Birkebeineren Ski Stadium and Lillehammer.

What to take: You'll need classic cross-country skis, poles, gloves, sunglasses, sunscreen, lots of warm layers (including a merino base-layer and a waterproof shell), a hydration pack, energy gels/bars and a backpack to carry the required 3.5kg (3.5lb) of weight.

Where to stay: Lillehammer has accommodation to suit every pocket. To really get into the atmosphere of the race and its backstory, book a room or apartment at the Birkebeineren Hotel (birkebeineren.no), right by Håkons Hall where the race finishes.

Essential things to know: Outside of the Birkebeinerrennet, sample classic cross-country skiing on the myriad magical trails at Sjusjøen resort (visitsjusjoen.no). You can rent ski equipment, including *pulks* (sleds for pulling children along) from Sjusjøen Skisenter (sjusjoenskisenter.no/en).

On pilgrimage to Santiago de Compostela

Pilgrims have been making their way to Santiago de Compostela since the 9th century. Follow in their footsteps to discover Spain – and yourself.

Pilgrimage is an ancient spiritual practice. Once prescribed as atonement for sins, it is now a voluntary undertaking – but the same purpose remains: letting the walker contemplate their life by separating from it. Millennia old, it is tried and tested.

Atheists and those of every faith undertake the Camino. There is far more that joins us than separates us.

Walking is deeply human, but our increasingly sedentary lives mean our bodies are no longer used to it. By the time you reach Santiago, covering 30km (19 miles) a day on foot will feel like the most normal thing in the world.

The Monasterio de Zenarruza has welcomed pilgrims for almost a thousand years, and the worn walls of the cloister here are carved with scallop shells and the Cross of St James. At dinner, a monk in a black cassock ladles out bowls of stew to the footsore and the weary; later, during the candlelit Vespers, your mind can drift as you wallow in the slow and soporific chanting. It could be this time, or it could be any other. Tomorrow

© JUSTIN FOLKES / LONELY PLANET

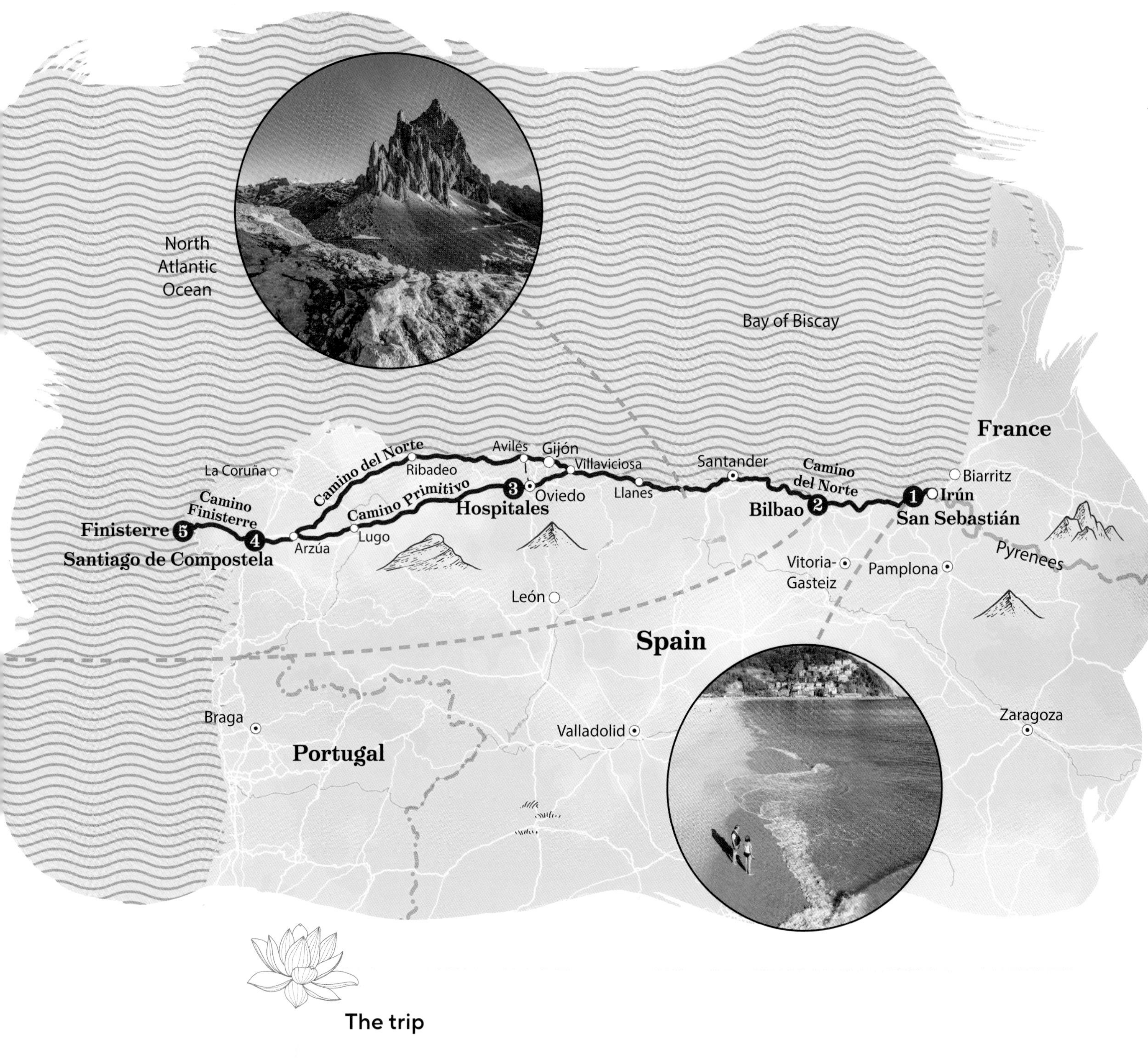

The trip

Distance: 908km (564 miles)
Mode of transport: On foot
Difficulty: Moderate

1 San Sebastián A beautiful coastal city, renowned for its beaches and delicious *pintxos*

2 Bilbao Take a break from walking in the architecturally stunning Guggenheim Museum

3 Hospitales Climb through carpets of wildflowers in the Cantabrian Mountains, passing medieval pilgrims' hospitals

4 Santiago de Compostela Don't just visit the cathedral – the whole city is a marvel

5 Finisterre The end of the road and, very nearly, the westernmost point in Europe

Clockwise, from left: walking the Camino through Galicia; Guggenheim Museum, Bilbao; Picos de Europa, Cantabrian Mountains; San Sebastián shores

Adam Weymouth, Camino de Santiago pilgrim

It was the first time I had done anything like this, but I was inspired by a friend who had done the Camino the year before, also with no experience. I still can't explain exactly what it was that kept me going, or how I changed as a result, but I'm not sure if that matters. I do know that ever since, when I've wanted time to reflect and think, I've turned to walking.

will be another long day, and there are many more days to go. But tonight, with a belly full of food and a bed waiting, you are once again at peace.

This is life on the Camino. Gruelling days punctuated by moments of deep rest. Periods of intense introspection, then a blessing from a stranger.

Choosing your Camino

To look at a map of the many different Caminos is like looking at the watershed of some vast river, with its many tributaries flowing across Europe and converging on Santiago. The Camino Francés (French Way) is the most popular route, but there are advantages to a path less trodden. There is more chance of a bed in the pilgrim hostels, and more opportunities for self-reflection. And then there is the scenery. The Camino del Norte (Northern Way) hugs Spain's Atlantic coast, and is said to be the most beautiful path. At Oviedo you can switch over to the Camino Primitivo, walked by Santiago's very first pilgrim, King Alfonso II the Chaste, in the 9th century. Climbing over the Cantabrian mountains, it not only offers unparalleled views, but ratchets up the hardship as well. Why else are you on a pilgrimage?

Walking makes us human

Pilgrimage is ancient. Australian Aboriginal people continue to practise the world's oldest religious rituals, of which walking is an integral part. Each of the major religions has its pilgrimage sites. There is something universal about walking that connects us, fundamentally, to what it is to be a human.

© JUSTIN FOULKES / LONELY PLANET

"Maybe, in these difficult times, we are hunting for meaning in the simplicity of walking"

During the Middle Ages, up to a quarter of a million people made the journey to Santiago each year, but by the 20th century the Camino had drifted into obscurity. Its revival was driven by Elías Valiña, a Galician pastor who revitalised the trails and marked them with yellow arrows. In 1985, 690 people made the newly signposted journey; in 2022, it was 438,000. The anthropologist Victor Turner suggested that pilgrimage grows in popularity during 'periods of destructuration and rapid social change'. Maybe once again, in these difficult times, we are hunting for meaning in the simplicity of walking.

Today, devotion is not the only motivation for a pilgrimage. There are teenagers on gap years and those starting retirement. There are artists, the bereaved, endurance athletes, families – and plenty who aren't quite sure what moved them to undertake something both so strenuous and so simple. The beauty of a shared journey is the chance to chat with your fellow pilgrims, on the trail or over dinner. Yet despite the diverse rationales, the common experience is that there is something radical in the simple act of putting one foot in front of the other. You cannot spend several weeks at walking pace without experiencing something profound.

Setting out

The route begins at Irún, just over the border with France. The footpath hugs the cliffs and descends to small, secluded beaches where you can eat a lunch of bread and cheese, wash off the morning's grime and siesta before carrying on. It passes through the great Basque cities of San Sebastián and Bilbao, famous for their food and their Txakoli, a white wine poured from height to carbonate it. It is exciting to be surrounded by crowds again, but after a day of trudging the tarmac it is a relief to be spat back out, onto the path again. Back to the scallop shells and the now-familiar yellow arrows that guide your way across Spain to salvation.

Climb through forests that smell of pine resin and salt. Leave the sea behind and scramble higher

Clockwise, from above: scallop-shell waypoints mark the route; Camino de Santiago pilgrim's passport; Camino on horseback; en route to Santiago in Galicia

Walking into freedom

For three decades, the Belgian organisation Oikoten – a Greek word meaning both 'away from home' and 'by one's own effort' – ran a programme for young offenders whereby, if they walked the Camino from Brussels to Santiago, they could avoid incarceration. The four-month journey, of around 2500km (1553 miles), was undertaken with a mentor, who was also doing it for the first time; and it gave the teenagers – many of whom had never left the places they were born – a vital opportunity to reflect on other ways of life. 'It was such a good experience,' said Marc Vangasse, who walked when he was 17. 'The big thing that I learned is that many people were very nice. It's amazing how good people are, and it's not about nationality. We all have it in our heart.' The programme's rates in preventing reoffending were remarkable, far more successful than institutions. 'A walk is medicine,' said Marc. 'Prison is not.'

still to mountain passes that look out over the ocean you have come from and the valleys where you're headed. Slither down the other side to a small village clinging to the hillside, and a single bar and a simple lunch with a glass of cheap red wine drawn from the *spina*. Griffon vultures soar overhead. On thousand-year-old bridges and cobbled lanes you can see where countless feet have trodden the stones before you. Those walkers felt the same as you; they even smelled the same. There is a powerful link in realising that both your suffering and your joys have been experienced in equal measure by other pilgrims down the ages.

Yes, it is hard work. Your thighs will hurt. Your back will hurt. At night you will examine your raw and blistered feet and wonder what on earth you were thinking. But there is something about walking day in, day out that begins to make a shift in both the body and the mind. You start to feel that this is what your legs are actually meant for. Your thoughts have the space to wander.

Journey's end

The cathedral of Santiago de Compostela is magnificent. To arrive at the end of a long day, after many weeks on the road, is to feel that same awe that pilgrims have experienced here for a millennia. During the Mass, a huge censer swings in great arcs across the transept, reaching almost to the roof (it was said to have been needed in times past to mask the stink of the pilgrims, although in truth the pong is not much better today).

> "Maintaining feelings of peace, of an active body, of an openness to strangers – that is the true challenge of pilgrimage"

For non-Catholics, Santiago may not be the most meaningful place to stop walking. Three more days west is Finisterre, so-named by the Romans for being at the very end of the (known) world. You walk carefully onto this spit of land surrounded by the sea on all three sides, and you literally can't go further. Traditionally you hurl your walking stick into the waves here, and make a bonfire of your threadbare clothes as you watch the sun tip below the horizon.

Of course, in the 13th century, this was not the end. Tomorrow, you would turn round and walk back. Certainly the train is tempting, but you might find that by now you've got used to this pace. Of waking up and walking and having nothing more complicated to think about than how much you will eat at lunch. The return can be much harder than the first journey. Maintaining those feelings when you stop – of peace, of an active body, of an openness to strangers: whatever your faith, that is the true challenge of pilgrimage.

© JUSTIN FOULKES / LONELY PLANET

©MATT MUNRO / LONELY PLANET

Clockwise, from left: Santiago de Compostela cathedral; Tarta de Santiago, the 'cake of St James'; sunrise over Santiago

Practicalities

Region: Basque Country/Cantabria/Asturias/Galicia, Spain
Start: Irún
Finish: Finisterre

Getting there and back: Regular ferries run from England's south coast to Santander and Bilbao in Spain throughout the year, and take about 24 hours. From the ports, trains connect to Irún and Santiago.

When to go: The Camino Francés can be unbearably hot at the height of summer, but on the coastal and mountain routes there should always be a breeze. It's a beautiful walk at any time, but winter might well mean snow on the Camino Primitivo – go prepared.

What to take: As little as possible! The kilos really count when you're carrying your bag all day on the trail. There's no need for a tent if you plan to reach a hostel every night. Hanging a scallop shell from your rucksack will identify you as a bona fide pilgrim.

Where to stay: Get your pilgrim passport (*credencial*) from the Confraternity of St James (csj.org.uk) before you go. It's your access to the network of pilgrim hostels that span the length of the route.

Where to eat and drink: In most towns there is a restaurant near the hostel offering a cheap set menu for pilgrims on presentation of a *credencial*.

Tours: There are dozens of tour operators who will carry your bags from hostel to hostel so that you don't have to stagger under the weight of your own gear – but this arguably isn't the true pilgrim experience.

Essential things to know: Everyone has their own tips and tricks for dealing with the inevitable blisters: talcum powder, rubbing feet in alcohol and bursting blisters with sterilised needles are favourites. But you can't beat good shoes and good socks, and blister plasters for emergencies.

Remembering D-Day on the Normandy coast

Motor along France's most emotive coastline, where peaceful swathes of sand and moving memorials provide a living lesson in WWII history.

The D-Day sites can trigger intense emotions: horror, sadness, respect, regret, humility. Learning about the conditions endured by soldiers and locals during WWII encourages empathy, gratitude and introspection, perhaps prompting you to question your own physical and emotional capacity in challenging situations.

Learning about D-Day, the Battle of Normandy and the ghastly price paid for France's liberation from Nazi tyranny reveals the profound legacy of these historic events. Superbly designed museums and memorials use film, audio, digital technology, photography, dioramas and wartime artefacts to make WWII history tangible and encourage critical thinking.

The powder-soft beaches and breeze-swept buffs of Normandy's northern coast are quiet today. Outside of summer, bar the odd wading bird and plump harbour seal, they are deathly quiet. The piercing solitude inspires inner stillness, and renders contemplation of the incongruous cacophony of cannons, gunfire and human terror that ripped through these quiescent beaches during WWII even more unsettling.

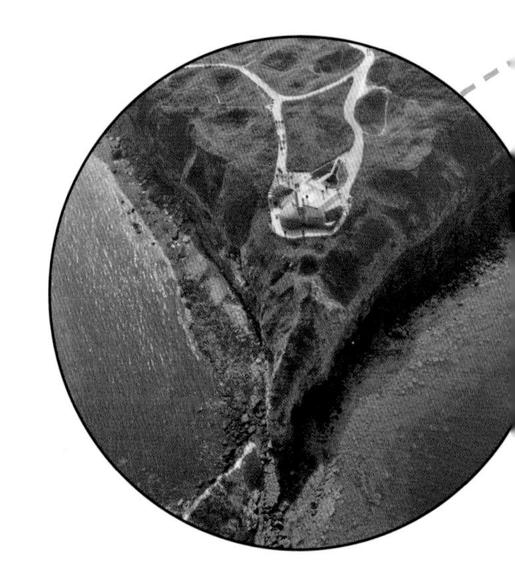

© SHUTTERSTOCK / PERESANZ

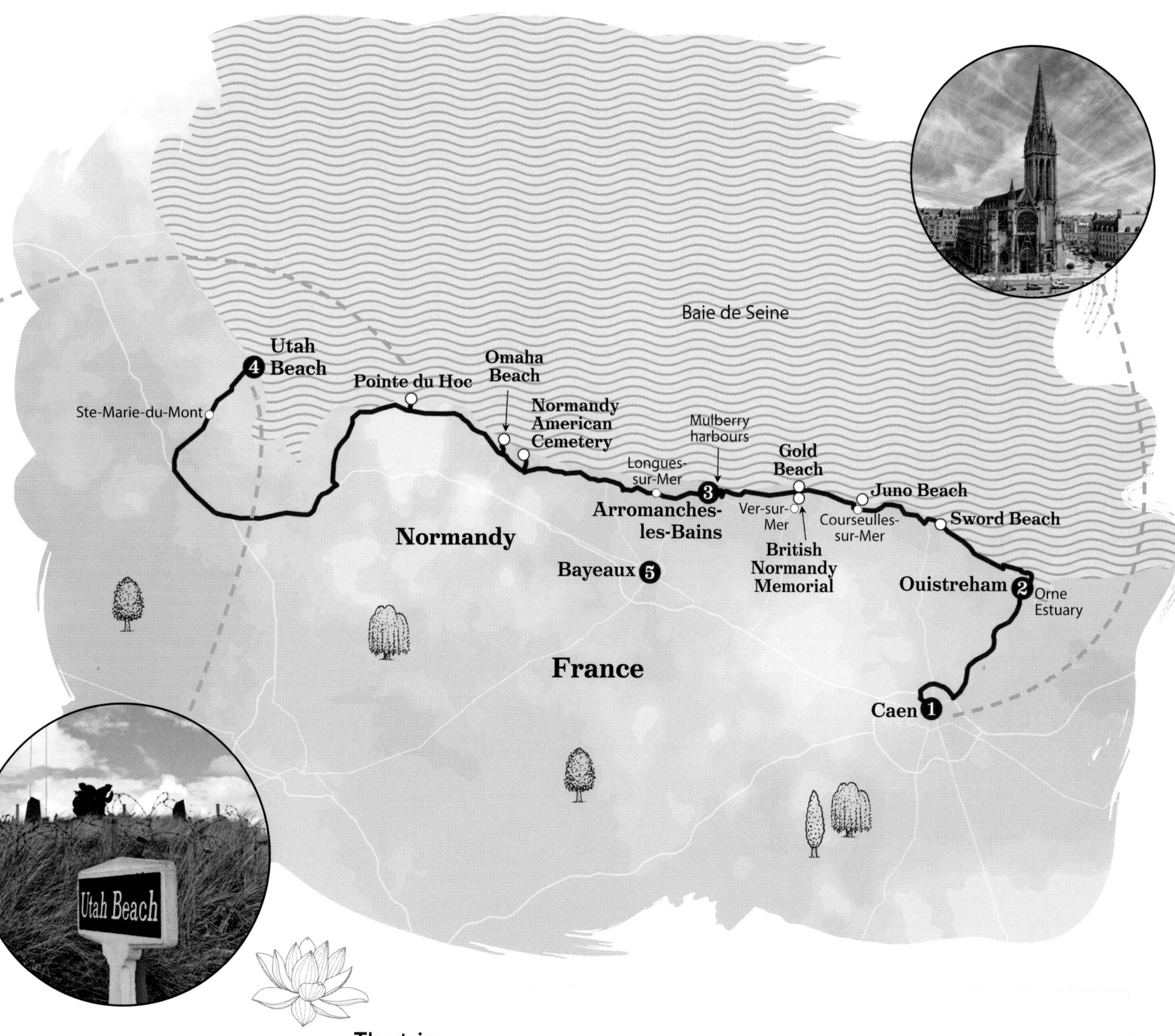

Clockwise, from left: Normandy American Cemetery; Pointe du Hoc; Église St-Pierre, Caen; Utah Beach

The trip

Distance: 140km (87 miles)
Mode of transport: Car, e-bike
Difficulty: Easy

1 Caen Handsomely rebuilt after the ravages of 1944, with a top-drawer WWII museum and medieval castle walls and abbeys

2 Ouistreham Experience a section of the German Atlantic Wall defences at Musée du Mur de l'Atlantique, set in a former bunker

3 Arromanches-les-Bains Learn more about the D-Day landings at the Musée du Débarquement

4 Utah Beach Explore an exceptional museum dedicated to the *débarquement* (landings), set on the sands stormed by Allied troops in 1944

5 Bayeux This inland town holds Normandy's largest Commonwealth war cemeteries, a fine Gothic cathedral and the world's most celebrated tapestry

Claire Lesourd, D-Day tour guide

I was with a D-day veteran on Omaha Beach, who'd been there with the first wave of the invasion at 6.30am on 6 June 1944. The tide was out, the sky was grey and it was misty, just like when he came ashore long ago. He looked at me and said: 'You know, some people think I'm a hero because I survived this place. The true heroes are those who rest in the military cemetery.' To me, that veteran and all the other ones I meet, are a lesson in humility and courage.

Code-named Operation Overlord, the D-Day landings were the largest seaborne invasion in history. Early on the morning of 6 June 1944, swarms of landing craft – part of an armada of more than 6000 battleships and destroyers – hit the beaches of northern Normandy. Allied soldiers, weighed down with heavy backpacks, life belts and weapons, poured onto French soil. Approximately 160,000 stormed ashore along five sandy beaches – code-named Sword, Juno, Gold, Omaha and Utah – that fringe the 90km (56-mile) stretch of coast between Ouistreham and Ste-Marie-du-Mont. British and US paratroopers were dropped inland.

Enemy perspective

Begin your D-Day journey in Norman capital Caen, today rebuilt after being almost completely destroyed in the fierce battles that followed the landings. Insightful exhibitions at the Caen-Normandie Mémorial, 3km (2 miles) northwest of the centre, offer a powerful overview of the entire war. During the 76-day Battle of Normandy alone, the Allies suffered 210,000 casualties, including 37,000 foot soldiers; a third of all the French civilians killed during WWII also died during the fighting here. Film, animation and audio testimony proffer food for thought on the realities of war, the trials of occupation, the joy of liberation – and the fragility of peace.

From the museum, a 20-minute drive north along the D515/D514 unveils a rousing first glimpse of the sea in the fishing port of Ouistreham. Beyond its lighthouse and ferry port is Le Grand Bunker – Musée du Mur de l'Atlantique. It occupies a vast and menacing former Nazi bunker, part of the Atlantic Wall German fortifications that were established to defend 4000km (2485 miles) of coastline from Norway to southern France. Climb to the telemetry post for a panoramic view of the Orne Estuary, the port of Le Havre, the Baie de Seine and the golden sweep of D-Day's five landing beaches. Imagining what flashed through the minds of German soldiers, as they looked out to sea from here on 6 June, is unfathomable.

> "Displays of humble everyday objects are stark reminders of the humanity of soldiers who left behind everything to fight"

Unsung heroes

A short drive or meditative seashore walk west brings you to Sword Beach. On these beautiful broad sands, at 7.25am on 6 June 1944, British troops – bolstered by significant units of Canadian, Commonwealth, Free French and Polish commandos – attacked. German resistance was quickly overcome and within hours the beach was secured. Among some 29,000 men who clambered from flat-bottomed landing craft, through deep water and gunfire, onto treacherous sands charged with booby-trapped stakes, barbed wire and mines, was British Commando Lord Lovat of the 1st Special Service Brigade. At Sword Beach's Lone Piper Memorial, pay your respects to Lovat's personal piper William 'Bill' Millin who – unarmed – played his bagpipes on the battlefield while shells exploded and soldiers fell around him.

Never forgotten

Holidaying families enjoying summer sun, sand and sea in pretty Courseulles-sur-Mer, 16km (10 miles) west of Sword Beach, are a world away from its sobering history. Canadian battalions landed quickly here, at Juno Beach, on D-Day, but then faced the bloody task of clearing German forces trench-by-trench before moving inland. Mines took a heavy toll. The 381 Canadian soldiers who died that day, along with 5500 killed in the Battle of Normandy,

Clockwise, from above: Mulberry harbours, Arromanches-les-Bains; Utah Beach's Musée du Débarquement; US Sherman tank above Utah Beach

War and peace in Bayeux

Two cross-Channel invasions – almost 900 years apart – gave Bayeux a front-row seat at defining moments in Western history. The dramatic story of the Norman invasion of England in 1066 is told in 58 scenes by the astonishing 70m-long (230ft) Bayeux Tapestry. Embroidered a few years after William the Bastard, Duke of Normandy, became William the Conqueror, King of England, it illustrates the dramatic, bloody tale with verve, vividness and incredible artistry, and offers insightful scenes of everyday life in 11th-century Normandy. Centuries later, Bayeux was the first French town to be liberated – on the morning of 7 June 1944 – after the D-Day landings, so survived WWII unscathed. In its vast War Cemetery, the Latin inscription on the memorial to soldiers whose remains were never found reads: 'We, once conquered by William, have now liberated the Conqueror's native land.'

are remembered in displays at Juno Beach Centre, overlooking the sand. Backed by wind-tangled grassy tufts and wild dunes, a traditional Inuit inukshuk stone sculpture commemorates the lives of Indigenous Canadians lost in the fighting.

Driving west along the coastal D514, or riding the Voie des Français Libres cycle path, brings you to the village of Ver-sur-Mer. Bucolic green fields chequered with *bocage* – the Normandy hedgerows that proved so troublesome for Allied troops to storm through – ribbon from the sea to the British Normandy Memorial, just inland. The stark simplicity and isolation of the site, etched with the names of 22,442 men and women under British command who were killed in summer 1944, is stirring; 350 died during the chaotic D-Day landing at nearby Gold Beach. Their personal stories, recorded by themselves or by relatives on the memorial's smartphone app, prompt further remembrance and reflection.

Mulberry harbours

Midway along the Normandy beaches, 7km (4 miles) west of Gold Beach in Arromanches-les-Bains, the Allies constructed one of two artificial harbours to disembark troops and supplies unimpeded by the tides. Explore the swathe of golden sand at low tide to uncover the scattering of prefab concrete pontoons, pierheads and floating breakwaters constructed in Britain and towed across the Channel in sections. These relics of the Mulberry harbours endure as silent witness to the Allies' determination, ingenuity and technical wizardry.

You can learn more about this high-risk operation, which took just 12 days to complete, at Arromanches' state-of-the-art new Musée du Débarquement on the beachfront. Immersive digital projections hurl you into the din, clatter and chaos of a wartime military harbour. Displays of humble everyday objects, from water bottles scarred with bullet holes to army-issue *Harry Potter*-esque spectacles, are stark reminders of the humanity of soldiers who left behind family, their civilian lives – everything – to fight and die.

Lest we forget

Drive west along the D514 to Longues-sur-Mer, where a German battery dug into the cliffs is an unsettling stop: the still-intact monster-sized cannons here fired shells weighing 45kg (100lb) onto both Gold Beach and Omaha Beach, 16km (10 miles) west.

The bloodiest D-Day battle haunts vast and magnificent Omaha. The attacking US 1st and 29th Infantry Divisions poured onto a beach defended by three heavily armed German battalions

> "No words do justice to the soul-wrenching sea of white crosses honouring soldiers killed in combat"

supported by mines, underwater obstacles and extensive trenches. Some US troops, overloaded with equipment, disembarked in deep water and drowned; others were cut to pieces by machine-gun and mortar fire from cliffs above the sand. Take in the expanse of empty beach, then wind your way up the footpath, past German bunkers, to the Normandy American Cemetery. No words do justice to the soul-wrenching sea of white crosses here, honouring 9386 American soldiers killed in combat – it's an incredibly emotional sight.

Shells also rained down on Omaha from German artillery guns on Pointe du Hoc to the west. At this chilling clifftop site, the earth still pitted with huge bomb craters, you can explore German bunkers and eerie casemates scarred by bullet holes and blackened by flame-throwers.

US troops landing at Utah Beach, 50km (31 miles) west, fared better than their Omaha comrades. Most landing craft came ashore in a relatively lightly defended sector, and by noon the beach had been cleared and 4th Infantry soldiers had linked up with paratroopers from the 101st Airborne. A B26 bomber plane and landing craft used in combat are displayed at Utah Beach's Musée du Débarquement. From the museum, you can look down on the sands where, by nightfall on 6 June 1944, some 20,000 more American troops and 1700 vehicles had arrived. Little did those men then know that in the three weeks it would take to reach Cherbourg, US forces would suffer one casualty for every 10m (33ft) advanced. Standing on Utah's wide swathe of golden sand, so peaceful today, it feels ever more important to remember the sacrifices that were made here in the name of freedom.

© SHUTTERSTOCK / ORXY

Left: sunset over Omaha Beach

Practicalities

Region: Normandy, France
Start: Caen
Finish: Utah Beach

Getting there and back: Cherbourg, with car ferries to Poole and Portsmouth, is 1hr 30min by car from Caen and 45min from Ste-Marie-du-Mont, just inland of Utah Beach. Le Havre ferry port is 1hr 30min by car from Caen, 2hr from Ste-Marie-du-Mont. Trains link Paris' Gare St-Lazare with Caen in 2hr 30min. Pick up a rental car at Caen train station.

When to go: Commemorative events are held during June's D-Day Festival. Spring (March to May) promises fewer crowds. Summer (July and August) is beach season.

What to take: A windproof jacket, umbrella, pocket knife (for picnics) and a big appetite.

Where to stay: Hotels and family-friendly *chambres d'hôtes* (B&Bs) pepper the cities, coast and peaceful villages inland.

Where to eat and drink: Buy picnic supplies – a baguette, regional cheeses, farm-fresh fruit, organic apple juice, cider and *poiré* (Norman pear cider) – at produce markets in Caen, Bayeux and Courseulles-sur-Mer. The choice of cafes, bistros and restaurants is dizzying: expect sublime scallops, oysters and other seafood, and Norman specialities like *boudin* (black pudding) and salt-marsh *agneau* (lamb).

Tours: Caen (caenlamer-tourisme.com) and Bayeux (bayeux-bessin-tourisme.com) tourist offices can recommend guided minibus tours; reserve in advance. Independent travellers can consult Liberation Route (liberationroute.com), or follow one of eight signposted circuits. Eolia Nomandie (eolia-normandie.com) runs guided electric fat-bike tours of Omaha Beach from Colleville-sur-Mer.

Essential things to know: Tourist offices sell detailed road maps. Caen-la-Mer tourist office rents umbrellas fitted with headphones, GPS and a D-Day-themed audioguide. When visiting the beaches and memorials, don't leave valuables in your car.

Hiking the Celtic Way from Glastonbury to Stonehenge

An ancient pagan way linking legendary mystical locations, this walk encompasses prehistoric, Roman and Christian sites as well as rolling scenery and expansive views.

Withdrawing from normal life is a central part of this walk, allowing you time to reflect in nature while challenging your body. That it is bookended by two great mystical sites with hints of the supernatural and the occult, and links to Celtic and Arthurian legends, means it encourages both a questioning of beliefs and a delve into the wisdom of those who went before us.

The route is awash with remnants of times past and offers ample fodder for further research and exploration into the lives of our ancestors over thousands of years.

A shroud of mystery cloaks Glastonbury, a town rooted in legend – it's said to be the birthplace of English Christianity, the burial place of King Arthur and the location of the Holy Grail. Much of this stems from a medieval campaign to attract pilgrims to its abbey, but Glastonbury has been a meeting place for thousands of years. When you venture beyond the main street's clutter of crystal shops, bong sellers and New Age bookstores, the surrounding landscape reveals itself as Glastonbury's real draw. The low-lying, often mist-covered plains, rolling hills and ancient forests here are drenched in history.

A legendary start

Hills rising above the plains became beacons for travellers and often took on sacred significance – none more so than Glastonbury Tor, said to have been home

© ADAM BURTON / AWL IMAGES

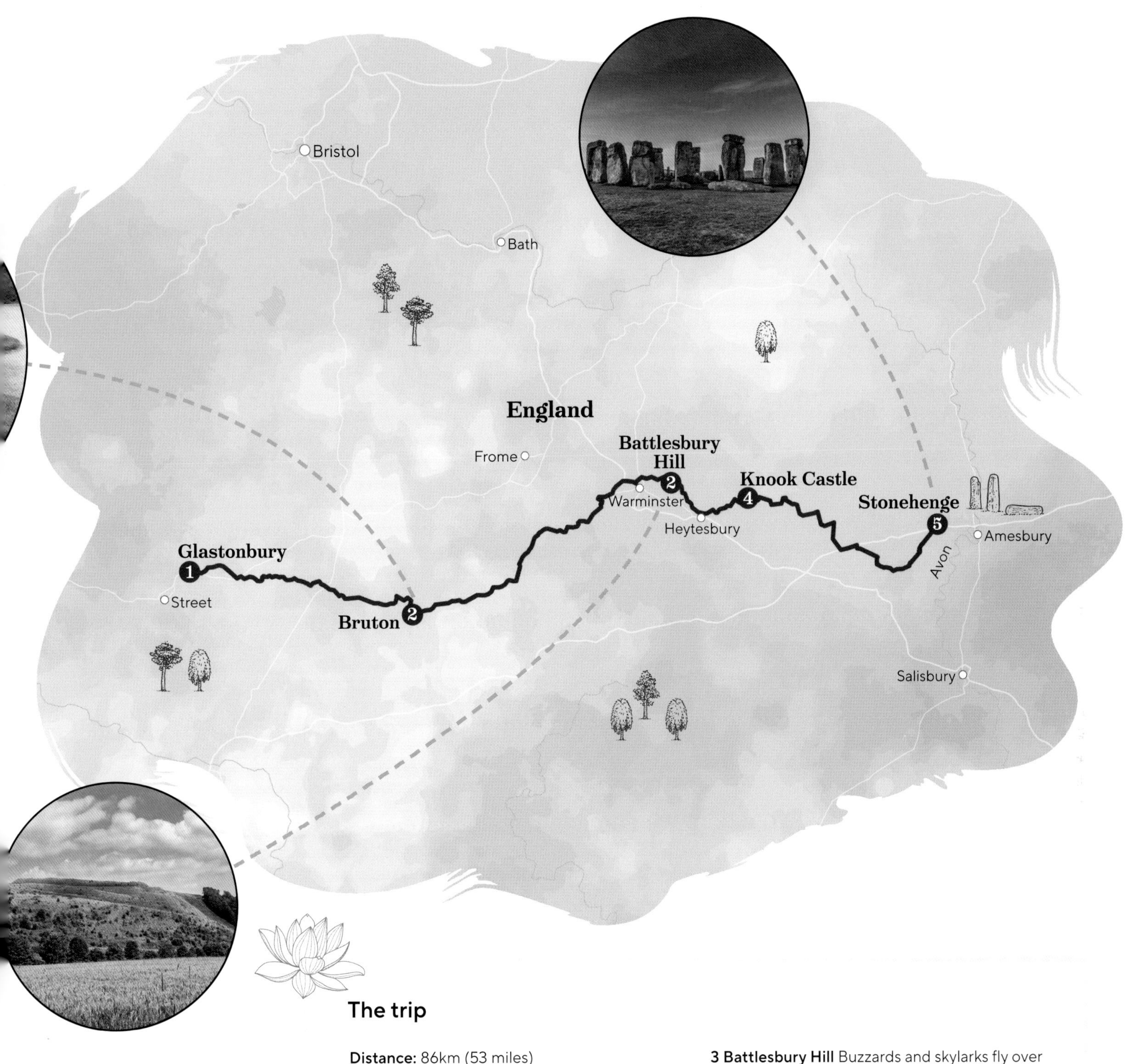

The trip

Distance: 86km (53 miles)
Mode of transport: On foot
Difficulty: Moderate

1 Glastonbury Birthplace of English Christianity, Arthurian legend, bohemian counter-culture and Britain's largest festival

2 Bruton A chichi town with an outstanding art gallery for quiet indulgence

3 Battlesbury Hill Buzzards and skylarks fly over one of Britain's finest Iron Age hill forts

4 Knook Castle Sweeping views and numerous earthworks on a hilltop inhabited for millennia

5 Stonehenge A 5000-year-old stone circle that is one of Britain's great archaeological mysteries

Clockwise, from left: morning mists below Glastonbury Tor; Hauser & Wirth art gallery, Bruton; Stonehenge; the Iron Age hill fort at Battlesbury

Maria Williams, hiker

I walked this route for charity with someone specific in mind and that changed the whole experience. It made me appreciate the good parts all the more, and gave me the determination to keep going when the trail got tough and the rain came down. The enduring legacy of the landscape, along with my reasons for being there, left me acutely aware of my own transience and helped me re-evaluate what I was doing in life and why.

to a fairy king, an entrance to the underworld, an intersection of ley lines and the mystical Isle of Avalon. It's a formidable vantage point that was of obvious importance to many cultures. The terraces on its slopes are thought to be an ancient ritual labyrinth dating to Neolithic times, while the tower on its summit is all that remains of the 15th-century St Michael's Church. For walkers contemplating the four-day hike to the great mystical stone circle at Stonehenge, it makes a fitting starting point, its enduring significance bestowing a sense of gravitas to the endeavour.

Returning to the past

Making your way down the far side of Glastonbury Tor, with kestrels keening overhead, the path to Stonehenge follows a section of the Celtic Way, a little-used trail meandering through quiet countryside that offers a striking contrast to the clamour of the landmark sites at either end.

The trail follows an old drovers' road through the somnolent villages of West and East Pennard before scaling Lamyatt Beacon, once the site of a Roman temple. Although nothing remains, Roman life doesn't seem quite so far removed as you stand on its summit surveying the landscape. After all, who wouldn't want a view like this from their window?

Continuing on, the trail crosses streams and rivers, lush fields and rural roads, and with few other walkers about, there's plenty of time to wallow in your thoughts before arriving at medieval Bruton, a well-heeled market town on the River Brue. A historic packhorse bridge crosses the river here and there's a handsome dovecote, but the town's quaint streets are more famous today for the Hauser & Wirth art gallery and its magnificent garden.

© SHUTTERSTOCK / TJOGR

"The trail crosses streams and rivers, lush fields and rural roads; there's plenty of time to wallow in your thoughts"

From Bruton, the trail winds along hills and ridges, through woodlands and past manor houses and meadows. The hills form a ring of defence over the lowlands, their summits topped by ancient forts that offered their inhabitants fair warning of invasion. From Penselwood to Gaer Hill and Cley Hill, the bowl barrows and terraced hillsides once used for cultivation speak of human habitation over countless millennia. There are portals to the past everywhere you look, and should you choose to explore them, the link between human consciousness and the landscape becomes ever stronger. Which is just as well, as progress is not always straightforward here, with overgrown paths and freshly ploughed fields contriving to hide the trail and test your patience. But wrong turns often lead to unexpected discoveries and rewarding chance encounters.

Exploring ancestral memory

Past the ancient forest at Longleat are Bronze Age and Iron Age burial mounds at Arn Hill and Battlesbury; Heytesbury has 14th-century almshouses and a 19th-century lock up. From here, the route skirts the Imber Range military training ground, and on through meadow and forest to Salisbury Plain. The climb to Knook Castle hillfort is rewarded with sweeping views and the earthworks of a Roman village at its summit.

The last stretch is the gentle climb to Stonehenge, the monolithic ring of standing stones that has acted as a meeting place for 5000 years. Arriving on foot offers a tangible connection to all those who have made their journey here in the same way, and allows time to contemplate this enigmatic, humbling and extraordinary site from afar. Stonehenge poses more questions than it answers – archaeologists still debate whether it was a celestial observatory, ceremonial site or burial place, but the mystery does nothing but add allure. And perhaps that is the significance of taking the time to walk here: to step aside from the everyday, to challenge mind and body, and to acknowledge that by opening ourselves up to opportunities that go beyond what we know, we make life far richer.

© AROUNDWORLD/ SHUTTERSTOCK

From top: Stonehenge, route's end for the Celtic Way; Longleat House and parkland

Practicalities

Region: Somerset and Wiltshire, England
Start: Glastonbury
Finish: Stonehenge

Getting there and back: From Bristol, bus 376 runs to Glastonbury in 1hr 20min. There are no public buses to Stonehenge; the nearest bus service is to Amesbury, about 3.5 km (2 miles) away, where you can catch the X4 bus for the 40min ride to Salisbury.

When to go: May and September are good times to take on this walk, with long evenings and a better chance of fine weather, but none of the summer crowds that swarm to Glastonbury and Stonehenge in the peak summer months.

What to take: Good walking boots and rain gear are essential as the route is quite hilly and the weather can be unpredictable, even in midsummer. The route isn't well marked and phone signals aren't always reliable. Carrying OS Landranger maps 183 and 184 is well advised.

Where to stay: You'll find plenty of choice for places to stay in Glastonbury and Salisbury. The walk is best done over four days with overnight stops at the Turk's Hall in Bruton, the Bath Arms in Horningsham and the Angel Inn in Heytesbury.

Where to eat and drink: You'll pass through numerous villages along the way so you'll only need to carry enough food and water for each day. There are pubs in each of the overnight stops for food.

Essential things to know: The route spans some busy roads, so take care when crossing. Close access to the stone circle at Stonehenge is only possible on Stone Circle Experience tours, which take place outside normal opening hours (book well in advance).

Stargazing on the Cambrian Mountains Astro Trail

Find space, hope and fresh perspective by wishing on stars and spotting planets in the dark skies of Wales' starkly beautiful Cambrian Mountains.

The expansive night skies show us that life is bigger. At one with the universe, with just the stars and sounds of the night for company, we are free from the chaos of daily life and very much in the moment. With no boundaries to space or time, stargazing can help us reconnect with nature, giving us peace and purpose.

With its roots in prehistory, astronomy is the oldest and most universal of the natural sciences, with mythological, cosmological, calendrical and astrological origins. Observing the stars teaches us a great deal: as well as identifying planets, constellations and galaxies, we can gain understanding of orientation, magnitudes and measuring distances.

When night falls in the Cambrian Mountains, the darkness is total. Nicknamed the 'Desert of Wales' for its vast expanses of nothingness (certainly not for its weather), this neck of the country is Wales at its wildest: rocky summits fall to wind-blasted, bracken-cloaked moors, river valleys and lichen-draped forests of oak and spruce where it's quiet enough to hear your own heartbeat. Drive the narrow lanes that whip through the drizzly heights to who-knows-where and the only traffic jam you'll encounter is unruly sheep on the road.

Eyes on the skies

The lack of people and light pollution is naturally a recipe for astonishingly dark skies, ripe with stargazing potential. The Bannau Brycheiniog to the south and Snowdonia to the north hog the limelight with their

© DAFYDD WYN MORGAN / CAMBRIAN MOUNTAINS INITIATIVE

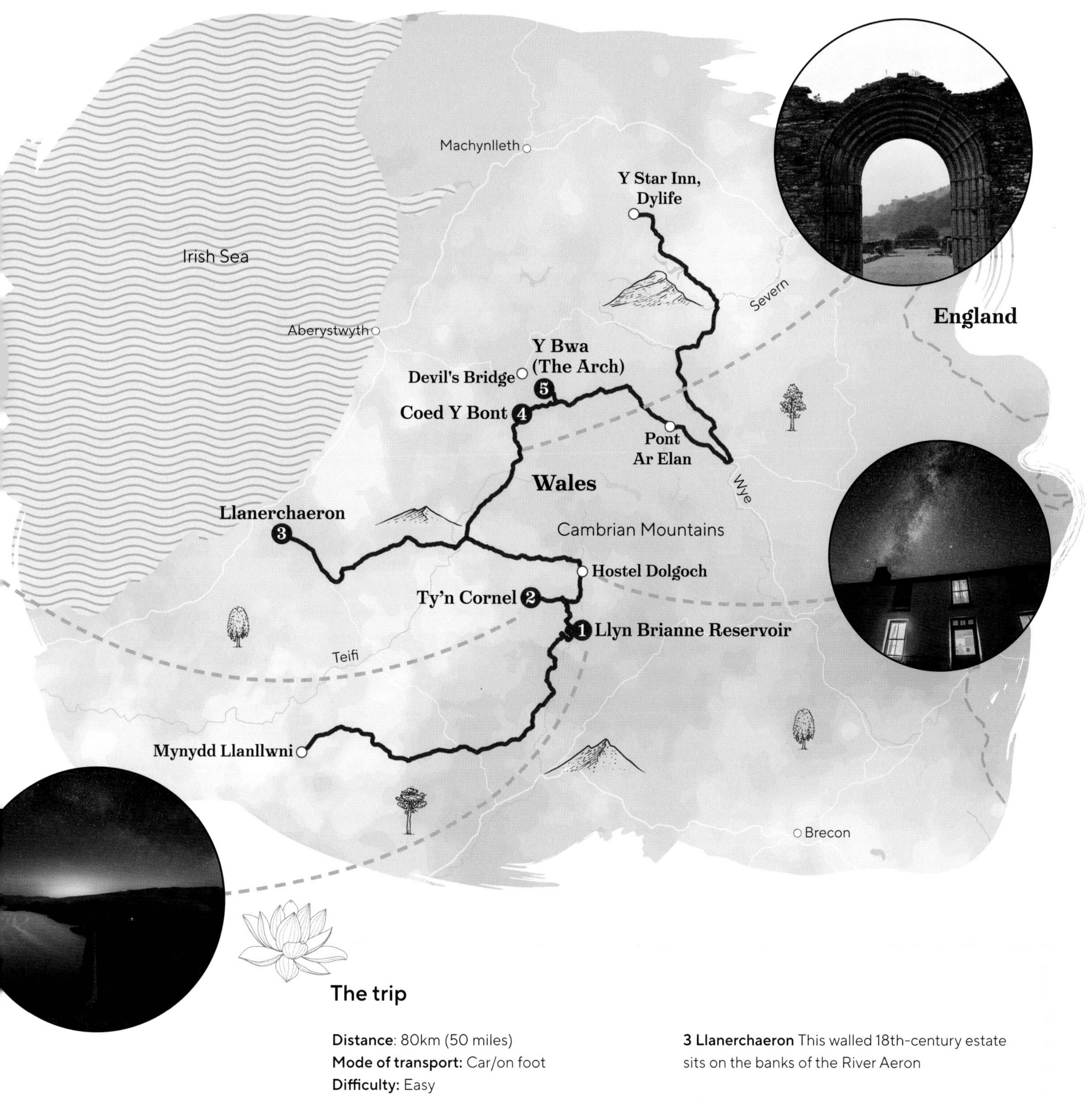

The trip

Distance: 80km (50 miles)
Mode of transport: Car/on foot
Difficulty: Easy

1 Llyn Brianne Reservoir Stars glitter brightly above this fjord-like, forest-rimmed reservoir, home to the UK's tallest dam

2 Ty'n Cornel This farmhouse hostel is Britain's most remote, with unrivalled views of night skies

3 Llanerchaeron This walled 18th-century estate sits on the banks of the River Aeron

4 Coed Y Bont A remote, serene woodland near the ruined Cistercian abbey of Strata Florida

5 Y Bwa (The Arch) Once the grand entrance to the Hafod Estate, now an astrophotographer's dream

Clockwise, from left: taking in the night skies on the Cambrian Mountains Astro Trail; Ty'n Cornel; Strata Florida Abbey; Hostel Dolgoch; Llyn Brianne Reservoir

John C Barentine, founder of Dark Sky Consulting

The Cambrian Mountains have a valuable resource that is rapidly vanishing right now: night-time darkness. Stargazing brings sensations of calm in the quiet contemplation of nature as well as the enthusiasm we all have as children when we want to learn and know everything. With no specialised knowledge or equipment, you can immediately connect with nature and ponder life's big questions: what is the universe? What is our role in it? And what does it all mean?

© DAFYDD WYN MORGAN

International Dark Sky Reserves, but the planets and constellations can also be observed with startling clarity in this off-the-radar corner of the country – and you'll likely have the universe all to your lucky self. Cue the Cambrian Mountains Astro Trail, an 80km (50-mile) self-guided journey ticking off nine locations that have been awarded Milky Way Class Dark Sky Discovery Site status, turning the cosmic light on this beautiful region.

'Witnessing the night skies in the Cambrian Mountains means you've arrived in one of the world's most amazing landscapes,' says Dafydd Wyn Morgan, project manager for the Cambrian Mountains Initiative. 'The Cambrian Mountains are home to some of the darkest, low-light communities in the world. The palette of colours is incredible by day, but by night it paints an entirely different picture. Here the skies are *bola buwch,* as we say in Welsh, or as dark as the belly of a cow. Experiencing the peace of night in the Cambrian Mountains on a mountainside or summit is so soothing. It can be so dark that you jump out of your skin if you hear footsteps or voices. I love being able to photograph things that are millions of miles and light years away. Once I was with a fellow astrophotographer who was setting up his camera at 1am when a meteor flew right across the night sky in front of us. He could sense my smile in the darkness.'

Celestial moments

Dafydd is right. These skies are sensational. The trail zigzags from the remote moorland hill of Mynydd Llanllwni in the south to the car park of Dylife's aptly named Y Star Inn pub in the north, but most stargazers cover it in stages, stopping in one dark-sky location for a night or two before moving onto the next. With a pair of binoculars or – better still – a telescope, you can spot the hazy spray of the Milky Way, meteor showers and the International Space Station, but some stars and constellations are visible with the naked eye alone: Polaris (the North Star) and the Plough (the Big Dipper), for instance.

North of Mynydd Llanllwni, the Llyn Brianne Reservoir twists and turns through spruce forests barely visible in the darkness. The only sound is the muffled bleats of a thousand sheep and little gasps escaping from would-be astronomers as the stars and planets blaze in vaulted skies. In summer here, you can see meteor showers. And autumn and winter are just as magical, opening up phenomenal views of the Milky Way, Taurus, Orion and Sirius.

The road to nowhere

The Astro Trail has one rule of thumb: if you're on the road to nowhere, keep going. Lit only by the stars, single-track lanes wriggle through the moors to Ty'n Cornel, a glorious farmhouse hostel tucked away in a valley near Soar-y-Mynydd, the most remote chapel in Wales. The moorland rolls on north through spectacular nothingness to another stargazing hotspot, Hostel Dolgoch, where, after a night peering up at galaxies, you can crawl into a cosy bunk in an off-grid 17th-century farmhouse hidden in a wild, wild valley near the source of the River Tywi.

> **"The only sound is the little gasps escaping from would-be astronomers as stars and planets blaze in vaulted skies"**

Moving west of Hostel Dolgoch brings you to the pitch-black skies of Llanerchaeron, where an 18th-century Georgian villa and parkland estate designed by architect John Nash hug the forested banks of the River Aeron. Swinging northeast instead takes you to Coed Y Bont, a pocket of remote woodland near romantically ruined Cistercian abbey Strata Florida, where you might spot a nocturnal pine martin or, as you look towards the horizon, a shooting star.

So much space

A quick spin north of Coed Y Bont reveals one of Dafydd's stargazing favourites, Y Bwa – or The Arch – a freestanding 18th-century stone arch that was once the stately gateway to the Hafod Estate. It's near Devil's Bridge, where waters crash deep into a gorge swirling in local myth. On clear, moonless nights here, the sky appears immense and infinite, lavishly embroidered with stars and planets.

East of here is the penultimate stop: the arched stone bridge of Pont Ar Elan, deep in the Elan Valley, a great fretwork of reservoirs, dams and forests spread across Wales' midriff in wondrous solitude. Far removed from any light pollution, this silver-tier International Dark Sky Park reveals its celestial magic after dark. And as the road edges north and Snowdonia's brooding peaks pucker up, you reach end-of-trail Y Star in in Dylife, a rambling old drover's inn where you can expect a view of distant planets with your pint.

On the Cambrian Mountains Astro Trail there is space. Space to escape. Space to think. Space to dream. Space above.

© SHUTTERSTOCK / DAVE KNIBBS

From top: Pont Ar Elan in the Elan Valley; stargazing hotspot Y Bwa (The Arch)

Practicalities

Region: Cambrian Mountains, Wales
Start: Mynydd Llanllwni
Finish: Y Star Inn, Dylife

Getting there and back: You'll need your own wheels to reach the trail's dark-sky locations, which can be driven south to north or vice versa. The nearest major car rental locations are in Aberystwyth (west), Brecon (east), Ammanford (south) and Machynlleth (north). Be mindful that roads can be narrow (sometimes single-track), winding, unlit and prone to ice/flooding in winter.

When to go: Stargazing is possible year-round. Winter months are darker but colder. Visit within a week of the new moon to maximise darkness hours; and try to time your visit for the biggest meteor showers, including the Perseids (August), Geminids (November) and Quadrantids (January).

What to take: Check the weather in advance and bring warm, waterproof layers, a blanket, boots or wellies and binoculars. Even in summer, nights can be cold. Let your eyes adapt to the darkness, then use a planisphere or an app like Stellarium Mobile, or bring a star chart for the month.

Where to stay: Stargaze and stay in historic farmhouses-turned-hostels Ty'n Cornel and Hostel Dolgoch, both Dark Sky Discovery Sites. Or glamp on a working farm at Wigwam Holidays Hafren in Staylittle, which arranges stargazing workshops and events. Y Star Inn offers lodgings in a terrifically isolated location.

Where to eat and drink: Bring a flask of something hot to help keep warm, and snacks if you intend on staying up all night.

Essential things to know: Cambrian Mountains Dark Skies (cambrianmountainsdarkskies.co.uk) should be your first port of call, with a downloadable Dark Sky Guide and interactive map, plus plenty of handy tips on how to stargaze and where to stay. There's more info on the Cambrian Mountains website (thecambrianmountains.co.uk).

Follow seafaring saints to Iona

Mystical Iona lies at the westernmost edge of Europe, and making a pilgrimage to the island is a tradition that dates back to the first centuries of Christianity.

Irish monks found that the solitude of Iona was a conduit to spiritual enlightenment, and anyone travelling to the island today – with or without faith – can also find quiet corners in which to refocus the mind.

Iona offers a history lesson on how Christianity spread inland from the Atlantic, with the island serving as a shining light in the so-called Dark Ages. The natural world of Iona – from seals to abundant seabirds – is also illuminating in its own way.

Bracing walks in stiff Atlantic winds will leave you feeling refreshed and rejuvenated.

In the northwesternmost reaches of the British Isles, the Inner Hebrides lies scattered across the sea like the broken pieces of a smashed plate. Some islands – Skye, Jura – are sizable chunks, while others – Rum, Eigg – are modest shards. Among them is Iona, a piece tiny enough to miss the bristles of a sweeping brush, but facing square against the cold leagues of the Atlantic. Though this out-of-the-way place stood at the boundary of the known world for centuries, it was one of the first footholds of Christianity in Western Europe – and ever since it's been a destination for generations of pilgrims. Today, getting to Iona entails travelling through some of Scotland's most magnificent landscapes and seascapes – a journey that often touches the souls of those who undertake it.

By rail through the Glens

In ages past pilgrims walked to Iona: now it's very

© SHUTTERSTOCK / REIMAR

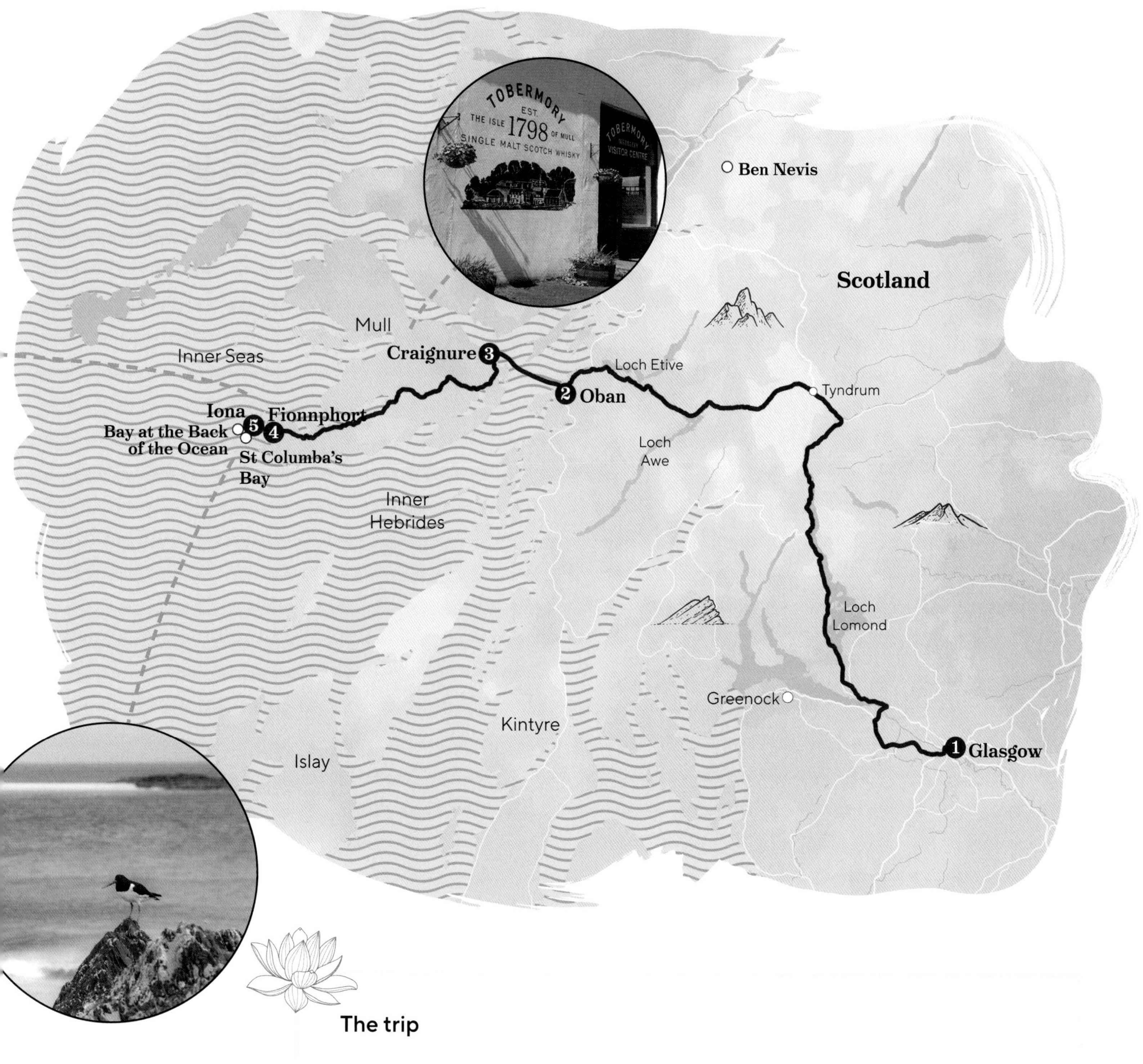

Clockwise, from left: Inner Hebrides homes on Iona; Tobermory's whisky distillery; oystercatcher on the Iona coast; the Bay at the Back of the Ocean

The trip

Distance: 233km (145 miles)
Mode of transport: Train, ferry, bus and on foot
Difficulty: Easy

1 Glasgow Scotland's biggest city provides an absorbing starting point

2 Oban This spirited port is a gateway to the islands to the west

3 Craignure The Isle of Mull's ferry port is a stepping-stone to Iona itself

4 Fionnphort This tiny village on the Ross of Mull is the last staging post before you reach Iona

5 Iona A tiny island which has stirred the imagination of millions

Arriving at the
Isle of Iona

Oliver Smith, Iona pilgrim

Though small, Iona has a tapestry of different landscapes – a windswept hill in the north, farmland in the middle and white-sand beaches on all sides. Perhaps the most memorable place for me was the west coast's evocatively named Bay at the Back of the Ocean. Here you could stand and feel the full force of gales inbound from the Atlantic, look at rags of white surf topping the waves – and imagine the faith that carried St Columba through these rough seas from Ireland long ago.

possible to drive there too. But perhaps the best way to get there is via train, ferry and bus. It's a complex undertaking which, like all true pilgrimages, requires effort, planning and persistence – but the rewards are window-seat views of Scotland's most majestic scenery, from blustery glens to shimmering lochs.

Start out at Glasgow Queen Street Station, a transport hub both for Scotland's biggest city and the densely populated Central Belt. But places and people tend to thin out once you board a northbound service to Oban. Here, the trains chunter along a section of the West Highland Line – one of the world's most celebrated railway journeys – passing the finger-like inlet of Loch Long and skirting bonnie, island-strewn Loch Lomond before turning westward, under the shadow of snowcapped mountains, at Tyndrum. Pretty little Oban is where the train meets the tides: its harbourside station is positioned so passengers need only make a few strides to board CalMac ferries bound for the Isle of Mull. If you've time to kill before making a connection, it's worth exploring Oban – head for the 19th-century McCaig's Tower folly on the hilltop, or the seafront St Columba's Cathedral, named after the saint who founded the abbey at Iona.

Over the water to the Hebrides

When the weather is clear, the views from the ferry deck are wildly beautiful, with the battlements of Mull's Duart Castle to the left and Lismore Lighthouse teetering on wave-lashed rocks to the right. For many on board, the Isle of Mull is the final destination, with the colourful fishing port of Tobermory and the west coast's white-sand beaches exerting a powerful lure. For pilgrims, however, Mull is a stepping-stone to its far smaller neighbour, Iona. From the port at Craignure,

"Step away from the hubbub of the wider world and into the sanctuary of this little landmass out in the waves"

buses follow a remote single-track road westward, passing penitently under the hulking mass of Ben More. Look out for standing stones by the roadside – ancient signposts erected to mark the way to Iona.

The tarmac runs out at Fionnphort; you can see Iona Abbey rising up across the strait from the village's lobster-pot-strewn quays. A second, smaller CalMac ferry makes a second, shorter sea crossing from here, clanking and crashing across the narrow Sound of Iona. These ferries are an ancient institution – medieval pilgrims summoned them by setting fire to lumps of heather. Then, as now, to make the passage is to cross a kind of threshold; stepping away from the hubbub of the wider world and into the sanctuary of this little landmass out in the waves.

Making landfall on Iona

It was an epic sea journey that first brought St Columba here from Ireland in 563 CE, sailing with his companions in a wicker coracle through the squalls. He founded his abbey in the shape of Jerusalem and used it as a base from which to spread Christianity eastward into the pagan mainland. Iona was home to a community of monks who farmed, fished and copied the gospels; Columba himself slept on a bed of stone, and walked into the cold morning sea to recite psalms. An easy 10-minute walk from the ferry dock brings you to Iona Abbey – a stout Benedictine structure, where Gothic arches frame the Sound's waters and seabirds swoop over the cloisters. In the grounds you can find the mound on which St Columba's simple timber hut may once have stood – a focal point of pilgrimage.

Celtic Christianity is rooted in nature, though, and the entire island has countless spots for contemplation in the company of the Atlantic elements. There could be nowhere better to end your journey than St Columba's Bay – a pebbly cove at the southern end of Iona, an hour's walk away – where it is believed the saint first came ashore. Listening to the rush of the wind, the squawking of gannets and the hush of the tides raking the stones, you have reached both the edge of Eurasia and your journey's end.

From top: looking over the Sound of Iona to Iona Abbey; McCaig's Tower in Oban

Practicalities

Region: Inner Hebrides, Scotland
Start: Glasgow
Finish: Iona

Getting there and back: Glasgow is well-served by rail links from across the UK, including direct services from London (around 5hr) and Edinburgh (50min). It's easy to change from Eurostar trains into London St Pancras on to Glasgow-bound services, as London Euston is just next door. Glasgow is also accessible on overnight sleeper services from London.

When to go: This part of Scotland can be visited year-round, but is arguably best experienced in early summer (May and June) and early autumn (September and October). High summer can mean bigger crowds and midges (biting insects), while winter brings shorter days and the risk of rougher seas.

What to take: Warm and waterproof layers are essential for any journey in the Inner Hebrides. Midge repellent is a good idea during the summer months.

Where to stay: Iona has a smattering of hotels and guesthouses – one of the largest is the whitewashed St Columba Hotel, where rooms look across the Sound of Iona to Mull.

Tours: You can arrange travel to Iona independently, booking trains within Scotland via ScotRail (scotrail.co.uk), ferries with CalMac (calmac.co.uk) and buses across Mull with West Coast Motors (westcoastmotors.co.uk). Iona Abbey is in the care of the Iona Community (iona.org.uk), an ecumenical Christian organisation which offers long retreats on the island.

Essential things to know: En route to Iona, it's worth stopping off to make the most of your time in Scotland. Glasgow is a vibrant metropolis with outstanding museums such as Kelvingrove Art Gallery; Tobermory on Mull is home to an excellent whisky distillery; and from Tyndrum you can detour northward by train to Ben Nevis, the UK's highest peak.

The Magna Via Francigena: hiking through the heart of Sicily

Experience the spirit of resilience and revival in Sicily's sparsely populated interior, where restored pilgrim pathways are bringing new life to remote communities.

The resourcefulness of communities along this route, through some of the most deprived parts of Italy, offers a lesson in resilience and innovation – not least in areas where organised crime has historically dominated.

This route has ups and downs, both literally and metaphorically. Many stages pass through remote countryside and melancholy abandoned villages; you'll need to rely on your wits and map-reading skills (phone signal isn't guaranteed).

Interwoven with Christian history and jewel-like churches, Sicily's rural landscapes offer the space for quiet contemplation and meditation, regardless of faith.

Sutera, so it's said, has been touched by the hand of God. This town, stacked theatrically around the steep peaks of Sicily's interior, is crowned by its 'broken rock' – a craggy precipice that, according to legend, was split in two by lightning that struck at the moment of Christ's death. Today, religion still shapes life in Sutera. Church bells bring people to worship several times daily; Semana Santa fills the streets with Easter processions that rival

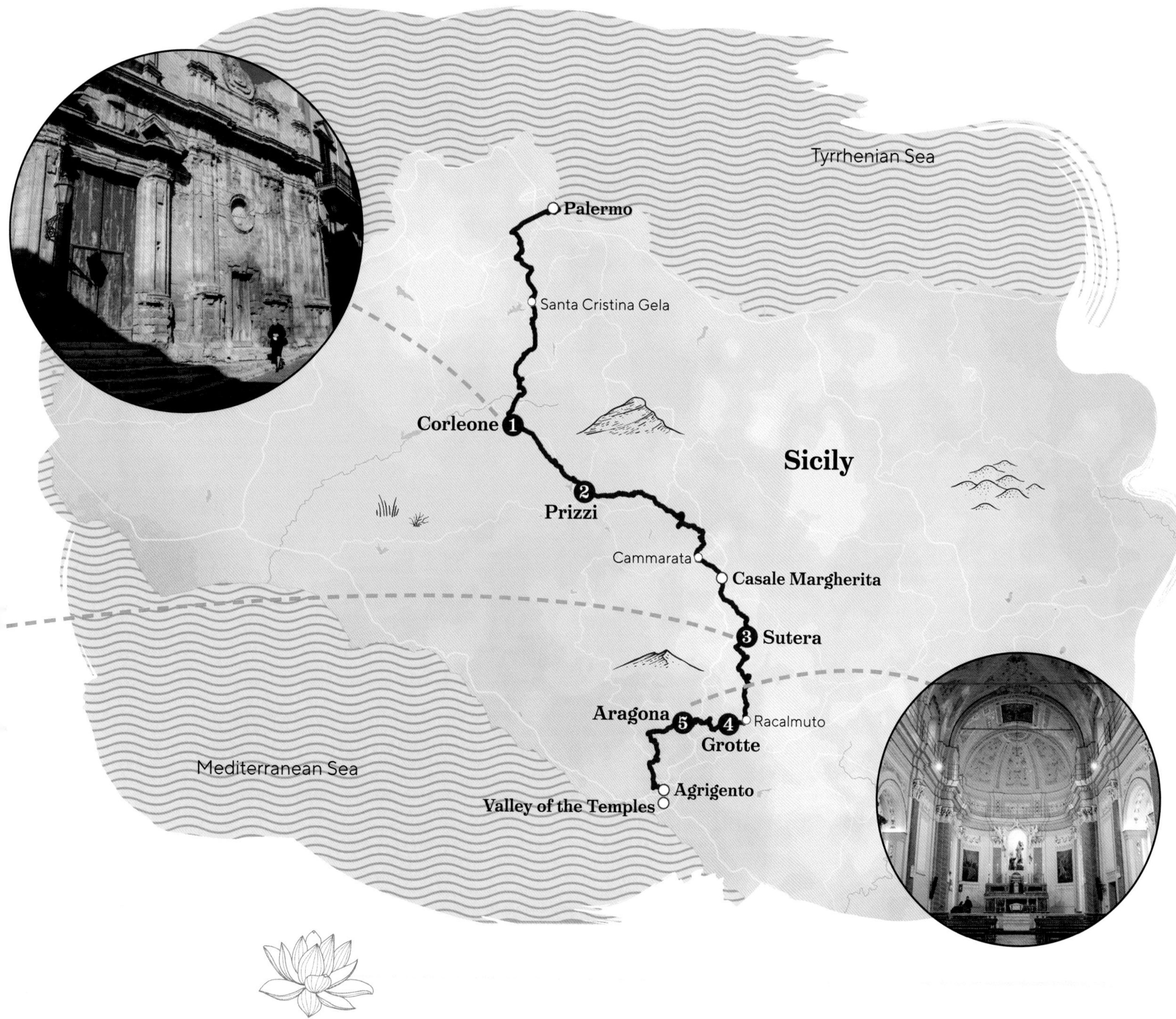

The trip

Distance: 180 km (112 miles)
Mode of transport: On foot
Difficulty: Moderate to difficult

1 Corleone At this spectacular hilltop town, hikers can learn about Sicily's Mafia history and the brave local resistance to organised crime

2 Prizzi Switchback trails bring hikers to this medieval hamlet, with a buzzing hostel and an archaeological museum

3 Sutera Another dramatically set hilltop town, with a fascinating ethnological museum and a mountaintop church bell for pilgrims to ring

4 Grotte The murals lining Grotte's labyrinthine Old Town are part of a vibrant street-art project that's reviving local culture

5 Aragona A handsome town where Gothic churches and the castle of Sicily's 18th-century prince display frescos and jewelled ex-votos

Clockwise, from left: Palermo Cathedral; looking out over the Rabato district, Sutera; Chiesa di San Domenico, Corleone; Chiesa del Carmine, Aragona

Marialicia Pollara, Corleone guesthouse owner

This part of inland Sicily has such a rich history and culture, yet in Corleone, we're known simply as a place of Mafia. But now, with the arrival of pilgrims, we can tell our real stories. There's more demand for hospitality, and locals are converting vacant buildings as lodgings for guests. We are so proud of the Magna Via – it brings people together and it's a route I always find joy in walking, again and again.

those in Spain for gilded, saint-worshiping splendour. But few outside Sicily's heartland know of this town which, like many in Italy's rural south, had seen its ageing, fast-shrinking population so depleted that any new birth was greeted like the Second Coming.

With its youngsters fleeing for opportunities on the coast and Italian mainland, Sutera didn't wait for divine intervention. Taking destiny into its own hands, the municipality has settled dozens of asylum seekers into the vacant stone buildings of its medieval Rabato neighbourhood over the past decade or so. 'Greeks, Arabs, Spanish: Sicily has seen thousands of years of immigration,' says local guide Miri Salamone. 'We're used to integrating. Italy has turned its back on rural Sicily, so we've had to find our own way to do things.' It's a spirit of resilience found across the hill towns of the island's interior, a region looking to put itself firmly back on the map with the revival of its old pilgrims' paths.

Laying the trail

In 2012, Sicily's abandoned, ancient trails, once part of a pilgrimage route from Canterbury via Rome to the Holy Lands, were re-mapped by a group of Italian volunteers – historians, archaeologists and naturalists. In a labour of love that took 10 years to come to fruition, the 1000km (621-mile) network of old paths, trading routes and *trazzere* (grazing tracks) now form waymarked hiking routes across the little-visited interior of the island. The main artery, the Magna Via Francigena, cuts 180km (112 miles) through the heart of the country from Palermo on Sicily's Tyrrhenian

© FRANCESCO LASTRUCCI

"The resourcefulness of communities along this route offers a lesson in resilience and innovation"

coast to Agrigento on the Mediterranean; those walking its length can have a pilgrim's passport (*credenziali*) stamped at participating cafes, chapels, shops and B&Bs at stages along the route.

South to Corleone

The Magna Via Francigena officially starts in Palermo, where you can pick up a pilgrim's passport at the cathedral, but most hikers skip the Sicilian capital's dreary outskirts and catch a bus into the hills above, bound for the Via Francigena's best start-point, Santa Cristina Gela, where the village cafe offers a passport stamp and packed lunch. Out on the trail from here, the route weaves through the herb-rich grazing land that gives a fragrant-sweet flavour to the milk used to make Sicily's exemplary ricotta-stuffed cannoli. Muddy livestock populate wheat fields patchworked around lakes, a peculiarly English vista that contrasts sharply with sudden canyons of subtropical green, thick with palms, aloes and prickly pear cactus.

A long day's walk concludes in the fabled town of Corleone, draped across several towering tabletop mountains. At CIDMA, an anti-Mafia museum set in a former orphanage, guide Federico Blanda wants to set the story straight. He passionately relates Sicily's complex history of organised crime as something far from the fictionalised romanticism of the *Godfather* movies. Tours explore harrowing images of Mafia shootings by photojournalist Letizia Battaglia, and court documents detailing the hundreds of Mafiosi tried in Palermo's Maxi Trial. 'The Mafia wasn't born in Corleone – it's all over Italy. In the 1980s and '90s, at the trial's height, we were a population of 10,000; just a handful were Mafia. Yet Corleone's people still live with the stigma.' It's a story of immense bravery and resistance that contrasts sharply with the town's incredible beauty.

Old traditions, new beginnings

The route rollercoasters across Corleone's mountains, passing churches and monasteries, and following the San Nicolò River as it rumbles through a chain

© MARZOLINO/ ISTOCK / GETTY IMAGES PLUS

Left: sample organic wine in the hilltop hamlet of Prizzi
Opposite: Corleone, home to the CIDMA anti-Mafia museum

Grotte's graffiti art

Since 2017, the tiny town of Grotte, just inland from Agrigento, has been on a mission to bring new life to its abandoned Old Town, with artists transforming its streets with murals. Friezes of mountain flowers and local herbs unfurl across flaking masonry, multiethnic angels hover above doorways, and 1950s ad posters are reimagined as statements against consumerism. Spot a snake-woman coiling across the facades of two houses, by local artist Andrea Casciu; it recalls the Sicilian legend of La Biddina, a heartbroken girl whose sorrow turned her into a serpent. Elsewhere, you'll find the Dutch-master-subverting *Boy with a Stone Earring*, by Venezuelan Luis Gomez de Teran, along with several clever optical illusions that shift with the angle of vision. Come for the art and linger longer for *arancini* rice balls and ricotta-filled cannoli, freshly made at family-run Bar Marconi on Piazza Marconi.

of waterfalls that once powered the region's flour mills, now largely defunct. In the nearby hilltop hamlet of Prizzi, far from any main highway, there are more stories of resistance and revival. 'We're remote. People didn't arrive here just passing by,' says Totò Greco, a local 'friend' of the Francigena. 'And there wasn't anywhere to stay. But with hikers, things are changing.' Totò has converted his grandfather's vacant house into a hostel and a hub for his Sikanamente association, which incentivises youngsters to stay in the area.

New opportunities in hospitality, and work reviving the region's neglected vineyards, have brought Totò and other natives back from jobs overseas. Prizzi is a fine place to sample new organic wines conjured from the region's ancient grape varieties, and to learn about its pre-Roman Sicani culture at the impressive Hippana Archaeological Museum. Agricultural communities, too, are feeling the benefit of the Francigena. The route continues through the remarkably green woodland of the Monte Carcaci nature reserve to arrive at Casale Margherita, an *agriturismo* on an organic farm just outside Cammarata, set around a restaurant and swimming pool. 'So many of us went abroad to find work,' says owner Carmelo. 'But I've returned home, to set up life here with my wife and kids. Hikers will spread the word about our region.'

From peak to sea

Upwards, onwards, over steeply rolling hillsides, the route reaches Sutera and its distinctive 'broken' peak.

© FRANCESCO LASTRUCCI

"We're remote. People didn't arrive here just passing by, and there wasn't anywhere to stay. But with hikers, things are changing"

The town's excellent Museo Etnoantropologico, set in a beautiful old convent building, is filled with medieval farming tools, exquisite icon paintings and pointy-hooded religious robes from the region's sizeable Semana Santa celebrations. Museum guides lead trekkers up the 157 steps to Sutera's bell tower, where pilgrims can ring their arrival across a panorama of peaks extending, in clear weather, to Mt Etna in the island's far east.

From here, the Magna Via Francigena offers glimpses of the Med's azure-blue glow, weaving through towns where narrow medieval streets are replaced by palm-shaded piazzas and grand Italianate *palazzi*. In the town of Racalmuto, topped by a Norman castle, the 1930s home of novelist Leonardo Sciascia conserves the typewriters, Modernist art and shelves of books that served its most celebrated native, with mountains views framed through elegant windows. 'All my books have not only been written in that place,' said the author of Sicily's heartland, 'but it is as if they are connected to it. To the landscape, the people, the memories, the feelings.'

As the route winds seawards, arts and culture abound. In the village of Grotte, named for the cave-houses carved into the hillside here, vast murals bring vibrant colour to abandoned buildings – an ever-expanding street-art project that demands more than a quick pause for breath. In Aragona, the 18th-century seat of Sicily's Spanish-appointed royals and birthplace of celebrated Sicilian playwright Luigi Pirandello, museums set in former castles and the crypts of Gothic churches promise rich treasures. Spare an afternoon for exploration; those that linger too long might skip the final 9km (6-mile) walk to the coast at Agrigento, instead hopping on the train for the 20-minute ride to the Valley of the Temples. Sicily's Unesco-listed Greek ruins make for a spectacular seafront terminus – but your pilgrimage isn't over until you complete a final climb to the city's hilltop cathedral, where a *testimonium* certificate awaits triumphant Magna Via Francigena trekkers.

Clockwise from left: street art in Grotte; Agrigento's Valley of the Temples; on the trail between Corleone and Prizzi

Practicalities

Region: Sicily, Italy
Start: Palermo
Finish: Agrigento

Getting there and back: Several European airlines serve Palermo, connected to Agrigento by frequent trains (2hr).

When to go: The optimum time to hike is between March to mid-July and September to mid-November, when the weather is most forgiving. Summer is sizzling, and the route has little shade or water-refill points. From mid-November onwards, storms can make trails hard going; there are detours above floodplains, but some sections can become impassable nonetheless.

What to take: Walking poles can help on rocky or muddy ground and deter the ubiquitous stray dogs. Pack light, even if your main luggage is being transported ahead of you, and take layers for all weathers, waterproofs in spring and autumn, plenty of sunscreen and sturdy walking footwear.

Where to stay: Locally run B&Bs, hostels and guesthouses are gradually popping up along the route, but book well in advance to guarantee a room for the night.

Where to eat and drink: Before you depart each morning, pack a lunch and as much water as you can carry (refill points are few to none).

Tours: Operators such as UTracks (utracks.com), SloWays (sloways.eu) and Walkers' Britain (walkersbritain.co.uk) offer accommodation bookings, maps and route notes, and can transport luggage ahead of you by taxi each day.

Essential things to know: This walk is graded as 'moderate', but you need a good level of fitness to enjoy it. Most take around six days to hike the route, covering 11km to 26km (7–16 miles) per day, and ascents and descents of between 400m and 1000m (1312ft to 3280ft). Local news, recommendations, discounted accommodation, tour guides and all manner of useful info can be found on the route's website (viefrancigenedisicilia.it)

Paddle Scotland's Great Glen

Discover the Highlands – and awaken your sense of adventure – on a coast-to-coast canoeing odyssey across Scotland.

The lochs that punctuate the Great Glen, especially Loch Ness, are mind-blowingly massive. And deep. And dark. Being out in their midst, sitting in a small canoe surrounded by all this vastness, is a humbling experience. Some people find it confronting, others liberating. Either way, it stays with you.

The lochs along the length of the Caledonian Canal are layered with legends, castles, ruins, ghosts, myths and monsters, offering a fabled float through centuries of history.

Canoeing for 3–5 days is physically and mentally demanding. Expect to dig deep and discover muscles you didn't realise you had.

Paddling the Great Glen is an aquatic quest that will remain in your brain forever. Oft-recounted anecdotes are born on this waterway and powerful bonds can be forged between people who share the experience – providing they're still speaking to one another by the time they reach the final take-out. Because doing the full route is no flippant float in the park: it takes several days, and demands stamina and a degree

© MATTHEW WILLIAMS-ELLIS/ GETTY IMAGES

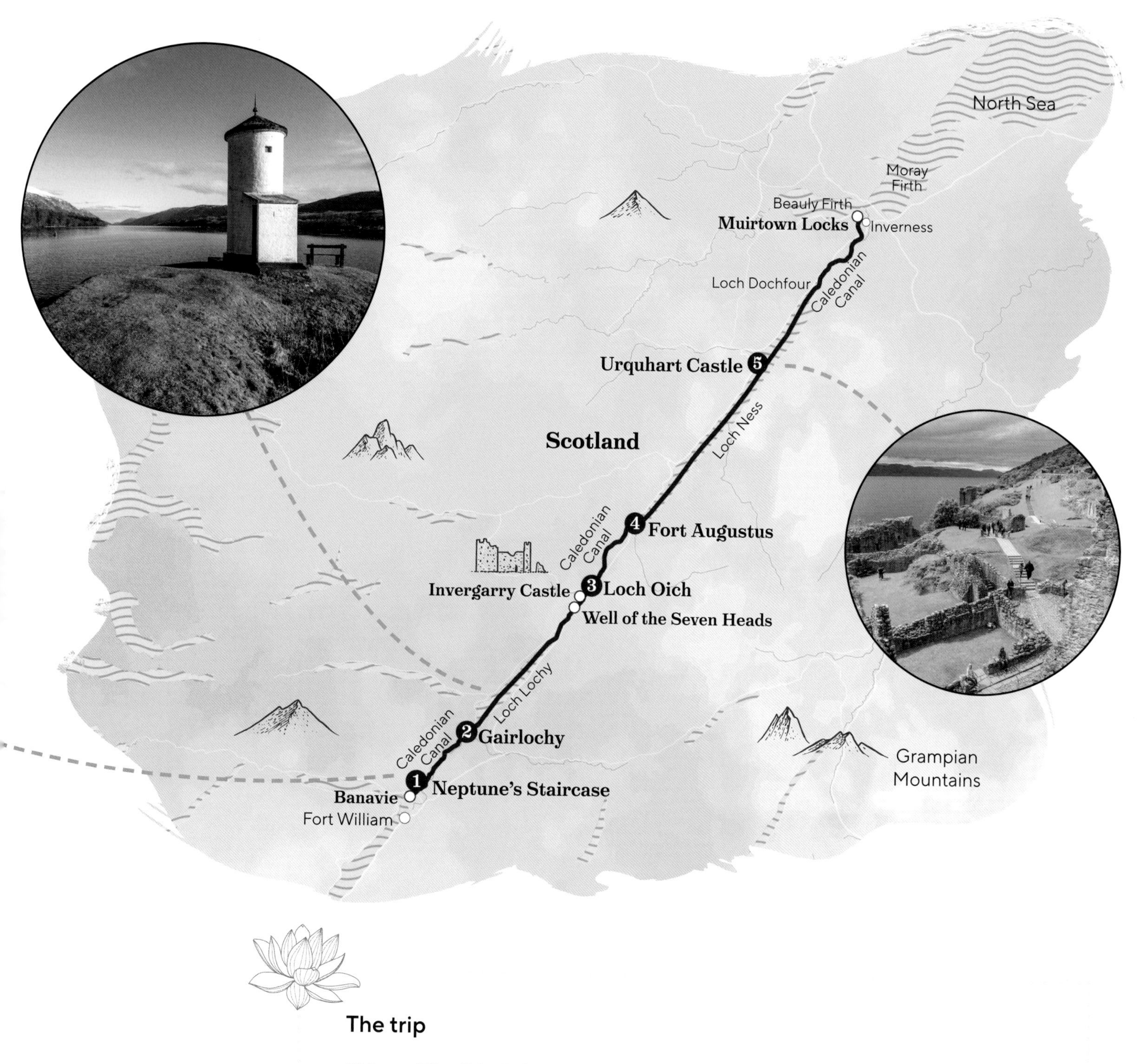

The trip

Distance: 95km (60 miles)
Mode of transport: Canoe/kayak
Difficulty: Moderate to difficult

1 Neptune's Staircase, Banavie Start at the top of Britain's longest liquid staircase; its eight locks connect Loch Linnhe to the Caledonian Canal

2 Gairlochy, Loch Lochy The first overnight stop for those spending five days paddling the route

3 Loch Oich Horrible history aplenty surrounds lochside Invergarry Castle ruins and the Well of the Seven Heads

4 Fort Augustus Time to start monster-spotting as you set out across leviathan-like Loch Ness

5 Urquhart Castle The evocative ruins of this 13th-century stronghold near Drumnadrochit overlook Loch Ness

Clockwise, from left: canoeing the Caledonian Canal; a Highland cow, Banavie; Gairlochy Lighthouse, Loch Lochy; Urquhart Castle, Loch Ness

Patrick Kinsella, Great Glen canoeist

I first paddled the Great Glen with my father in 2011 (a year before the canoe trail officially opened), when I'd been living on the opposite side of the planet to my family for 12 years. Dad had distilled in me a love of the outdoors, but the tyranny of distance had robbed us of many experiences. Spending four days in a canoe, wild camping on loch shores, was the most concentrated amount of uninterrupted time we'd shared for decades. I'll never forget it.

of courage. Think carefully about who you set out on this expedition with, because you'll be making a whole reel of enduring memories with them.

Locks to lochs

The Caledonian Canal stretches for some 95km (60 miles), but the challenge for canoeists and kayakers attempting to paddle it isn't found in the manmade waterways that flowed from the quill of engineering genius Thomas Telford when he designed the navigation in 1803. These sedate sections only account for a third of the waterway's total length. The crux of a coast-to-coast canoeing expedition along the Great Glen lies in the negotiation of the immense lochs, gouged out by a geological fault that's been ripping the Highlands in half for millennia.

Loch Ness is the biggest of the Great Glen's four freshwater lochs, and it's the scariest, even if you don't believe the stories about what lives in its depths. Along with Ness, Loch Lochy is also classified as open water. And when the wind whips through the Great Glen – as it frequently does with great fury, after roaring in from the Atlantic and becoming trapped in the Highland valleys – the waves on these two inland seas get alarmingly large.

And yet, on calm days – of which there are many, too – these same lochs can be incredibly peaceful places to paddle, with secluded spots along the banks that are almost somnambulant in their serenity. When conditions are fine, paddlers can pause and put down their oars, to drift and let the elements decide the direction of travel while they take in the surroundings, observe the wildlife or just disappear into their own heads for a while.

© MATTHEW WILLIAMS-ELLIS / GETTY IMAGES

> "Powerful bonds can be forged between people who share the experience – if they're still speaking by the final take-out"

On a multiday trip, there are myriad opportunities to moor and explore the banks, hunting for treasure, both tangible and ephemeral, between ruined castles and verdant woods. When wild camping along the shore, the evenings and early mornings are especially tranquil times to gaze out across the lochs. And once snug in your sleeping bag, you can listen to the lapping of the water while sailing off to sleep.

Way to go

The Great Glen can be explored in either direction, but most opt to paddle with the prevailing wind on their backs. This means starting in the west, near Fort William, which looks out over the Atlantic, and travelling east to Inverness, where the North Sea runs into the Moray Firth on Scotland's east coast.

At its western extremity, the Great Glen is soaked in the salt water of Loch Linnhe, but the first proper put-in for paddlers is at the top of Neptune's Staircase in Banavie, where the Caledonian Canal and the Great Glen Canoe Trail truly start. The 10.5km (6.5 miles) of easy canal paddling between here and Gairlochy is the perfect warm-up before you enter the first of the large lochs, 15km-long (9 mile) Loch Lochy, where conditions can be much more challenging.

Those intending to spend five full days on the water often make Gairlochy their first overnight stop, but other paddlers prefer to continue along Lochy's shores, where wilder camping can be enjoyed. If the elements allow, it's better to stay close to the western shore, which has a more natural feel since it doesn't have a road running parallel to it.

Historic castles & karma drama

A short section of canal connects Loch Lochy to Loch Oich: much smaller, but awash with bloody history. Paddlers pass 17th-century Invergarry Castle, which was the stronghold of the MacDonnells (a branch

© 4KCLIPS / SHUTTERSTOCK

Clockwise, from above: the eight locks of Neptune's Staircase connect Loch Linnhe to the Caledonian Canal; Urquhart Castle, overlooking the western shore of Loch Ness; paddling Loch Lochy

© WONDERLUSTPICSTRAVEL / SHUTTERSTOCK

Discovering monsters

Loch Ness is famed for the mysterious monster that allegedly lives in its depths. Sightings of this beast – variously described as a plesiosaur-like dinosaur, a whale or a humped serpent – began in earnest in the 1930s, but earlier accounts exist, including one from the 6th century. Several serious scientific attempts to locate the enigmatic animal have been conducted, but so far it has remained elusive. Nevertheless, a sizeable industry has grown around the monster, and 'Nessie' is a major tourist drawcard. Some Loch Ness maps even highlight the 'monster zone', where most sightings have happened; paddle over these if you're brave (or cynical) enough. In fact, Celtic folklore has it that most Scottish lochs have a resident monster: Loch Lochy has a kelpie called Lizzie and a *tarbh uisge* (water bull), while Loch Oich boasts a two-humped beast with a dog's head. So keep your wits (and camera) about you.

of the Clan Donald) until government troops blew it up after the Battle of Culloden in 1746, to punish the MacDonnells for supporting Bonnie Prince Charlie.

Nearby is the Well of the Seven Heads monument, above a spring named after a spectacularly gory story that began in 1663, when Alasdair MacDonald and his brother Ranald were killed by rivals within their clan. Two years later the brothers were avenged when Iain Lom (Bald John) found, killed and decapitated the seven murderers. On his way to Invergarry Castle to present the heads of the slain men to the clan chief, Lom stopped to wash his trophies at this spring, and it was thereafter known as Tobar-nan-Ceann – the Well of the Heads. Seven stone renditions cap the monument today.

From the eastern end of Loch Oich, the calm canal continues through bucolic farmland to Fort Augustus, where a flight of five locks drops boats down to the start of Loch Ness. Paddlers, of course, can simply portage (carry their boats) around the locks.

Lonely Ness & happy Ness

When beginning the paddle across Loch Ness, it's not an encounter with a monster that most sensible people are fearful of. It's the immensity and the tempestuous nature of the loch itself. Ness is 37km (23 miles) long and up to 230m (755ft) deep. It holds more water than all the lakes of England and Wales combined, and it doesn't suffer fools gladly. Even in a benign mood, Loch Ness is often lumpy. Small ripples scooting across the surface can play tricks with your eyes, and mind too, creating illusionary sights in the water – especially when you start dwelling on the demonically dark depths below your boat.

If conditions are calm, most paddlers strike out across the middle of the loch, but you should always check forecasts carefully before setting off. The safest approach is to hug the banks, although this adds quite a bit of distance to the journey. The east bank is quieter; the A82 runs along the western shore, but here the popular and picturesque ruin of Urquhart Castle, which dates to at least 1296, is well worth a paddle-past.

To journey's end

Bona Lighthouse heralds the end of Loch Ness, and here paddlers pass into little Loch Dochfour. Just before Dochgarroch, at the Ness Weir, the waterway splits. Most Great Glen canoeists and kayakers, laden with gear, stay left and cruise into Inverness along the canal, but experienced whitewater paddlers fork right to run the wild River Ness, where rapids include Fast Eddy and the Dragon's Tail.

The canal empties into Beauly Firth at Clachnaharry Sea Lock, where it meets the cold salt water of the North Sea – but just as the majority of paddlers begin the trail at Banavie, most also finish at Muirtown Locks, because beyond here you are simply portaging around locks for the sake of it. And by this stage, you have earned the right to go and enjoy a celebratory dram of the life-affirming local whisky: *uisge beatha* (the 'water of life').

Left: wild camping on the shore of Loch Ness

© MATTHEW WILLIAMS-ELLIS / GETTY IMAGES

Left: striking out across the open waters of Loch Ness

Practicalities

Region: Scottish Highlands
Start: Banavie
Finish: Muirtown Locks

Getting there and back: The overnight Caledonian Sleeper train from London terminates at Fort William. Banavie, on the West Highland Line, is served by trains from Fort William. Buses between Fort William and Inverness run alongside the lochs on the A82.

When to go: April to October are best. In summer, midges can be a serious annoyance, so come prepared.

What to take: Always wear a PFD (personal flotation device) when you're on the water. Independent paddlers will need all their own gear, but if you're hiring a boat or joining a tour, the following will be useful: several dry-bags, peaked cap, sunglasses, insect repellent, warm layers, waterproofs, water-shoes or neoprene booties, head torch, grippy gloves. Camping gear if needed.

Where to stay: There are serviced campsites along the route, and you can wild camp on the loch shores (follow leave-no-trace ethics and observe Scotland's Outdoor Access Code); note that you're not allowed to wild camp beside the canal sections. There are B&Bs dotted along the trail at places such as Gairlochy, South Laggan, Invergarry, Fort Augustus and Drumnadrochit.

Where to eat and drink: Villages and towns have pubs, cafes and shops selling food.

Tours: To paddle the Great Glen independently you need some kayaking/canoeing experience. All the necessary kit can be hired from Explore Highland (explorehighland.com) and Snow Goose (highland-mountain-guides.co.uk), which also offer shuttles and fully guided trips.

Essential things to know: Before starting out, register with Scottish Canals (scottishcanals.co.uk); it's free and gets you a canoe licence, waterproof map, access to facilities such as toilets and lots of up-to-date information. Visit greatglencanoetrail.info for more on the route.

Hiking the Wexford–Pembrokeshire Pilgrim Way

Tread in the footsteps of Celtic saints, reconnect with nature and rediscover lost rhythms on this over-the-sea pilgrimage trail from Ireland to Wales.

Pilgrimage provides a clear physical path for an inner journey – like a moving meditation. Intention is at its heart, whether you're resolving a problem, letting go or changing direction. As St Augustine said: *solvitur ambulando* ('it is solved by walking').

The route dives deep into Celtic history and folklore, replete with saintly legend, fairy forts and dolmens (megalithic tombs), druidic temples, Iron Age hillforts, holy wells and mighty cathedrals.

At one with the elements and your own thoughts, you'll build strength and stamina on this multiday walk, especially on the steep, rocky coast path. You'll typically walk 15km to 20km (9–12 miles) a day.

To walk the twisting, verdant country lanes of County Wexford is to feel the spirit of Celtic saints who once crisscrossed the storm-tossed Irish Sea. Over lichen-clad rock, mossy root and through muddy field, the Wexford–Pembrokeshire Pilgrim Way from Ferns to St Davids forges the saintly connection between Ireland and Wales, shining a light on the trail St Aidan took to see mentor and friend St David. And it all starts with a well.

Saints alive

Lore has it that St Mogue's holy well in Ferns miraculously sprang up when the monastery was built in the 6th century. When the monks needed clean drinking water, St Aidan told them to chop down a tree: the well miraculously burst forth from its split trunk. With the evocative medieval ruins of its Augustinian abbey, castle and Gothic-revival cathedral, Ferns itself punches high historically. It might be a quiet town

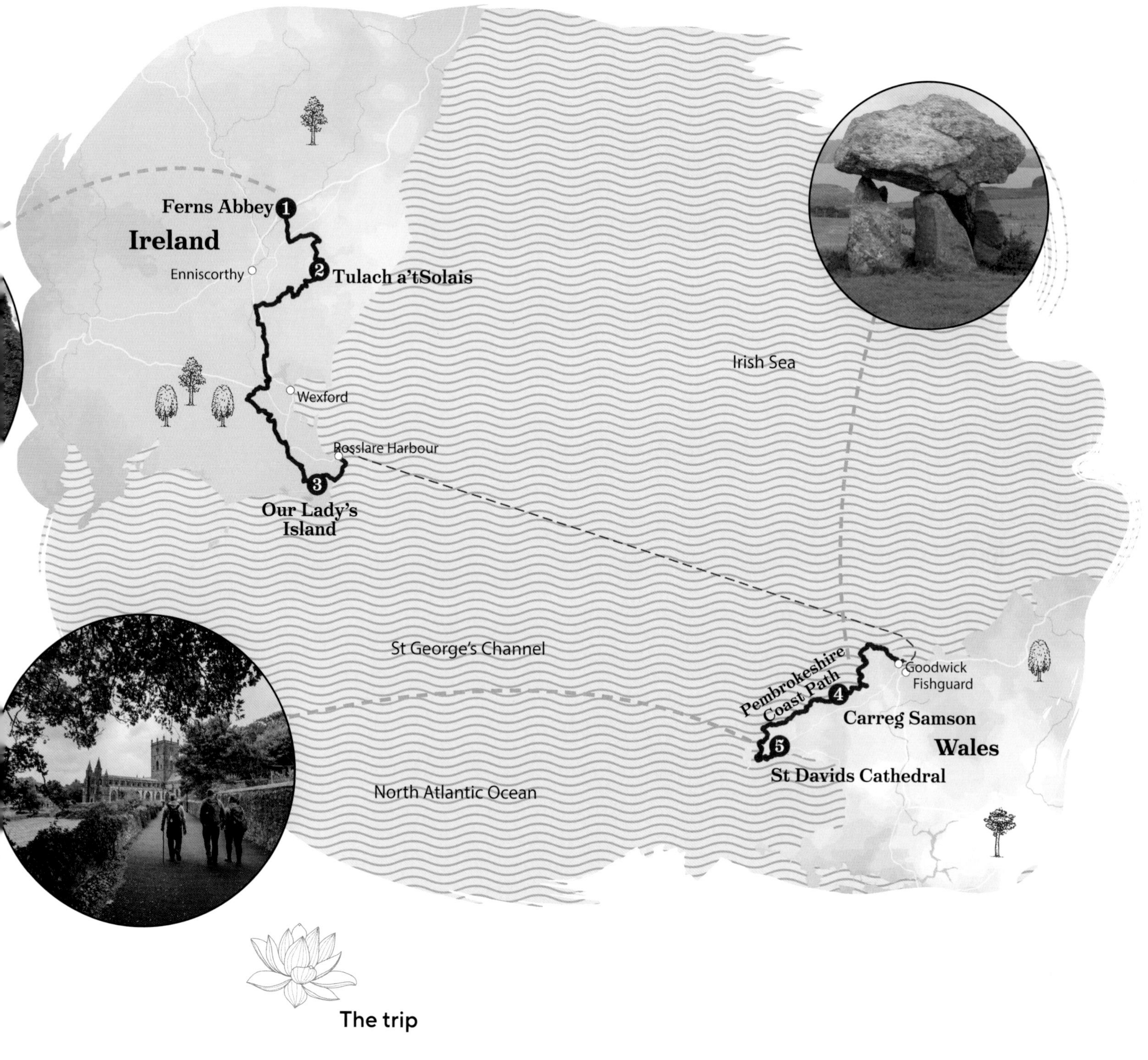

The trip

Distance: 260km (161 miles)
Mode of transport: On foot, ferry
Difficulty: Moderate

1 Ferns Abbey A ruined Augustinian abbey on the spot where St Aidan built his 6th-century monastery

2 Tulach a'tSolais This poignant 1798 battlefield memorial beautifully captures the solstice sun

3 Our Lady's Island Pilgrims come from afar to Ireland's oldest Marian pilgrimage site

4 Carreg Samson Topped by a massive capstone, this clifftop Neolithic burial chamber whispers of Wales' ancient past

5 St Davids Cathedral A medieval marvel and pilgrimage magnet, home to the shrine of St David

Clockwise, from left: hiking Wales' Pembrokeshire Coast Path; the castle at Ferns, Ireland; Carreg Samson, Wales; approaching St Davids Cathedral;

Iaian Tweedale, guide

Here you feel the spiritual connection with the living landscape – what the Celtic saints would have called a 'thin place', where the gap between Heaven and Earth is small. We pass holy wells, Celtic chapels, druidic temples and great cathedrals. But most important of all is the path itself. After walking for a few days, your mind calms and you observe things differently – seeing rocks, flowers and birds as if for the first time. It is when the outer journey from place to place becomes an inner journey from head to heart that self-discovery really begins.

today, but back when St Aidan became its first bishop and went on a monastery-building frenzy, it was a hotbed of spiritual activity. But St Aidan had itchy feet. Waving his disciples goodbye, he travelled on foot to the coast, crossing the Irish Sea to Pembrokeshire to study under St David. When he returned in 570 CE, he brought with him hives of bees, now remembered with a striking art installation: a trio of giant medieval beehives where bees once again make honey.

Heaven & Earth

Treading the path south, Tulach a'tSolais monument in Oulart commemorates the United Irishmen's uprising of 1798. Aligned to capture the solstice sun, it's a place of contemplation, with an inner chamber hiding two shrine-like curving tablets carved from Irish oak that open the view to the sky. The same peaceful air hangs over Oylgate, further south, where pilgrims linger at St David's holy well and in Raphael's Healing Garden, where sculptures of Tobias from the Old Testament remind us to accept life's challenges with grace. Spreading across wood and marshland, the Irish National Heritage Park near Wexford invites reflection with its recreated Celtic monastery and High Cross from St Aidan's time; a short hike south to Forth Mountain lifts spirits with soul-stirring views.

The final stretch on Irish shores threads along a seemingly never-ending dune-flanked beach, where the remains of an 8000-year-old petrified forest appear at low tide. Eventually, you reach Our Lady's Island, Ireland's oldest Marian pilgrimage site, with a sacred history predating Christianity (it is believed female druids gathered here). In August and September, it draws pilgrims in their thousands, who make nine circuits of the island – sometimes barefoot.

© COURTESY WEXFORD-PEMBROKESHIRE PILGRIM WAY

"Over lichen-clad rock, mossy root and through muddy field, the trail forges the saintly connection between Ireland and Wales"

Over the seas to Wales

Braving the choppy seas between Ireland and Wales must have been a proper adventure in the 6th century, but today it's a simple ferry ride. The trail gathers scenic momentum as it hooks onto the Pembrokeshire Coast Path, where ragged cliffs dive to a booming Atlantic. History is writ large here, too. Just out of the port in Goodwick, you arrive at Llanwnda's 6th-century chapel, where St Aidan had a fistfight with St Gwyndaf over who got the right to name the holy well.

Stride on south and the coast will likely hold you in its thrall – just as it did those early Celtic saints – as you hike above remote coves where seals pup in autumn, and round wind-battered Strumble Head, where storm petrels and Arctic skuas wheel above a lighthouse, and seals and dolphins splash in a sea full of rusting wrecks. On a clifftop near Abercastle, the haunting Neolithic dolmen of Carreg Samson reverberates with all the mysteries of its 5000-year history.

St Davids in sight

St David's Head ups the drama further, as the path threads along wildflower-freckled cliffs and past castaway bays. History is etched in an Iron Age hillfort, the Neolithic burial chamber of Coetan Arthur; legend has it that the volcanic outcrop of Carn Llidi was where an angel told St Patrick to go on his mission to Ireland. All of this is the drumroll for the dinky city of St Davids, where a mighty cathedral sits where St David founded his 6th-century monastery. By the 12th century, when Pope Callixtus II decreed that two pilgrimages here equalled one to Rome, St Davids was a destination to rival Spain's Santiago de Compostela. Pilgrims still flock in to pray at the saint's gilded shrine.

But it is nearby St Non's that really pulses with the spirit of those early saints. With its chapel ruins and holy well, this wild headland is where St David, patron saint of Wales, was born during a fierce storm in 500 CE – or so the story goes. It's an apt place to end, peering out to a quicksilver sea, with only your own footsteps and thoughts for company – just like those penance-seeking, blister-footed saints and pilgrims.

© COURTESY WEXFORD-PEMBROKESHIRE PILGRIM WAY

From top: route's end at St Davids Cathedral, Wales; commemorating the 1798 United Irishmen's uprising at Tulach a'tSolais

Practicalities

Region: County Wexford, Ireland/ Pembrokeshire, Wales
Start: Ferns, Ireland
Finish: St Davids, Wales

Getting there and back: Stena Line ferries run twice-daily between Fishguard in Pembrokeshire and Rosslare in Ireland (3hr 30min). From Rosslare's port, trains run to Wexford, from where the 740 bus goes to Ferns. From St Davids, buses serve the train stations in Fishguard and Haverfordwest.

When to go: Conditions can change fast year-round – from dazzling sunshine to fog, wind, storm and rain. Spring through to autumn are optimum: May and June bring a riot of wildflowers on the coastal path, and in September and October you can spot seal pups on Pembrokeshire beaches.

What to take: Bring wind- and waterproof clothing, hiking trousers and boots and a lightweight rucksack. Carry at least 1L of water each day, snacks and, in summer, sunscreen and insect repellent. Compact binoculars increase your chances of spotting wildlife.

Where to stay: Accommodation en route is plentiful, from basic campsites to B&Bs and hostels. For something with true pilgrim spirit, try the converted stone stables at Old Deanery estate in Ferns; and the lovely Old School Hostel and budget B&B in Trefin.

Where to eat and drink: The trail passes through many towns and villages, with pubs, restaurants and shops.

Tours: Guided Pilgrimage (guidedpilgrimage.co.uk) and Journeying (journeying.co.uk) offer guided tours on the route.

Essential things to know: For the inside scoop on the trail, visit the official site (wexfordpembrokeshirepilgrimway.org), with a day-by-day route breakdown, an audioguide and reflections on pilgrimage focused on self-discovery. The path is largely well-signposted, particularly on the Pembrokeshire leg, and fully digitally waymarked (download the map at outdooractive.com).

Walk France's Robert Louis Stevenson trail with a donkey

The rural trek is a challenge, but you won't be alone – follow in the footsteps of Robert Louis Stevenson and hike it alongside the ambling gait of a donkey.

You may end up reshaping how you show persistence without losing your temper or your resolve: donkeys don't carry the cliché of being stubborn without merit.

This trip may open the door for a more 'we-centric' worldview. When you can share victories and losses with a donkey, it's not a far jump to begin feeling camaraderie with pine trees under whose shade you rest, or with the gurgling streams from which you drink

Two weeks of hiking alongside a donkey, covering 15km to 20km (9–12 miles) per day on hilly terrain, should certainly add definition to your calves.

A few kilometres out of Le Puy-en-Velay, and the fantasy of lolling along the GR70 – aka the Robert Louis Stevenson Trail – is buried in the dust somewhere below the donkey's hooves. The donkey who is currently refusing to budge another step. Seemingly immune to your tugs, it walks two steps forward, only to veer off the trail to snatch a mouthful of grass. Whose idea was this again? Wasn't there something about creating a bond with the donkey and seeing the dormant volcanoes, rolling hills, vast prairies and rocky canyons of central France? The suspicion that it's impossible to reach Alès in two weeks seems more and more certain by the second.

How did Robert Louis Stevenson do it? He didn't have a GPS or modern hiking boots – but he did write a book about it: *Travels with a Donkey in the Cévennes*. And yet, in the present day, when the donkey does finally grant its accord to walk forward, a feeling of

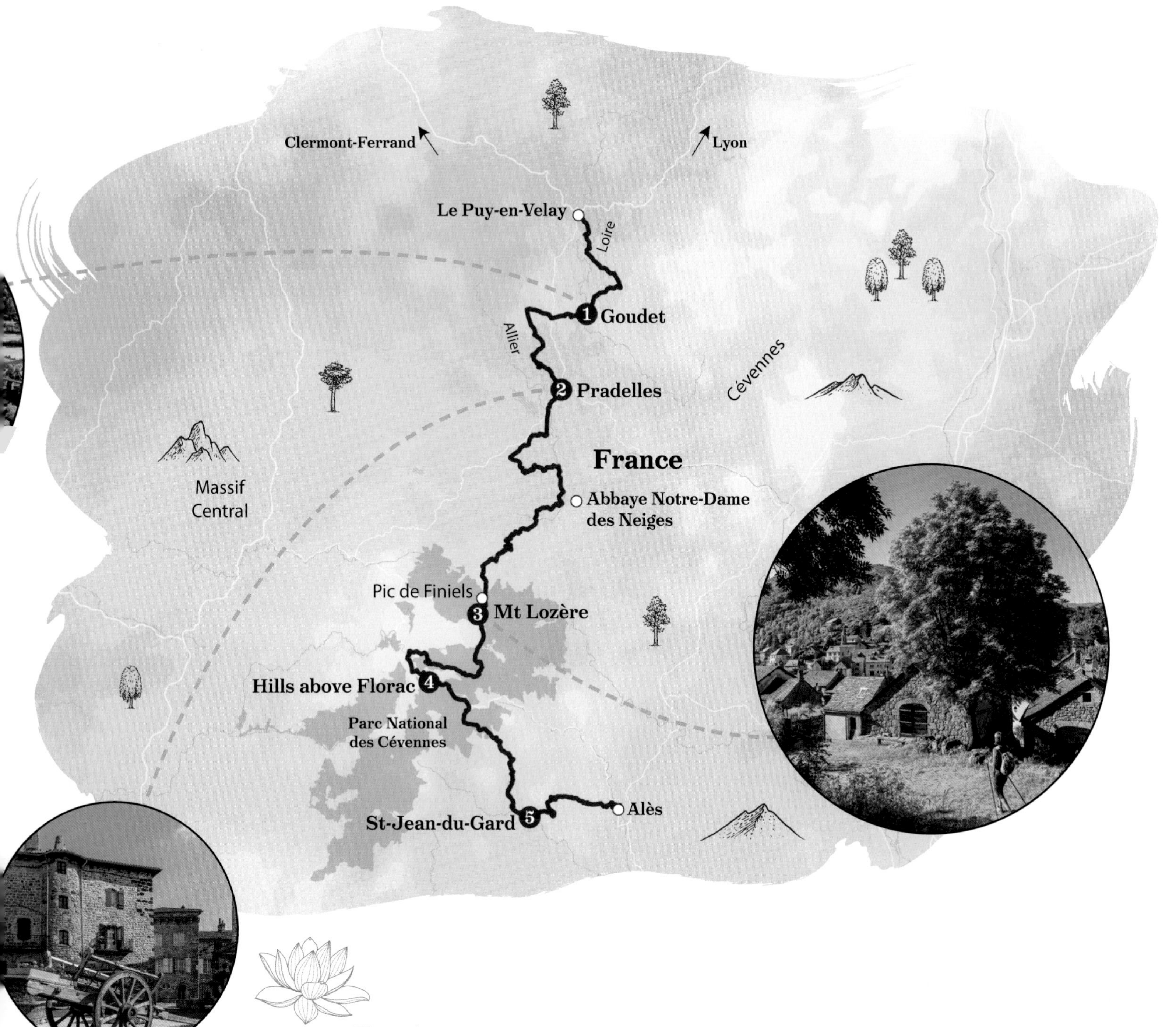

Clockwise, from left: approaching Pradelles in tandem; Goudet, in the Haute-Loire; on the trail in the Lozère; pretty-as-a-picture Pradelles

The trip

Distance: 272km (169 miles)
Mode of transport: On foot
Difficulty: Medium

1 Goudet Flanking the village is a thin ribbon of river that grows into the turbulent Loire

2 Pradelles Take a slow amble through the paved streets of this medieval village

3 Mt Lozère Try to observe the often-elusive wildlife here: beavers, bats and birds of prey

4 Hills above Florac Take a deep breath of thyme-scented air – you're in the Cévennes!

5 St-Jean-du-Gard Here Stevenson and his donkey Modestine parted ways, with Stevenson completing his journey solo

Elise Gravel, Stevenson Trail hiker

We didn't really plan our GR70 in advance; when we arrived to rent a donkey, only one remained. Allegedly, he was the worst of the herd. But we immediately got along. Within a day, he felt like part of our family. And we even grew to love his slow walk; it gave us the chance to truck along slowly, appreciate the journey and finish the days without being too tired.

quiet confidence creeps in. Is it a feeling of connection with some forgotten part of the rural experience? This is travelling in an ancient way, sharing the experience with a smart, sometimes stubborn animal.

Volcanoes & valleys

The hardest part of the Robert Louis Stevenson Trail is setting up a rhythm. But once both traveller and donkey have found their (hoof)beat, there seems to be no better way to explore this landscape. The first of the route's four distinct regions, the Velay, reveals green-topped dormant volcanoes, which spread lava across the region some three million years ago. Near an invisible regional line that separates the Velay from the Gévaudan sits the gorgeous village of Pradelles, classified as one of the 'Plus Beaux Villages de France'. Travellers here might encounter the story of the 'Beast of Gévaudan', a wolf-like creature that's said to have terrorised the region in the 18th century.

Back on the road, take a page out of the donkey's book and observe the journey with all your senses. In the Gévaudan, the sound of streams and birds is predominant. Wild prairies, pine forests and pastures dotted with moss-covered rocks seem to soak up any other noise – except your donkey's brays in the morning. Those brays are more powerful than a strong cup of coffee as a means of waking drowsy hikers.

See the forest for the trees

With high white walls surrounded by forests and fields, the Abbaye Notre-Dame des Neiges welcomed Stevenson. Maintained by monks until 2022, a chapter of nuns cares for the domain today. For hikers looking for spiritual reflection, or simply quiet repose, the sisters uphold the tradition of hospitality for travellers.

Below: ups and downs as the trail winds through the Lozère
Right: soaking away the miles near Cassagnas in the Cévennes

But halls of worship and meditation are not the only way to gain clarity. After leaving the abbey, Stevenson spent the night under the sky in a pine forest. He so appreciated the accommodation of the forest that in the morning he thanked nature in his own way: 'And so it pleased me, in a half-laughing way, to leave pieces of money on the turf as I went along, until I had left enough for my night's lodging.'

Big views & dark skies

By the time you reach the Lozère, you'll likely have adapted your pace to the rhythmic hoofbeats of the donkey. Usually, this is slightly slower than a sporty hiker might choose on their own. But when climbing up to the summit of Pic de Finiels, it's not a bad idea to take it easy. The highest point in Lozère awaits at the top: 1699m (5574ft). On clear days, it's possible to see the Alps in one direction and the Mediterranean Sea in another. And on clear nights, the sky is full of stars: the Parc National des Cévennes carries an International Dark Sky Reserve label, denoting the superb possibilities for observing the heavens in the region.

Over all too soon

When the village of St-Jean-du-Gard comes into view, hikers may find themselves walking even slower than their donkey: this village marks the beginning of the end. Thinking back, you might mix your memories of an event with how the donkey reacted to the same occasion. You may remember a particular B&B with fondness – not because of the quality of the bed or tastiness of the dinner, but because the field where the donkey spent the night was full of lush grass.

The accomplishment of walking the Stevenson Trail with a donkey might come with a bittersweet taste. Robert Louis Stevenson himself was not immune to the camaraderie of travelling with a donkey: 'Father Adam wept when he sold her to me; after I had sold her in my turn, I was tempted to follow his example… I did not hesitate to yield to my emotion.'

Don't be surprised if you shed a few tears of your own when saying farewell.

Practicalities

Region: Haute-Loire/Ardèche/Lozère/Gard, France
Start: Le Puy-en-Velay
Finish: Alès

Getting there and back: Le Puy-en-Velay SNCF station has trains to/from Clermont-Ferrand (direct) and Lyon (one change). Alès SNCF station has trains to Nîmes (direct), Marseille (direct) and Lyon (one change).

When to go: Walk the GR70 between May and October, but note that high summer can be sweltering, and the trail may be crowded.

What to take: Essentials include a rain jacket, walking shoes, a hat and sunscreen. Lightweight sandals for the evenings are a comfy luxe, but avoid walking with your donkey while wearing them – those tiny hooves can break toes.

Where to stay: *Gîte d'étapes* (B&Bs) are your best option. Book in advance, and confirm that the lodging is equipped to welcome your donkey. Expect a supplemental cost of €5–10 for your donkey's board.

Where to eat and drink: Most *gîtes* offer *demi-pension* (dinner and breakfast) or a *full-pension* option which also includes preparing your picnic for the next day. Typical regional dishes might include local green lentils and charcuterie, as well as *aligot*, a filling dish of potatoes and stretchy cheese; and *coupétado*, a prune-laced bread pudding.

Tours: The Association Sur le Chemin de Robert Louis Stevenson (chemin-stevenson.org) is an excellent resource for trip planning and finding agencies to help organise a tour. Rando-Âne (rando-ane.com) is a trustworthy outfit from which to rent your donkey. Some agencies provide baggage transport.

Essential things to know: Wild camping is forbidden in many parts of the Lozère and Cévennes regions, due to wildfire risk. Inform the person from whom you've rented your donkey two days before you reach your arrival point so they can plan the pickup.

A Tóchar Phádraig pilgrimage – walking St Patrick's Causeway

Follow an ancient road and pilgrim path on a spiritual, cultural and historical journey through the raw beauty of the west of Ireland.

Whatever your beliefs, reflection is a major element of tackling this route and walkers are asked to abide by the pilgrims' rules: penance (no complaining); community (welcome those you meet along the way); faith (light a candle at the abbey before setting out); mystery (silence should be observed at particular points); and celebration (find the joy in the moment).

Passing fairy forts, famine graves, ancient ruins and mystical sites, the trail is a blend of pagan mystery and Christian pilgrimage, and offers walkers a chance to learn about local history from pre-Christian times to the present.

Medieval monks were not averse to a little cultural appropriation: taking pagan rituals and celebrations and repackaging them as Christian feast days; embellishing local lore into tales of miraculous endeavours; peddling legends to encourage an increase in pilgrims. Ireland's most famous myth tells of St Patrick casting out the island's snakes from the summit of Cruachán Aigle, a conical peak on the west coast, some time in the 5th century.

Cruachán Aigle became Croagh Patrick and the pilgrims followed, to a mountain sacred in pagan times and linked to Cruachan Aí – the seat of the high kings of Connaught – by a chariot trail now known as Tóchar Phádraig (St Patrick's Causeway). Walking this pagan-cum-pilgrim trail offers a glimpse into Ireland's complex history, but it is the walking with intent – a key part of any pilgrimage – that elevates this route from the scenic to the profound. Along with withdrawing

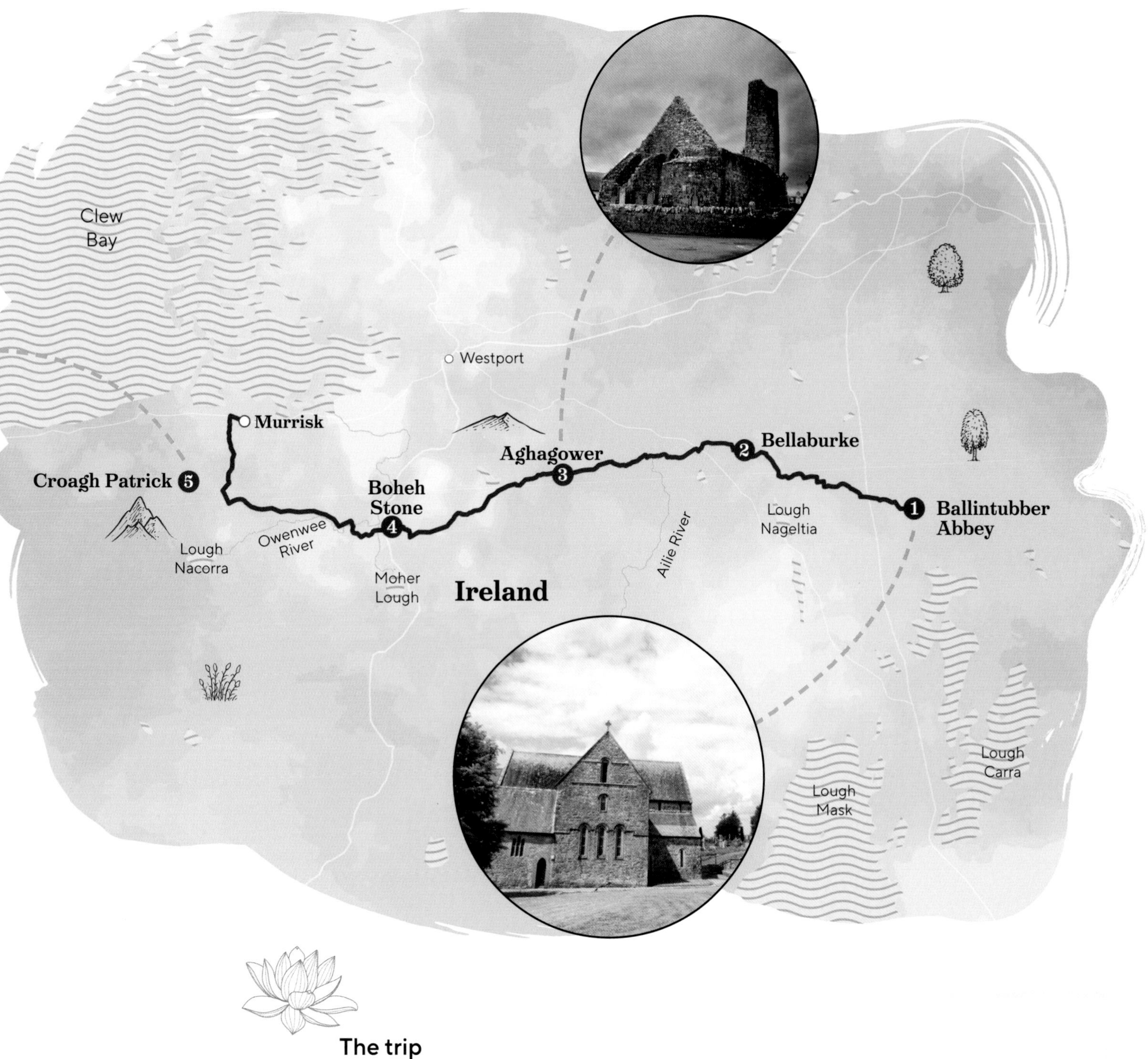

The trip

Distance: 35km (22 miles)
Mode of transport: On foot
Difficulty: Moderate

1 Ballintubber Abbey This 13th-century abbey has been in continuous use for over 800 years

2 Bellaburke A cluster of famine graves marks a tragic period of Irish history

3 Aghagower Historic village with an early Christian church, round tower and holy well

4 Boheh Stone Neolithic carved stone aligned with the setting sun on Croagh Patrick

5 Croagh Patrick Ireland's most sacred mountain, overlooking the islands of Clew Bay

Clockwise, from left: summiting Croagh Patrick; St Patrick surveying his snowcapped mountain; Aghagower's medieval chapel; 13th-century Ballintubber Abbey

Carmel Needham, Tóchar Phádraig pilgrim

In Ballintubber we were invited to pick up a rough stone whose sharp edges symbolised whatever was on our minds. To feel it as you walked was a reminder to give space to that worry, so there was something very symbolic about throwing the stone into the sea at the end of the walk, then returning to the abbey to be given a smooth stone to carry through life. I keep mine with me all the time.

into a landscape that creates space for reflection, walkers are asked to abide by the pilgrims' rules: to focus on finding joy and avoiding complaint, and to combine moments of welcome with times of silence. As the route winds through boggy fields and hazel forests, over trickling streams, up steep slopes and down mountains, it is this purposeful progression that makes this walk so special.

History's mark on the landscape

This trail sees only a trickle of walkers and, other than waymarking, little has changed since medieval times – leaving those who come here alone with their thoughts, the elements and the wild beauty of the west of Ireland. Leaving the grounds of 13th-century Ballintubber Abbey, you'll see the cobblestones of the ancient royal chariot route in the first field before continuing on into Mayo's hinterlands. The trail passes an abandoned mill at Killawullaun and along the banks of the Aille River for a first glimpse of Croagh Patrick. The 'Reek', as it is known locally, stays in view for much of the rest of the walk, shadows scudding across its slopes, quietly drawing the walker forward.

Crossing flower-filled meadows, streams, bogs and fields of sheep, there's a strong sense of those who have gone before. It's often a painful glimpse of the past, as the ruins of a small church and a series of famine graves at Bellaburke attest. A tragic period in Irish history, the failure of the potato crop in the mid-19th century led to the deaths of over a million people, while another million emigrated.

In this Ireland of struggle and strife, folklore flourished and nearby, at Creggaun 'a Damhsa (the 'Hillock of Dancing'), fairies were said to party close to a place where the river disappears beneath a series

© FREESKYLINE / SHUTTERSTOCK

"It is the walking with intent – a key part of any pilgrimage – that elevates this route from the scenic to the profound"

of cliffs. Mysteries like these were matched by those of the faithful, and at Aghagower (the 'Field of the Springs'), where St Patrick baptised his first converts, a large ash tree is said to have healing powers. Nearby, a medieval chapel, 10th-century tower and a series of wells all have associations with the saint.

However, it is the carved stone at Boheh which may be the most curious of all. Decorated with a series of circles and dating from around 3000 BCE, it aligns with the setting sun as its rays roll down the north face of Croagh Patrick each 18 April and 24 August.

The final climb

Crossing the Owenwee River, the landscape becomes increasingly wild, and a switchback trail leads up the slopes of Croagh Patrick. The Tóchar route doesn't take in the summit, but it's hard to resist. Rising directly from the sea, it's no surprise that this distinctive peak became a place of pilgrimage: its allure is unmistakable. Should you choose to hike up to its 764m (2507ft) summit, the steep scree-slopes of the final climb will test your mettle. It is only here, where you join the main trail up the mountain, that you are likely to encounter other walkers – though arrive on Reek Sunday, the last Sunday in July, and you'll join a stream of up to 25,000 faithful, many walking barefoot in the night to reach the top at dawn.

But whatever time of year you make the ascent, the reward is substantial, with the wide expanse of Clew Bay and its myriad islands laid out before you in a patchwork of blues and greens. For weary walkers it is a feast for the eyes and soul, and marks the culmination of a route that requires determination and perseverance, but which calms and fortifies in return.

Descending from the summit, while drinking in more views of Clew Bay, the Sheeffry Hills and the Connemara Mountains, the lure of a warm pub and a place to rest weary feet becomes ever stronger. Just spare a thought for the medieval pilgrims whose rite included walking back to Ballintubber to wash in ritual baths, a cleansing that symbolised a change of heart and their desire to return home renewed.

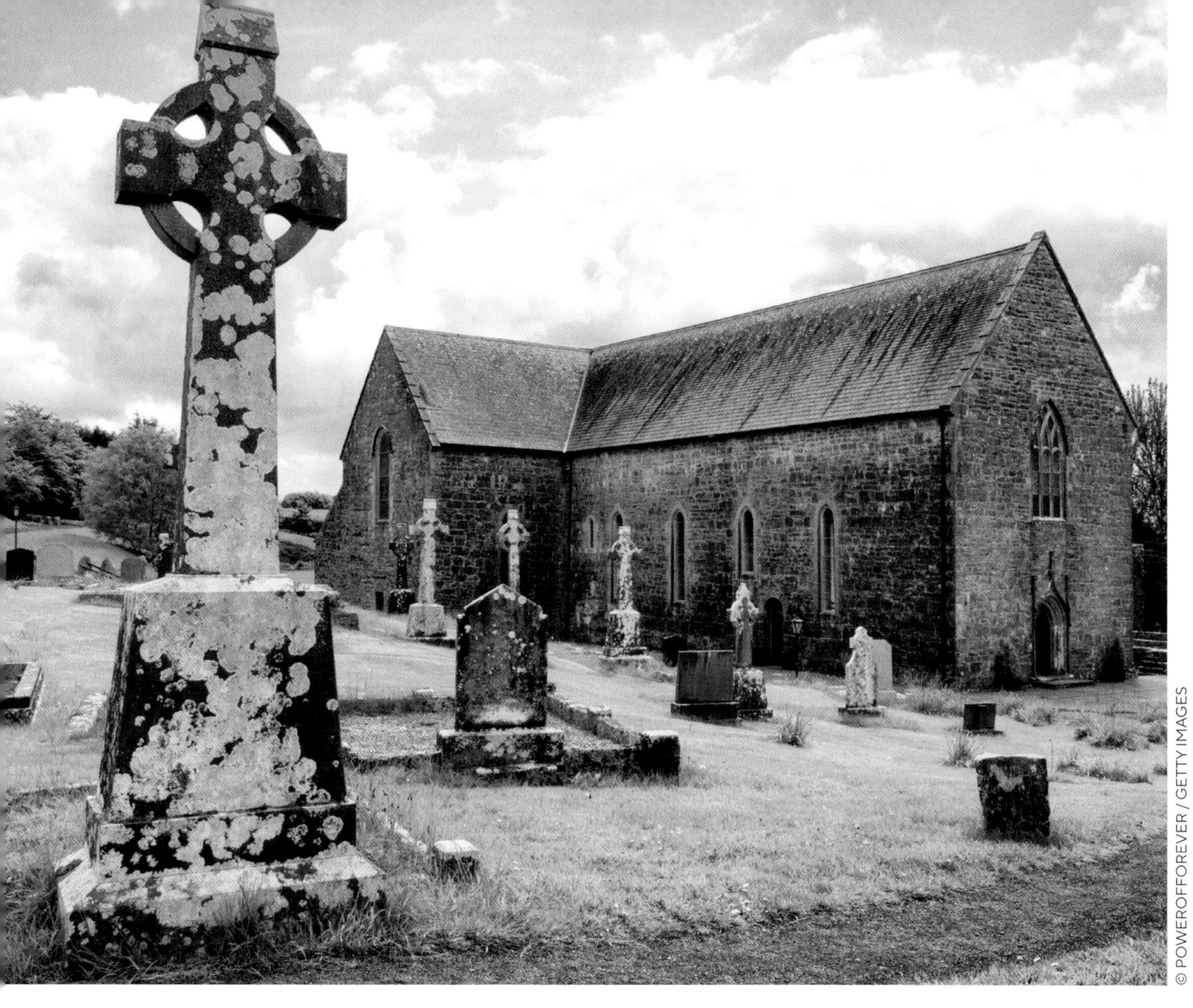

© POWEROFFOREVER / GETTY IMAGES

From top: Ballintubber Abbey, the start of the Tóchar Phádraig route; Clew Bay from Croagh Patrick

Practicalities

Region: County Mayo, Ireland
Start: Ballintubber Abbey
Finish: Murrisk

Getting there and back: From Castlebar, the closest train station to the start, take bus 422 for the 20min ride to Ballintubber village, from where it's a 20min walk to the abbey. From Murrisk, bus 450 takes 10min to reach Westport, from where you can catch a train to Dublin or bus 456 to Galway.

When to go: The summer months of June to September offer the best chance of dry weather, although you should be prepared for rain at any time of year. In winter, the route can be very boggy and wet underfoot, and conditions on the mountain can change dramatically in minutes.

What to take: Good walking boots and rain gear are essential – the route crosses numerous streams and can be wet and muddy year-round.

Where to stay: You'll find hotels in Castlebar, Murrisk and Westport, and B&Bs in Ballintubber and along the way at Aghagower. Booking in advance is recommended.

Where to eat and drink: Options for food and water are limited along the way, with the small shop and pub in Aghagower your best bet en route. Otherwise, you need to carry all you'll need. At the end of the walk, Campbell's in Murrisk is a popular country pub serving hearty food.

Tours: Ballintubber Abbey Trust (ballintubberabbey.ie) runs four guided group walks each year.

Essential things to know: Walkers must register with the abbey before departure and pay a small fee to cover insurance. The route is well marked and you'll be given a map and guide on departure. Walking the route alone is discouraged, and dogs are not allowed. OSI Discovery Series maps 30 and 38 cover the route.

Walking in the footsteps of Warsaw's Jewish ghosts

Historic landmarks, memorials and museums on this route through Poland's capital provide insight into the city's Jewish community, both past and present.

For those with Jewish and/or Polish roots, a visit to Warsaw can be an emotionally transformative, even cathartic experience. Standing in places directly impacted by the Holocaust is powerfully moving; and the Jewish community's contemporary revival is an inspiring example of the human capacity to overcome trauma.

Warsaw's Jewish archives and museums offer a wealth of information on family roots, age-old traditions and a community whose connections to Poland stretch back 600 years. This is an opportunity to engage with and gain a deeper understanding of Polish Jewish heritage.

Walking through Warsaw today, with its glittering new office blocks, meticulously recreated Old Town and soaring Stalinist Palace of Culture and Science, it's hard to imagine how broken the city was 80 years ago. 'Nowhere have I been faced with such destruction,' said General Dwight Eisenhower, Commander of the Allied Forces in Europe, after witnessing the ashes of Warsaw in September 1945.

Nazi Germany did its upmost to wipe Poland's capital off the map. In 1943, as if in a dress rehearsal, it liquidated Warsaw's Jewish Ghetto – about 2.5 sq km (1 sq mile) of the city, into which some 400,000 Jews had been crammed. Disease, execution and starvation in the ghetto, or later in concentration camps, claimed the lives of the vast majority.

Beneath the cold facts of the Holocaust lie the individual stories – heroically brave and harrowingly

© KAROL KOZLOWSKI / AWL IMAGES

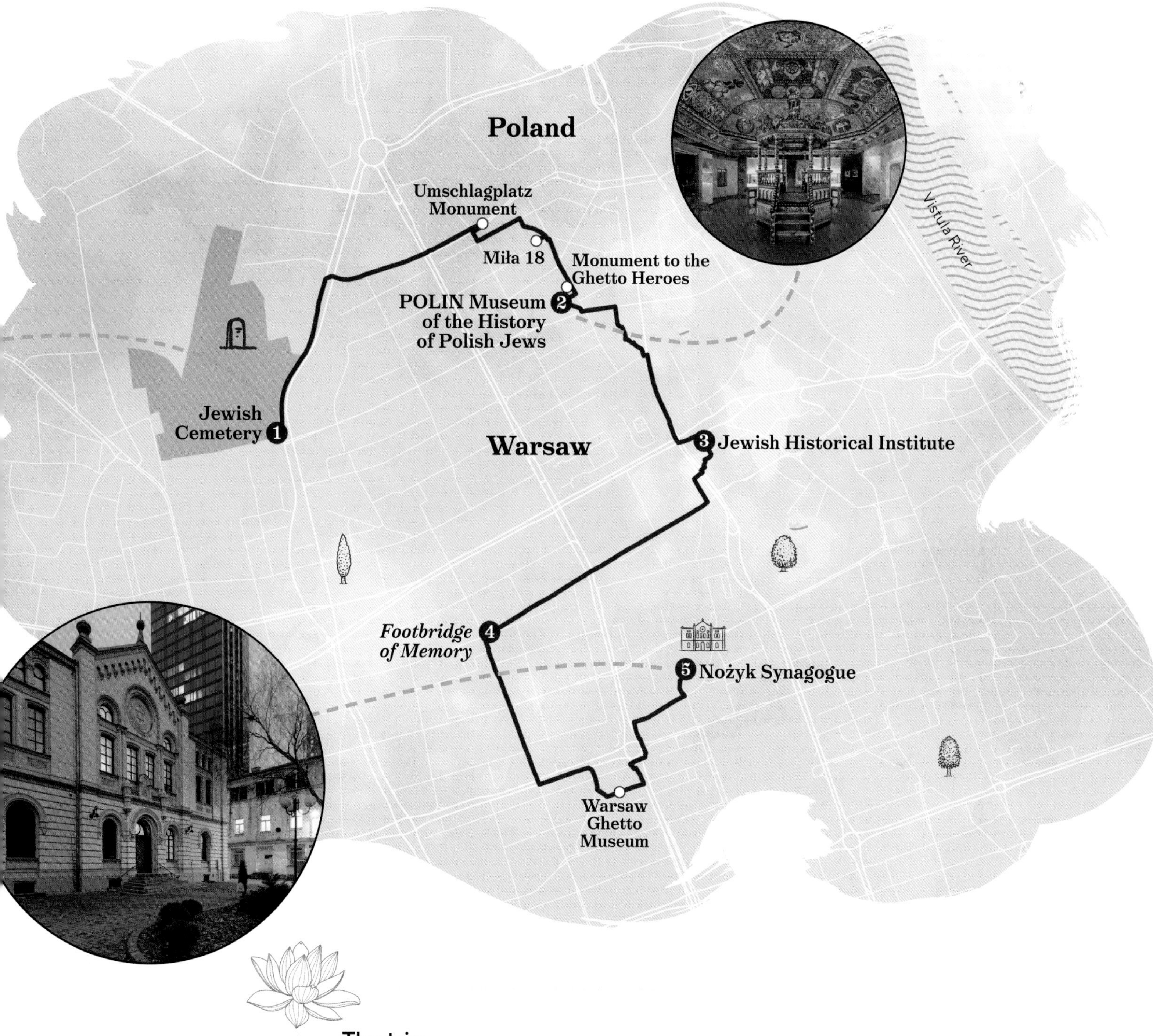

Clockwise, from left: Warsaw at dusk; memorial stone in the city's Jewish Cemetery; inside the POLIN Museum of the History of Polish Jews; Nożyk Synagogue

The trip

Distance: 7km (4 miles)
Mode of transport: On foot
Difficulty: Easy

1 Jewish Cemetery Wander this 19th-century burial ground, with its moving memorials to those lost in the Holocaust

2 POLIN Museum of the History of Polish Jews Explore the imaginative exhibitions at this beacon of education and collective memory

3 Jewish Historical Institute Discover a unique collection of direct testimonies about the extermination of Polish Jewry

4 *Footbridge of Memory* Imagine what it was like to traverse the Ghetto's wooden bridge at this contemporary installation

5 Nożyk Synagogue Marvel at the beautifully restored, functioning Jewish prayer hall, the only one in the city to survive WWII

David McClane, Warsaw visitor

I had been reluctant to take a Jewish tour of Warsaw, knowing that what I would hear would be harrowing, but I had no idea just how much these stories would affect me. Listening to our guide describe the horror, as we stood in the exact spot that it took place, already brought tears to my eyes. In my mind, you cannot understand a city without knowing about its history, no matter how disturbing.

tragic – that you can discover on a walk through Warsaw's Muranów and Mirów neighbourhoods.

Memorials to Jewish heroes

In Warsaw's Jewish Cemetery stands a bronze statue of Dr Janusz Korczak, the pen-name of Henryk Goldszmit, a children's book author and director of the city's Jewish orphanage. Korzcak was repeatedly offered sanctuary, but remained with his tiny charges in the Jewish Ghetto, accompanying them in 1942 when they were deported to Treblinka, never to be seen again. The death camp trains departed from Umschlagplatz. The platform is long gone, but the site is marked by a rectangular memorial symbolising the cattle wagons into which thousands of men, women and children were herded. The monument's marble walls are etched with over 3000 Jewish forenames, from Aba to Zygmunt, of those who perished.

At least 13,000 Jews fought back in the heroic but doomed Warsaw Ghetto Uprising of 1943, which led to the area's destruction. A memorial mound, Anielewicz's Bunker – also known as Miła 18 – stands above the hidden shelter used by Jewish resistance group ŻOB. Alongside several others, uprising leader Mordechai Anielewicz chose to commit suicide here rather than surrender; he is remembered on the 11m-high (36ft) Monument to the Ghetto Heroes, a carved frieze designed by Polish-Jewish sculptor Nathan Rapoport and unveiled in 1948, when much of Muranów still lay in ruins. In December 1970, Willy Brandt, then Chancellor of West Germany, spontaneously fell to his knees in front of this monument. 'Under the weight of German history and carrying the burden of the millions who were murdered,' he later recalled, 'I did what people do when words fail them.'

> "Beneath the cold facts of the Holocaust lie the individual stories – heroically brave and harrowingly tragic"

The Monument to the Ghetto Heroes stands on the western side of a park, dotted with several more memorials and dominated by the award-winning POLIN Museum of the History of Polish Jews. Along with the nearby Jewish Historical Institute (JHI), it's an essential stop, and well worth returning to for further research and contemplation.

Fragments of the ghetto

The JHI houses a precious collection of around 6000 documents, including diaries, drawings and photographs, that provide direct testimony on the atrocities of the Holocaust. Known as the Ringelblum Archive, it is named after the historian Emanuel Ringelblum, who created the organisation Oneg Shabbat in November 1940 to gather the collection. The archive was secured in metal boxes and milk cans, buried in batches, and unearthed after the war.

As you continue on, take note of the cast-iron slabs along the ground, and their associated 22 wall plaques. These shadow the Jewish Ghetto's original borders: 18km (11 miles) of brick walls that enclosed the 73 streets of the 'small' and 'large' ghettos, the two linked by a wooden bridge over ul Chłodna. Small sections of the wall also remain, such as at the corner of ul Żelazna and Grzybowska and in the courtyard of ul Sienna 55. At the junction of ul Chłodna and ul Żelazna is the *Footbridge of Memory* installation. Peer into the cavities in the metal support-poles to see grainy photographs of the original bridge. At night, you can gaze up at glowing fibre-optic cables strung between the poles to recreate the spectral shape of the Jewish Ghetto bridge. In his *Diary of the Warsaw Ghetto, October 1940 – January 1943*, Henryk Makower wrote that 'from the bridge you can see the "Aryan" Warsaw, to us seemingly so free. But such small pleasures are not allowed. The policemen on the bridge politely but firmly ask you not to stare and keep moving.'

As a final stop, step inside the Nożyk Synagogue to remember or say a prayer for those who perished in the Holocaust, and to celebrate the endurance and revival of Warsaw's Jewish community.

© W. KRYŃSKI / POLIN MUSEUM OF THE HISTORY OF POLISH JEWS

From top: the POLIN museum faces the city's Monument to the Ghetto Heroes; Warsaw's Jewish Cemetery

Practicalities

Region: Warsaw, Poland
Start: Jewish Cemetery
Finish: Nożyk Synagogue

Getting there and back: Warsaw has two airports as well as multiple international bus and train services.

When to go: You can do this walk year-round. Memorial events are held on 19 April, the anniversary of the Warsaw Ghetto Uprising; and 1 August, for the Warsaw Uprising of 1944. The Singer's Warsaw Festival (shalom.org.pl) of Jewish-related theatre, music and film, is also held in August.

Where to stay: Accommodation in Warsaw ranges from backpacker hostels to luxury boutique hotels. There are also plenty of centrally located rental apartments.

Where to eat and drink: The Warsze restaurant at POLIN Museum serves mainly kosher-friendly dishes, and can provide certified kosher food if you call ahead. For traditional treats like bagels and rugelach pastries, head to the Charlotte Menora cafe on Plac Grzybowski; the MenoraInfoPunkt here is an information hub where you can engage in contemporary Jewish life through activities such as cooking workshops.

Tours: Jewish-themed walking tours of Warsaw are offered by Orange Umbrella Free Tour (orangeumbrella.pl), Stancja Muranow (stacjamuranow.pl), the Jewish Historical Institute (jhi.pl/en), and Taube Jewish Heritage Tours (taubejewishheritagetours.com).

Essential things to know: Taube Philanthropies (taubephilanthropies.org) offers a free downloadable PDF of its Field Guide to Jewish Krakow and Warsaw. The Warsaw Ghetto Museum (1943.pl), in the former Bersohn and Bauman Children's Hospital, is scheduled to open in late 2025.

Artistic awakening on the French Riviera

Journey along the Côte d'Azur to embrace the *joie de vivre* of 20th-century artists lured to this fabled coastline by the luminosity and intensity of its light.

The art narrative that unfurls on the French Riviera offers an alluring invitation to interpret objects and landscapes from an alternative perspective.

Learn about Pointillism, Impressionism and other watershed 'isms' in art museums and galleries. In artist home-studios and gardens, observe intimate everyday details to understand French *joie de vivre* as a philosophy of life.

Admiring artistic masterpieces exercises your powers of observation and critical thinking; it can be a source of strength and a stimulus for cultural exchange and expression, perhaps inspiring you to create your own works of art.

Colour maestro and modernist Henri Matisse was bewitched by the intoxicating play of light spangling through the shutters of his hotel room in Nice, which he observed for days on end before signing off plans for his Chapelle du Rosaire in nearby Vence. But Matisse was not the only artist attracted to this corner of southern France. A journey along the coast, from Marseille to queen-of-the-Riviera Nice, offers a masterclass in modern art and an intimate insight into the lives of the movement's artists.

Art rocks

Begin with the backstory: a menagerie of bison, auks and ibex, etched into the walls of a cave near Marseille using flints, charcoal, red pigment and a herculean dose of patience by Palaeolithic artists. Submerged by the rising ocean over subsequent millennia, Grotte

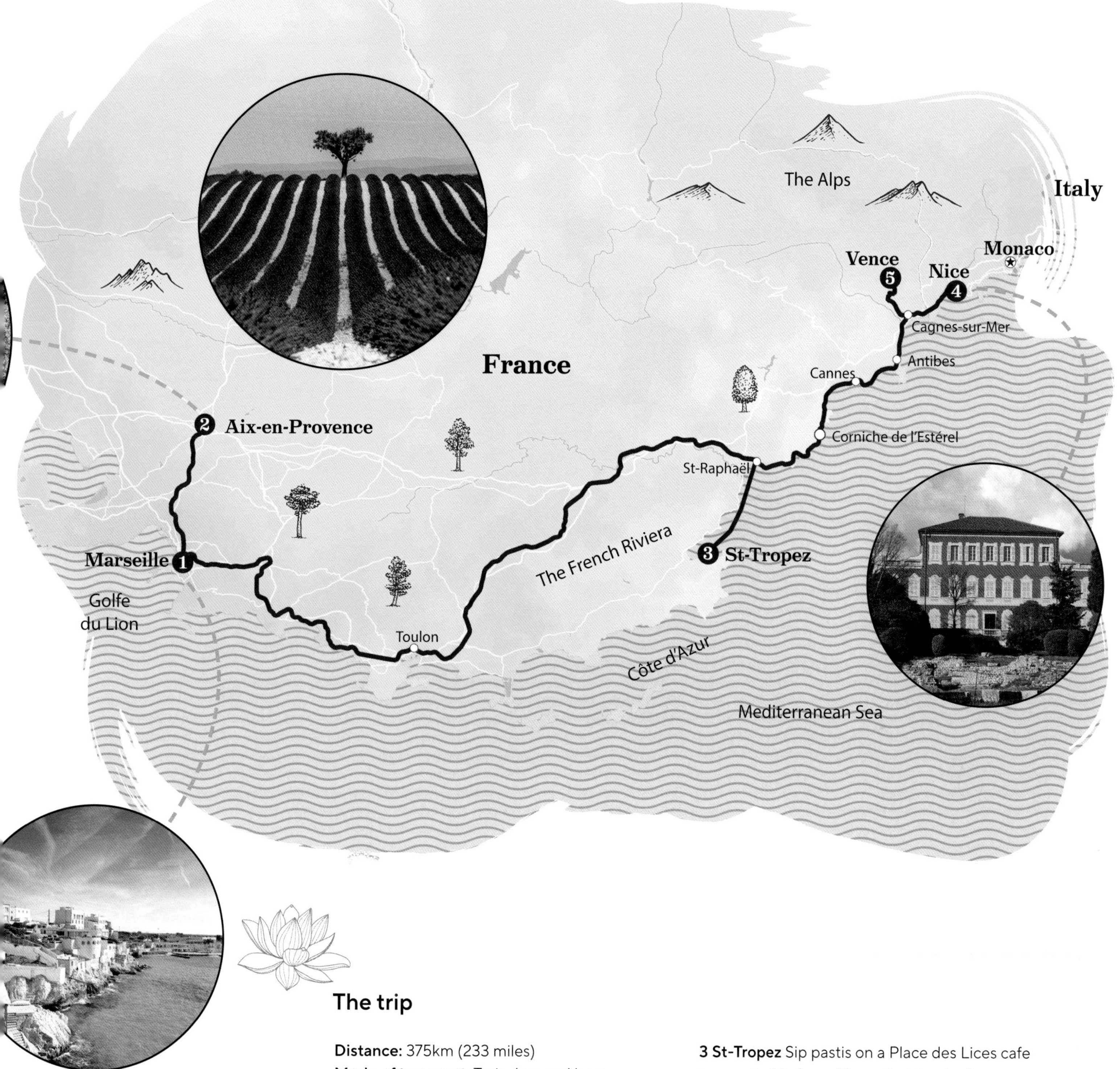

The trip

Distance: 375km (233 miles)
Mode of transport: Train, bus and boat
Difficulty: Easy

1 Marseille Vieux Port is the beating heart of this vibrant city, captured on canvas by numerous artists over the centuries

2 Aix-en-Provence Visit the Musée Granet to enjoy Cézanne masterpieces and a collection spanning antiquity to the 21st century

3 St-Tropez Sip pastis on a Place des Lices cafe terrace in this famed town that inspired so many artists

4 Nice Trace the evolution of Matisse's artistic career, from paintings and sculptures to paper cut-outs, at the Musée Matisse

5 Vence Pay your respects with a moment of serenity at the modern, Matisse-designed Chapelle du Rosaire

Clockwise, from left: Nice's Promenade des Anglais; Cours Mirabeau, Aix-en-Provence; rural lavender fields; Musée Matisse, Nice; Marseille

A bird's-eye view of St-Tropez

Nicola Williams, art lover

Avant-garde masterpieces fill the flagship art museums (MAMAC in Nice, Fondation Maeght in St-Paul-de-Vence) on this coastal Côte d'Azur trail. But no artwork haunts me like Chapelle du Rosaire. Learning the backstory behind the weeny white chapel and its spellbinding stained-glass 'light show' – designed by an aged, near-crippled Matisse for a Dominican nun who cared for him from 1943 – is a lesson in courage and our extraordinary artistic capacity to morph challenge into opportunity.

Cosquer remains underwater and inaccessible. On the Marseille waterfront, Cosquer Méditerranée plunges visitors, via floating explorer vehicles, into a near-perfect replica of this dazzling rock-art cave. Knowing these primeval artists had no rules or critics, and used little more than dirt and rocks to lay their creative souls bare, might serve as encouragement to unleash your own inner artist.

Back on dry land, hop on a bus or train bound for genteel Aix-en-Provence. Bronze pavement plaques trail Aix-born post-Impressionist master Paul Cézanne along elegant avenues and bijou squares beaded with sculpted fountains and polished *hôtels particuliers* (mansions). Several Cézanne works hang in Aix's Musée Granet – but you'll get a more immediate sense of what inspired him by taking a guided e-bike ride (or a tour in the sidecar of a vintage motorbike) to admire the limestone massif of Montagne Ste-Victoire, which he immortalised in a series of paintings.

Stirring sunsets & creative fire

Back in Marseille, the most seductive way to reach jet-set St-Tropez is a train to St-Raphaël, then an hour-long cruise along the coast. As the ochre-and-buttermilk bell tower crowning the fishing port inches closer, you might reimagine the arrival of Pointillist painter Paul Signac, who sailed into the harbour aboard his own boat in 1892. At the Vieux Port, ditch the A-lister *flaneurs* for modern art – infused with that illustrious Côte d'Azur luminosity – at Musée de l'Annonciade. The collection here shines a light on the late 19th- and early 20th-century painters who worked in St-Tropez, including Matisse – who summered here in 1904, drawing preliminary sketches for *Luxe, Calme et Volupté*. At sunset, seek out Paul Signac's sublime *St-Tropez au soleil couchant*, to join his glowing dots and fathom just what it actually was – is – about honeypot St-Trop that enamoured so many artists.

> "In the incandescent sunlight and dazzling vivacity of Nice, Matisse found his true *bonheur de vivre* (joy of life)"

The train from St-Raphaël to Antibes swooshes past a masterpiece of the natural kind, Corniche de l'Estérel, where needles of red rock swoop and soar among eucalyptus trees, turquoise inlets and a perennially bluebird sky. Stumble off in boat-bedecked Antibes and brace yourself for the azure-blue canvas of the Med, as admired from the seaside terrace of Château Grimaldi, where Pablo Picasso set up studio in 1946. He was already in his mid-60s by then, but this part of France had an immeasurable influence on his work. Of the dozens of paintings and drawings displayed here, no single piece captures the creative Côte d'Azur fire unleashed in this unorthodox Spanish artist quite like *La Joie de Vivre*. A dancing faun, centaur, goat and tambourine-playing nymph prance across the canvas, evoking the mythology of the Ancient Greeks who founded Antibes as Antipolis.

© FRANK LUKASSECK / 4CORNERS

Life & death

Fauvist Matisse shocked Paris with the explosive application of colour and distortion in his scandalously sensual and modern *Le Bonheur de Vivre*. But it was in the incandescent sunlight and dazzling vivacity of palm-fringed shores around Nice that he found his true *bonheur de vivre* (joy of life). Explore his phenomenal creativity at the Musée Matisse; then pay your graveside respects in the next-door cemetery of Monastère Notre Dame, where Matisse is buried.

Inland, Vence is an alluring footnote to the trip. Break the short bus ride up from Nice in Cagnes-sur-Mer, where the family home of Pierre-Auguste Renoir evokes the final struggles of the Impressionist genius, who worked with a paintbrush bandaged to arthritic fingers in order to capture the citrus groves and silvery olive trees framing his Provençal *mas* (country house) one last time. In Vence, the peerless Chapelle du Rosaire, crafted with passion and pain by an ailing 81-year-old Matisse, is the ultimate epitaph to the transcendent life-cycle of the modern artist.

© ALAIN COURTILLAT

From top: follow Cézanne to the Montagne Ste-Victoire on a guided tour by vintage sidecar; coastline along the Corniche de l'Estérel

Practicalities

Region: Provence-Alpes-Côte d'Azur, France
Start: Marseille
Finish: Nice

Getting there and back: Direct SNCF trains link Paris with Marseille (3hr 30min). Returning from Nice, high-speed TGVs zip to the capital (6hr); the luxury Le Train Bleu service no longer runs, but the Nice–Paris night train (12hr) offers a whisper of old-school romance.

When to go: Spring (March to June) and autumn (September and October) are optimum, pleasantly warm and with fewer crowds. April is particularly lovely, with flowering wisteria and bougainvillea blazing purple and pink across gold-stone facades.

What to take: Sunhat and shades, mosquito repellent, a pocket sketchbook and a corkscrew – chilled rosé is a sacrosanct aperitif at sundown.

Where to stay: Marseille and Nice overflow with hotels, from budget to luxury. Independent boutique hotels and family-run *chambres d'hôtes* (B&Bs) are often the most atmospheric. Book well in advance.

Where to eat and drink: Sample the day's catch at epicurean beach club Tuba in Marseille; Flaveur, in Nice, promises high-end Niçois gastronomy. The Côte d'Azur's smorgasbord of cafes, bistros, restaurants and beach bars leaves you spoilt for choice; for snacks on the go, look out for *socca* (chickpea pancakes) or a *pan bagnat* (tuna and veg roll).

Tours: Tourist offices in Marseille (marseille-tourisme.com) and Nice (explorenicecotedazur.com) have info on Riviera tours. For guided trips to Cézanne's Montagne Ste-Victoire from Aix-en-Provence, try La Belle Échappée (labelleechappee.fr) for a vintage sidecar, or the tourist office (aixenprovencetourism.com) for e-bike trips.

Essential things to know: Bring a water bottle to fill at fountains spouting *eau potable* (drinking water), and a scarf – make that two – to cover bare shoulders and thighs in Matisse's chapel and other churches.

A pilgrimage of faith through Germany, Switzerland & Austria

This contemplative journey takes in scenic and spiritual highs, from inspiring places of worship and the home of a historic Passion Play to a stretch of the Romantic Road.

Oberammergau's Passion Play dates back to the 1600s, when townspeople pledged to perform a staging of the Crucifixion if God would spare them further infection by the plague. That this tradition endures to this day is an inspiring testament to the power of faith.

Digging deep into the idea of the 'pilgrim's voyage' – the human urge to travel long distances to find meaning, purpose and, perhaps, divinity – this Alpine journey will likely offer plenty of inspiration for self-reflection.

On a route that has the Alps in full view at all times, there are many opportunities to stretch your muscles by hiking some superlative trails.

The idea of pilgrimage resonates with people of all ages, nationalities and life experiences; it is the ultimate journey of self-discovery, offering the chance to ponder on faith and the core of what makes us who we are. Circling through Germany, Switzerland and Austria by road, this contemporary pilgrimage journeys through the secular and the spiritual, taking in city sights, verdant valleys and marvellous mountain views alongside

© SHUTTERSTOCK / FOOTTOO

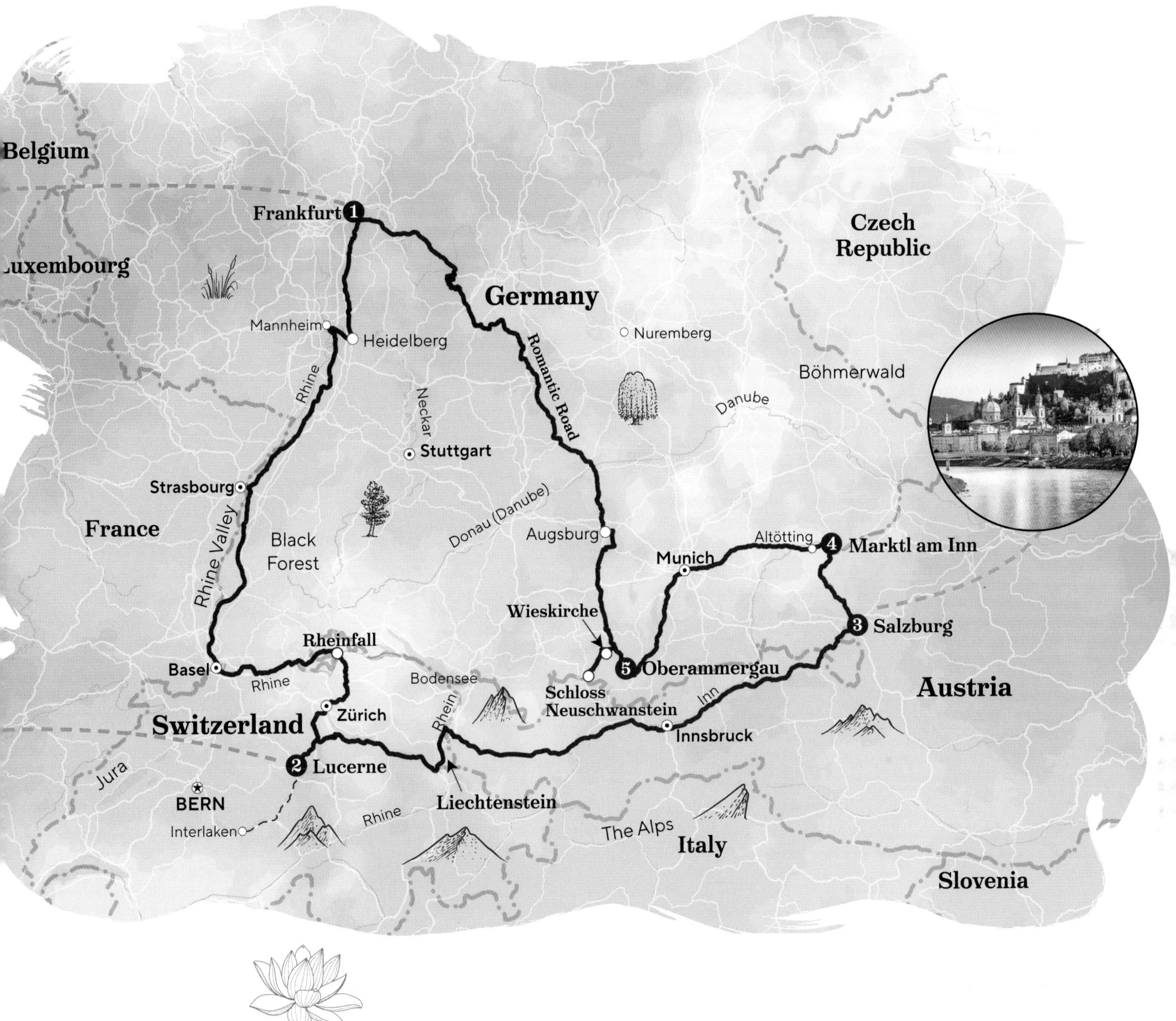

Clockwise, from left: Oberammergau's Passion Play; Lucerne's covered bridges; Frankfurt's Altstadt (Old Town); summer in Salzburg

The trip

Distance: 1916km (1190 miles)
Mode of transport: Car
Difficulty: Easy

1 Frankfurt, Germany This lively city with a medieval heart is a vibrant place to start and end the journey

2 Lucerne, Switzerland Soak up lovely lakeside views and a history-rich Old Town

3 Salzburg, Austria Beyond Mozart and *The Sound of Music*, this baroque beauty of a city offers a skyline of graceful domes and spires

4 Marktl am Inn, Germany This quiet village was the birthplace of the late Pope Benedict XVI

5 Oberammergau, Germany Appealing Bavarian town that's best-known for its world-renowned Passion Play

Harmony Diffo, Christian pilgrim

I found that this pilgrimage through central Europe was transformative, particularly because the spiritual road to Oberammergau to see the Passion Play is such an ancient, well-trodden path. I thought often about exactly how many pilgrims had followed my same path to find meaning, to learn more about their own spirituality, identify their unique gifts, and recapture the spark that makes one realise that we were all born for our own beautiful, unique reason.

inspiring places of worship and a village that's been staging a Passion Play for nearly 400 years.

Frankfurt to the Rhine Valley

The journey begins in Frankfurt am Main, where the gleaming tower blocks of 'Mainhattan' coexist with a handsome half-timbered Altstadt (Old Town). Head to the old centre to explore two fine shrines to the German Romantic tradition (of which you'll learn more when travelling the famed Romantic Road route on the last leg of this journey). Goethe-Haus is dedicated to author Johann Wolfgang von Goethe, who pioneered the German romantic novel and who was born at this spot; close by, the Deutsches Romantik-Museum (German Romanticism Museum) is the first of its kind, focused solely on the artistic achievements of the Romantic era.

South of Frankfurt, the route winds to Heidelberg, home to a ruined hilltop castle and Germany's oldest university. Take in the medieval Rathaus (Town Hall) as well as the University itself, founded in 1386. From the Altstadt, the ancient arches of the 1786 Alte Brücke (Old Bridge) span the Neckar River. A tributary of the Rhine, the Neckar flows through the Black Forest and meets up with the Rhine again in Mannheim, to the northwest. The Rhine Valley is the perfect place to make the first overnight stop before crossing over into Switzerland. It can be hard to tear yourself away from this alluring landscape of forested hillsides, near-vertical terraced vineyards and idyllic half-timbered villages, but console yourself with the knowledge that, after you've travelled through

© CHALABALA/ ISTOCK / GETTY IMAGES PLUS

"Innsbruck's rococo Basilika Wilten offers a respite for quiet contemplation, self-reflection or prayer"

Austria and Switzerland, you can linger longer en route back to Frankfurt at the end of the journey.

Waterside wonders in Switzerland

The Rhine serves as the border between Germany and Switzerland hereabouts; just over on the Swiss side at your first stop, the glorious Rheinfall, you'll hear the sound of roaring water as soon as you step out of the car. Walk down a steep incline for a close-up view – on clear days, the water is an incredible blue-green and the mist sprays twinkle with rainbows. Though just 23m (75ft) high, this is nonetheless the most powerful waterfall in Europe; on a hillock overlooking the cascade, Schloss Laufen adds to the charm.

There's more wonderful water – and a fine baroque church – on your next Swiss stopover, the lakeside city of Lucerne. With its stunning waterfront, medieval walls and ancient covered bridges, exploring Lucerne is a little like walking directly into a fairy-tale. The winding Old Town streets could absorb for hours: cross the Reuss River via the 14th-century Kapellbrücke; and take in the poignant Lion Monument, its recumbent big cat carved to commemorate Swiss soldiers who died in French Revolution battles. Back towards the waterfront, St Leodegar is one of the finest Catholic churches in the region, built on a foundation begun by the Romans in 735 CE.

Your stay in Lucerne may be a night or two, but could easily stretch to a week or more: there's a lively restaurant scene, excellent shopping and the promise of a scenic sunrise hike up nearby Mt Pilatus (you can also ascend by way of a cogwheel railway). Lucerne is also the perfect base from which to explore the breathtakingly beautiful Bernese Oberland. A favourite route is to follow the Brünig Pass to swoonworthy Interlaken, dwarfed by the postcard-famous peaks of the 3967m (13,015ft) Eiger and the 4158m (13,642ft) Jungfrau. Other picturesque small-village stops include Lauterbrunnen and Grindelwald.

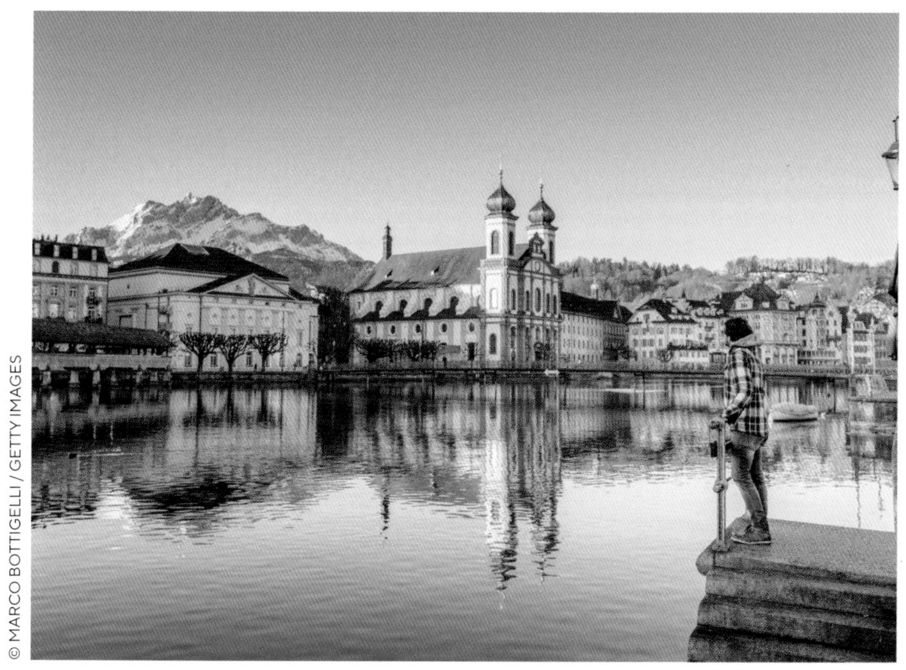

Clockwise, from above: castle-topped Heidelberg on the Neckar River; lakeside in Lucerne; hiking up Mt Pilatus, near Lucerne

An enduring passion

Both a creative and a historical marvel, the Passionsspiele (Passion Play) of Oberammergau (passionsspiele-oberammergau.de) has long drawn Christian travellers to Bavaria. It is rehearsed for years in advance, and almost all of the townspeople participate, with the play itself performed on an open-air stage and accompanied by a dramatic musical score and a supporting cast of live animals: Roman centurions ride real horses onto the stage, and Christ enters Jerusalem on Palm Sunday before the Crucifixion on a donkey. Since its first staging in 1634, the play has run every 10 years, though the pandemic saw the 2020 staging postponed until 2022; the next performance is scheduled for 2030. During the Passionsspiele season, which runs from mid-May to early October, Oberammergau is transformed, with busloads of theatregoers descending each day and hotels and B&Bs fully booked.

Into Austria

Head into the heart of the Alps for the next stop, historic Innsbruck, capital of Austria's gorgeous Tyrol region. As well as being a jumping-off point for summer and winter pursuits in the surrounding mountains, Innsbruck has a rich history to explore. For those seeking spirituality rather than sports, first stop might be the magnificently baroque Dom St Jakob, built between 1717 and 1724. 'Jakub' translates as 'James', and the cathedral lies on one of the myriad legs of the ancient Way of St James pilgrimage route to Santiago de Compostela in Spain. Elsewhere in town, seek out the Goldenes Dachl (Golden Roof), a late-Gothic oriel festooned with 2657 gold-plated tiles; it was built as a royal box over the town square for Emperor Maximilian I. Rococo Basilika Wilten offers a respite for quiet contemplation, self-reflection or prayer before heading to headline city sights like the stately Hofburg and the grand boulevard of Maria-Theresien-Strasse.

It would be hard not to be inspired by the cinematic mountain scenery as you travel on east through the Kitzbühel Alps en route to Salzburg, easily one of the most beautiful cities on the planet. Salzburg is best-known as the setting for *The Sound of Music* and as Mozart's birthplace, but for those of a more spiritual bent, it is best exemplified by the

© SHUTTERSTOCK / LIANEM

> "It would be hard not to be inspired by the cinematic mountain scenery as you travel through the Kitzbühel Alps"

Franziskanerkirche, with a late-Romanesque nave dating to 1208, a rib-vaulted Gothic choir and a florid baroque altar.

Looping back through Germany

Back into Germany, the route winds toward Frankfurt to complete the pilgrimage with a contemplative and cathedral-heavy finish. In Altötting, the Shrine of Our Lady is venerated for its Black Madonna image of Mary, said to have been the source of many healings. This 'Lourdes of Germany' has drawn pilgrims for over 500 years; many also seek out the town's stunning Catholic church, neo-baroque Basilika St Anna.

There's another focus for Catholic pilgrims a short way northeast in Marktl am Inn, birthplace of the late Pope Benedict XVI, Joseph Ratzinger. His former home, Geburtshaus, holds exhibits on his life and work, and has a reflection room for quiet contemplation. The nearby Pfarrkirche St Oswald, which houses Benedict XVI's baptismal font, offers another place for peaceful reflection before heading west to Munich. Bavaria's impressive capital was ruled for over 650 years by the dukes and kings of the Wittelsbach dynasty, who left their creative stamp on hundreds of pieces of divinely inspired architecture. Explore their history at the Residenz, then head to the soaring Frauenkirche, Munich's spiritual heart.

The route back to Frankfurt follows part of the famed Romantic Road, but before heading north, detour southwest to Oberammergau where, every 10 years, around 500,000 people descend over five months to take in the town's storied Passion Play. There's a small display on the history of the production in the foyer of the Passionstheater.

En route to Frankfurt, make a secular stop near Füssen to view Ludwig II's spectacular Schloss Neuschwanstein, a long-term architectural dream of the 'Mad King' and the contemporary model for the Walt Disney castles; and then at the Wieskirche, one of Bavaria's best-known baroque churches and a Unesco World Heritage Site.

© SHUTTERSTOCK / CANADASTOCK

Clockwise, from left: Schloss Neuschwanstein; looking down on Lake Lucerne from Mt Pilatus; Goldenes Dachl, Innsbruck

Practicalities

Region: Germany, Switzerland & Austria
Start/Finish: Frankfurt am Main, Germany

Getting there and back: Frankfurt is serviced by high-speed ICE train connections to major German cities like Stuttgart (1hr 10min) and Cologne (46min), as well as international hubs like London (5hr); the city's airport has flights from many European and international hubs. You can pick up rental cars from Frankfurt's train stations and airport.

When to go: May to October offer the best conditions, avoiding winter snowfall and spring rain; meadow wildflowers are in full bloom from mid-May through to mid-June.

What to take: A camera, a sunhat and, if you plan on doing off-route walks, hiking boots.

Where to stay: There are hotels and B&Bs in all the towns and cities along the route; book in advance.

What to eat and drink: Inviting options await all along the route. For traditional delights, the apple strudel (Apfelstrudel) at ConditCouture in Frankfurt is perfection, best warm with a scoop of vanilla ice cream. For a classic German beerhall experience, head to the Hofbräuhaus Brewery in Munich, which dates back to 1589.

©SHUTTERSTOCK / JAROMIR CHALABALA

Follow saintly footsteps on the Via di Francesco

From Florence to Assisi to Rome, follow in the footsteps of St Francis through statuesque hill towns with views so serene they're downright saintly.

St Francis renounced his privilege, material possessions and comfort to dedicate his life to helping animals, nature and the poor; many pilgrims take this journey to reflect on what comforts they could give up in their own lives.

While St Francis suffered from ill health most of his adult life, he still travelled throughout Umbria and beyond to spread his message of love and tolerance. But he listened to his body. He learned how to pace himself, stop and rest when needed, and ask for help. Many use this walk to assess how they can do the same – on this journey and in life back home.

In the stillness of the Italian countryside, there remains a constant stream of sounds and sights: the swoosh of a rushing river, the grunts and squeals of wild boar, the gentle sway of cypress trees, the trilling of birdsong. All of the senses are awakened on a walk along the Via di Francesco. Hikers might see a stag flit through an olive grove one day, feel the warmth soaked up by the pink stone of Assisi's Monte Subasio the next, and then drink in a *nocciola* (hazelnut) gelato frappè (and the town gossip) on the steps of a town square the day after. And those are just the external senses. What brings most pilgrims – secular and religious – to walk the Via di Francesco is a journey into the internal.

The man who became a saint

Originally a loose collection of places where St Francis walked during his lifetime – from his home in Umbria's

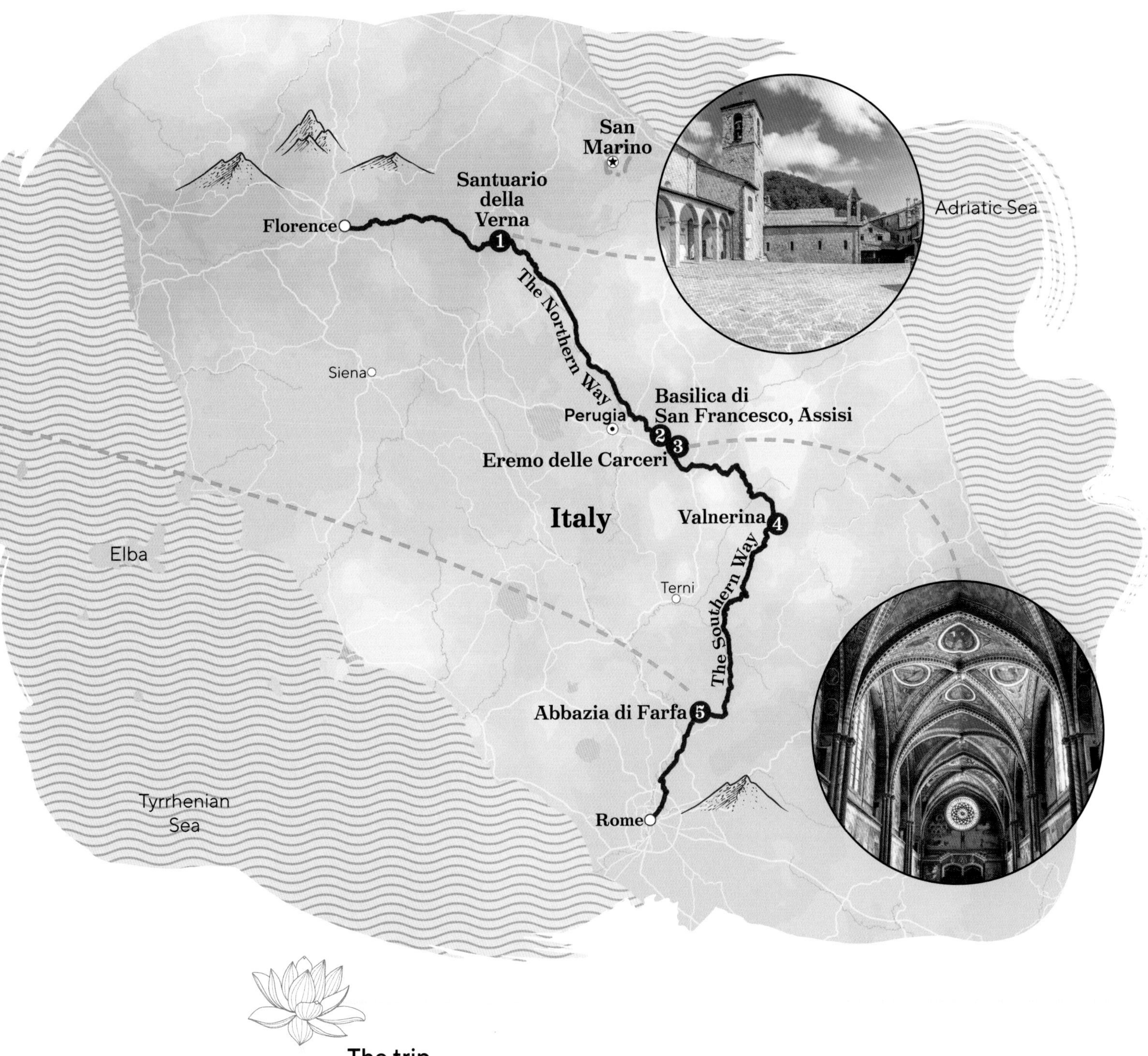

© ALESSANDRO SAFFO / 4CORNERS
© GIACOMO AUGUGLIARO / GETTY IMAGES
© SMARTSHOTS INTERNATIONAL / GETTY IMAGES

Clockwise, from left: Basilica di San Francesco, Assisi; Abbazia di Farfa, Lazio; Santuario della Verna, Tuscany; the frescoed interior of the Basilica di San Francesco

The trip

Distance: 550km (342 miles)
Mode of transport: On foot or by bicycle
Difficulty: Difficult

1 Santuario della Verna, Tuscany The cliffside monastic complex where Francis received the stigmata marks the start of the Northern Way

2 Basilica di San Francesco, Assisi The apex of the walk is the saint's eponymous fresco-filled cathedral and his much-venerated crypt

3 Eremo delle Carceri, Umbria This isolated hermitage, in the Assisian foothills of Monte Subasio, was built by St Francis himself

4 Valnerina, Umbria Imposing mountainous terrain along the rushing Nera river, and a centre of both spirituality and gastronomy

5 Abbazia di Farfa, Lazio Ornate monastic structures, dating back to 554 CE and surrounded by a medieval village

Mauro Cappelletti, hiking guide

I walked La Via di Francesco with my donkey, Luna, because she is the most gentle creature in the world. She can teach without speaking, she never complains and, you cannot believe, she has the biggest sense of humour. We walked 270km (168 miles) together, sharing all the effort of the path. The journey is like our life – downhill, uphill; rain, sun, fog. You just keep going and are happy to have someone at your side. Even better a donkey!

Assisi through neighbouring Tuscany and Lazio – the modern route has coalesced to form one single 550km (342-mile) journey that is quickly gaining status as a world-renowned pilgrimage.

To understand the journey, it helps to understand the beginnings of St Francis himself. The patron saint of Italy was born in Assisi around 1181 and died in 1226. The young son of a wealthy cloth merchant, he spent a year imprisoned during a war with nearby Perugia and fell gravely ill (most likely with malaria); he never truly recovered his health. While bedridden upon his return, Francis began an emotional and spiritual transformation. Having once been quite the *bon vivant*, he experienced doubt and spiritual confusion. On visiting the small church of San Damiano near Assisi (now one of the stops on the Via di Francesco), he had a vision that God asked him to fix his church.

Francis spent the next 25 years on a new spiritual path. He renounced all his prestige and wealth, and chose to live a life of poverty. No matter his ailments, he walked to all of the nearby towns – and even ventured abroad, to Syria and Egypt – preaching a new respect for the natural world, and for those living in poverty.

The magic on the Way

Many of those who walk this path today are also on a transformative journey. Though not all are religious believers, almost all are inspired by the life of the saint. At the many hermitages, churches, convents and monasteries along the route where Francis stopped to preach or do works of service, modern-day pilgrims often keep a reverent silence.

The journey is physically demanding. In and around the foothills of the Italian Apennines range, some days see elevation gains of 500m to 1000m

© VALERIOMEI / SHUTTERSTOCK

“In the stillness of the countryside, there remains a constant stream of sounds and sights: all of the senses are awakened”

(1640ft to 3281ft). And the urban settings aren't any different (they're not called hill towns for nothing). But something happens along the route. Many who walk – or bicycle – are not natural athletes or hikers, and it's not unusual to see pilgrims in their 70s and 80s. Like Francis, all who walk are encouraged to take rest days or go at a pace that works for them, and to ask for help if required. You never know what will happen when you tell the universe – and the route – what you need. 'We call it the "magic on the Way",' says Gigi Bettin, trail manager for the Francisco's Umbria region, 'when beautiful coincidences happen while you walk.'

Blaze your own trail

While there are dozens of books and maps about Spain's Camino de Santiago – walked by some 400,000 pilgrims in 2022 – the Via di Francesco hosts a mere fraction of those numbers. There are now a handful of guidebooks and a thoroughly detailed website (and advice groups on the usual social media sites), but the route should not be taken lightly.

The Via di Francesco's two sections – the Northern and Southern Ways – converge on Assisi's Basilica di San Francesco. The Northern Way to Assisi starts from either Florence or the Santuario della Verna in Tuscany; the Southern Way begins in Rome, though many walk it southwest from Assisi as a continuation of the Northern Way. Hiking the full 550km (342-mile) route takes about 28 days, and the trail is broken up into stages: daily walks of between 12km and 28km (7–17 miles). Each stage is fully detailed on the Via di Francesco website, and rated as 'easy', 'medium' or 'challenging'. Many stages have optional detours – to see the Cascata delle Marmore Roman waterfalls near Terni, for example.

But no matter the route, this is a personal journey about personal transformation. You might not give up all your worldly possessions after walking, but perhaps you'll opt to volunteer with animals, or just resolve to take an after-dinner stroll each evening. During the pilgrimage, you'll have plenty of time to contemplate, amid a backdrop of natural serenity and grandeur.

From top: the Via di Francesco offers a slow-travel route through the Tuscan landscape; explore Abbazia di Farfa on the Southern Way through Lazio

Practicalities

Region: Tuscany/Umbria/Lazio, Italy
Start: Florence or La Verna
Finish: Rome

Getting there and back: Trains and buses serve the larger towns along the way, and Rome, Florence and Assisi all have international airports.

When to go: The Basilica di San Francesco in Assisi hosts a Pilgrim's Mass between April to October, when the weather is most agreeable. July and August can be exceedingly hot (and crowded).

What to take: Hiking gear, sunscreen and a very large water bottle. Ensure that your backpack doesn't weigh more than 10% of your own body weight when it's full.

Where to stay: For devotees of Francis as the 'Poor Man of Assisi', spartan structures (churches, sacristies, convents and so forth) take in pilgrims devoting their walk to poverty. For others, the Umbrian tourism website (umbriatourism.it) offers a list of B&Bs, hotels and *agriturismos* along each stage.

Where to eat and drink: This is Italy, so the route never strays more than a few kilometres from a town with excellent restaurants, bars, pizzerias, gelaterias and porchetta trucks. Until more drinking fountains spring up along the way, carry a day's worth of water with you at all times.

Tours: The route's website (viadifrancesco.it) has comprehensive information. Pilgrim Paths (pilgrimpaths.com) offers self-guided, paced packages. For group tours, try Hike and Bike Italy (hikeandbikeitaly.com) or Ecologico Tours (ecologicotours.com).

Essential things to know: Read about St Francis before you go: *Trekking the Way of St Francis* by Rev Sandy Brown is invaluable. Ask for help on the way, and stay silent in all Franciscan hermitages.

Asia

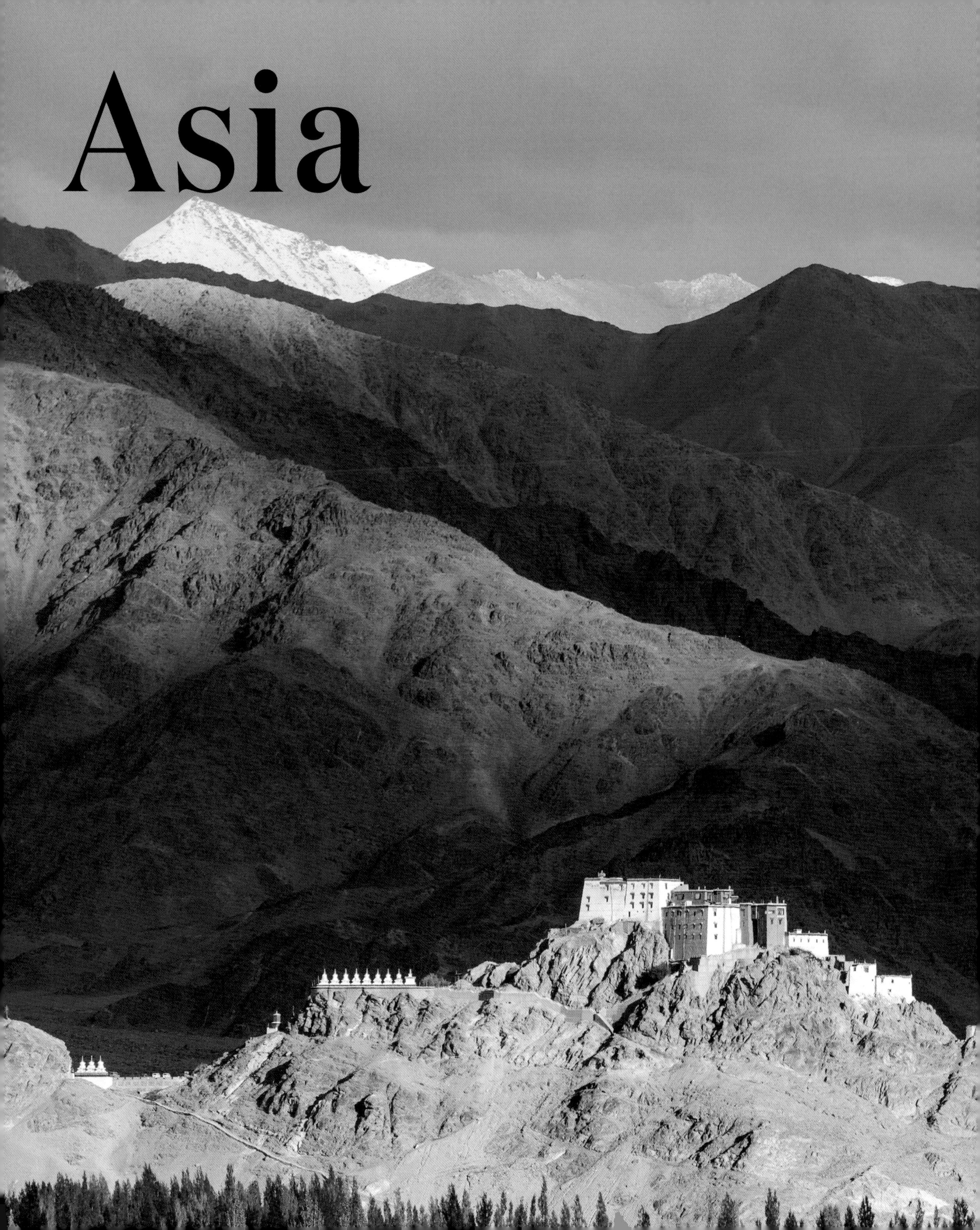

Thiksey Gompa
in Ladakh, India

Across India by train

Nothing will open your eyes to the complexities of India quite like travelling the length of this continent-sized country by train.

When you board an Indian train, the volume dials down and the breakneck pace of the subcontinent eases to a crawl – perfect for slow contemplation of where you are going, both geographically and philosophically.

Train travel in India is a great leveller – chatting to fellow passengers as the miles slip by, you'll learn how India ticks from the people who follow its rhythms all year round. You'll also get a crash course in how to deal with crowds, navigate bureaucracy, order street food on the hoof – and yes, get over your fear of India's public toilets.

India is big, the way continents are big. States come the size of countries, and a single train ticket can whisk you from the foothills of the Himalaya to the palm-shaded fringes of the Arabian Sea, taking in a continent's worth of cultures and climatic zones along the way. Unsurprisingly, this means journeys are measured in days and nights rather than hours and minutes.

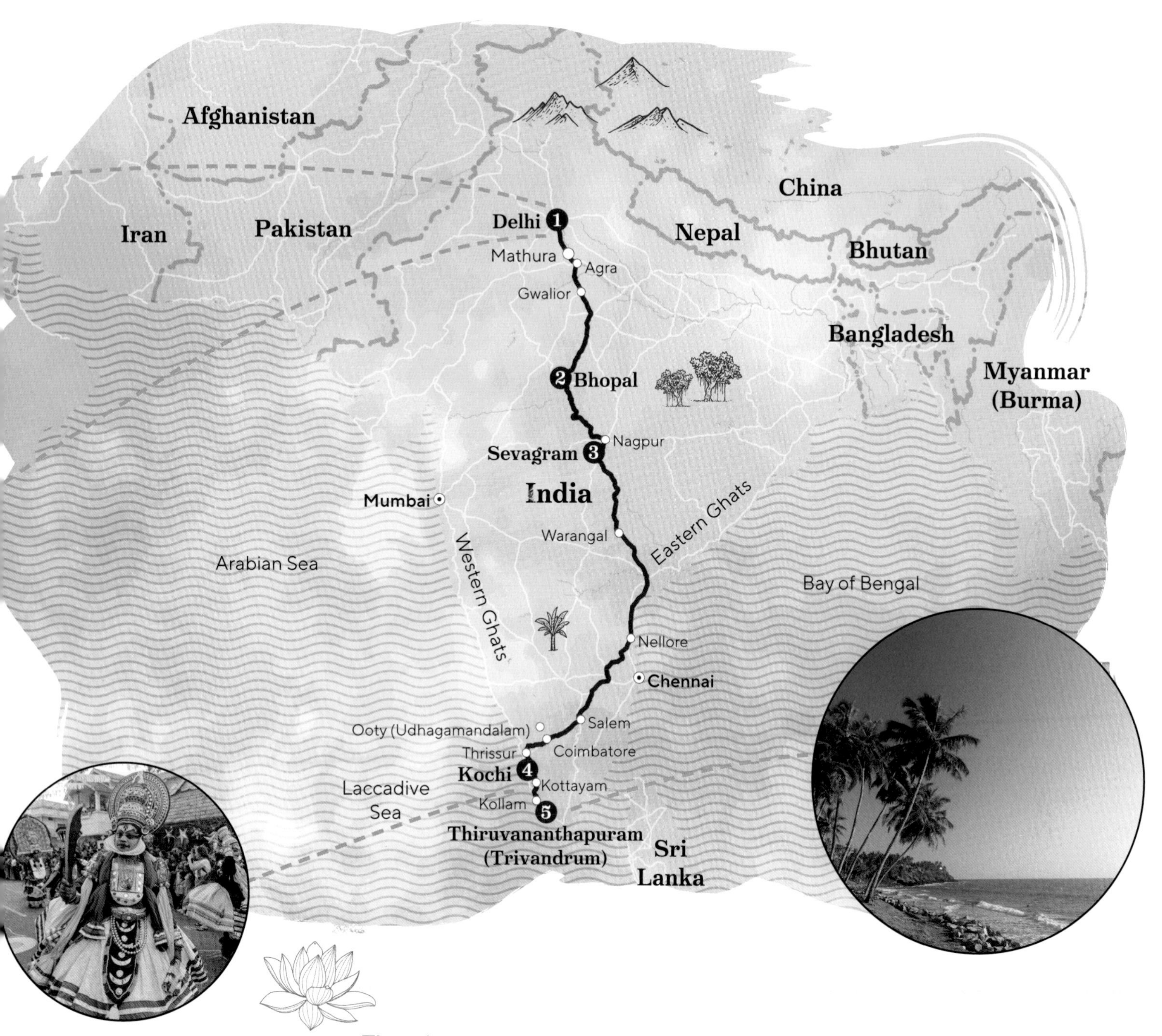

Clockwise, from left: Nilgiri Mountain Railway; Delhi's Jama Masjid; the Red Fort's Diwan-i-Am; Varkarla, Kerala; Kathakali dancer, Kochi

The trip

Distance: 3031km (1883 miles)
Mode of transport: Train
Difficulty: Easy

1 Delhi The Indian capital is exhausting but exhilarating, steeped in 2500 years of history

2 Bhopal Mosques, monuments and lakeshore views define this ancient, often overlooked city

3 Sevagram Detour to the ashram where Gandhi planned the struggle for Independence

4 Kochi Disembark at Ernakulam to explore this famous melting pot of cultures

5 Thiruvananthapuram (Trivandrum) Museums, markets, temples and churches – and easy access to Kerala's best beaches

Anirban Mahapatra, train traveller

For me, long-distance train rides in India are about getting into the heads and hearts of strangers sitting across the aisle, and becoming friends over endless rounds of tea. As a slideshow of ever-changing landscapes plays outside the window, I like to open myself to the plethora of real stories lived by real people in the many corners of this mind-bogglingly diverse country. And every journey makes me wiser about cultures, customs, beliefs, and life in general.

Every day, some 23 million passengers travel on India's railways, navigating 68,103km (42,317 miles) of train tracks, and joining them in the melee of boarding and disembarking is a joyous journey to the soul of the subcontinent. Every train ride is an education, revealing more of the country's secrets with every clickety-clacking mile.

Best of all, a train journey provides space to pause and think, a luxury rarely available at street level in this phenomenally crowded nation. In between memorable conversations with fellow passengers, there'll be long hours of gazing out of the window, lost in your thoughts, while India slips by like a documentary film behind the glass.

Getting the measure of India

This is one trip where the journey is definitely the destination, so plot a course right across the subcontinent to maximise your time on the rails. For a train ride with drama, it's hard to beat the Kerala Superfast Express – a 50-hour, 3031km (1883-mile) epic linking India's chaotic, colourful capital and Thiruvananthapuram (Trivandrum) on the Kerala coast.

Travelling the length of the country, you'll see the landscape swirl and change, as the dry plains of north India slowly give way to the jungles and beaches of the tropical south. En route, you'll mingle with people from every walk of life – students, teachers, families, holy men – and learn a little more about India with each passing mile.

© SEAN3810 / GETTY IMAGES

> "Every train ride is an education, revealing more of the country's secrets with every clickety-clacking mile"

But first, you'll need a sleeping-berth ticket – this means diving into the frantic booking queues at New Delhi railway station or navigating India's perplexing online train booking system. Either way, you'll get your first taste of the subcontinent's byzantine bureaucracy – and the sense of achievement that comes from completing all the forms correctly without losing your cool.

Out of Delhi, into the plains

With ticket in hand, prepare to be pleasantly surprised at how well things actually work in practice. New Delhi Railway Station is mobbed by crowds to make Woodstock balk, but once you board your carriage, the noise and chaos subsides and you can sit back and enjoy the ride.

Before you embark, take a stroll along the platform to stock up on sweet and savoury snacks from station vendors (there'll be opportunities to replenish your stash at longer stops along the route) and grab a copy of the *Times of India* newspaper, for a quick primer on the issues currently animating the subcontinent.

You may feel a moment of transcendental calm as the cacophony of New Delhi station recedes into the distance, but any hopes of easing back with a good book will soon be dashed as your fellow passengers seize the opportunity to make your acquaintance. Don't resist – train conversations are a crash course in the customs of the subcontinent. By the end of the trip, you'll be navigating India's cultural conventions like a seasoned pro.

After sharing your life story with half of the carriage, and answering dozens of questions about politics, sport, your religion, your home country, your marital status and the relative prices of essential items back home, the initial excitement will fade, and you can relax and settle in for the night.

Watching India roll by

Perhaps the greatest pleasure of Indian train travel is seeing ordinary life unfold beside the tracks, away

Below: Emperor Shah Jahan's Taj Mahal, Agra; **Left:** hawkers at Delhi's India Gate

Indian Railways 101

Travelling by train in India means navigating layers of bureaucracy, but it doesn't take long to learn the ropes. Every train has a number and a name, and every station is identified by a three- or four-letter code – you'll need all this information to make a booking, so use the 'Reserved Train Between Stations' search engine on the Indian Railways website (indianrail.gov.in) to scope out your journey. Online bookings can be tricky – it's easier to use the agency 12GoAsia (12go.asia) than the official Indian Railways booking site (irctc.co.in). When it comes to train classes, 2nd class is an unreserved free-for-all, but fan-cooled sleeper class and two-tier and three-tier AC sleepers are good for trips by day or night, with an in-seat meal service and beds that fold away by day to become seats. Skip the various 'chair classes' and 1st class – you'll be isolated from other passengers without a huge increase in comfort.

from the tourist circus. The plains where the Hindu god Krishna cavorted in the *Mahabharata* and the Taj Mahal town of Agra will slip by in darkness, but dawn brings Bhopal into view – an ancient Muslim city with a rich history that's often eclipsed by the industrial disaster of 1984. Consider breaking the journey here to feast on a buffet of street snacks in the Old City and admire the Taj-ul-Masjid, once the world's largest mosque.

Alternatively, enjoy a chapati-and-chai breakfast in your seat, then glue yourself to the windows for a live screening of Indian daily life in off-the-tourist-track towns such as Sevagram, where you can disembark to visit Mahatma Gandhi's Independence-era ashram. The sequence of railway crossings, temples, village houses and swathes of greenery is hypnotic – you'll soon be so lost in your thoughts that you'll barely notice the miles whisper by. With 50 hours at your disposal – or more if you break the journey – there'll be ample time to thrash out some of the big metaphysical debates: the meaning of life, your place in the universe, and whether you'll ever fully get the hang of this massive, magnificent country. Fellow passengers taking time for *puja* rituals, morning mantras and prayers facing west towards Mecca only add to the contemplative air.

A sample of the steamy South

By the second night, you should have your bedtime routine down to an artform, converting your daytime seat to a berth with the minimum of fuss. With the rhythm of the rails to lull you to sleep, expect to feel well rested as morning brings a first glimpse of the Western Ghats, India's jungle-choked southern hills.

There's a marked cultural shift as you move into the tropical south. Meat starts to vanish from menus, languages become more lilting, and *shikhara*-style temple towers give way to towering *gopurams* – skyscraper-like temple gateways covered in rainbow-coloured depictions of Hindu deities.

Around lunchtime, the Kerala Express swings into Coimbatore, offering a chance to break the journey and detour to the Raj-era hill station of Ooty (Udhagamandalam) by miniature locomotive on the narrow-gauge Nilgiri Mountain Railway. Alternatively, just settle back and watch tropical vegetation take over the scenery.

From Coimbatore, the tracks cross into tropical Kerala, where the views become increasingly verdant and waterlogged as you trundle towards the coast. You'll start to see Christian churches dotted among the palms – a religious tradition reputedly brought to Kerala by the apostle Thomas in 52 CE.

© SURESHKEGE / GETTY IMAGES

From top: emerald peaks in the Western Ghats; departing Mettupalayam, Coimbatore, on the Nilgiri Mountain Railway

Enticing stops abound – consider festival-tastic Thrissur (Trichur) or the historic port of Kochi at Ernakulam, or disembark at Kottayam or Kollam for a riverboat ride through the Kerala backwaters. Then, at last, you'll arrive in the Kerala capital, after a longer journey than that of the Paris–Istanbul Orient Express.

After two days and two nights of rattling and rocking, with only a train toilet for personal ablutions, celebrate your arrival in Thiruvananthapuram with a day at the beach. You can reach Varkala by train or Kovalam by bus in less than an hour, and be walking barefoot in the surf minutes later, followed by a slap-up dinner of Arabian Sea prawns or coconutty *molee* fish curry with lemon rice.

Practicalities

Region: Delhi to Kerala
Start: Delhi
Finish: Thiruvananthapuram (Trivandrum)

Getting there and back: While both cities have international airports, Delhi and Thiruvananthapuram are connected to almost everywhere by bus and train.

When to go: October to March, avoiding the summer monsoon rains and the stultifying pre-monsoon heat. If you come in midwinter, expect morning views to be obscured by fog in the northern plains.

What to take: For a comfortable train ride, bring ear plugs, hand sanitiser, a woollen shawl (as an impromptu blanket), a padlock and chain (for securing your bag to the baggage rack) and a music player with earphones (popping in the earpieces will buy you some privacy when you need it).

Where to stay: The train is your bedroom for two nights of travel – in air-con berths, you'll get crisp, clean bedding; in fan-cooled sleeper class, bring a blanket or a sleeping-bag liner.

Where to eat and drink: Snacks and drinks are available at every train platform in India; when the train stops in larger cities, you may have time to leap out onto the platform to buy supplies. For more substantial meals, order from the pantry car, or book an onboard delivery via the RailRestro app (railrestro.com).

Essential things to know: Reserve train tickets at least a few days before you travel at the station, or book online via 12Go (12go.asia) or the Indian Railways website (irctc.co.in). If you use a padlock to secure your luggage to the baggage racks, keep the key handy – you don't want to be scrabbling through your pockets as the train pulls out of your station.

Spiritual awakening in Bhutan

A huge national effort helped resurrect the Trans Bhutan Trail in 2022; sample a section of this cross-country hiking journey full of culture, spirituality and environmental lessons.

Bhutan's Buddhist principles – which are carried over into local attitudes to wildlife and the environment – may encourage a spiritual awakening, inspiring visitors to reassess their own lifestyles and beliefs.

Hiking past community farms and remote villages gives insight into Bhutan's unique national self-sufficiency; interactions with local people help explain the strong sense of community here.

The Trans Bhutan Trail is a technical hike, with steep ascents and descents, rock scrambles, stepping-stone river crossings, obstructive tree roots and tracts of leech-ridden forest. Each day feels like an achievement.

As far back as the 16th century, trail-runners called garps used the Trans Bhutan Trail (TBT) to pass important messages between the chain of strategic *dzongs* (fortresses) running west to east through the heart of the country. Of those original *dzongs*, Simtokha is the only one that remains completely intact, and it's a nationally important starting point for a multiday foray along a section of the TBT to Rukubji, covering around 55km

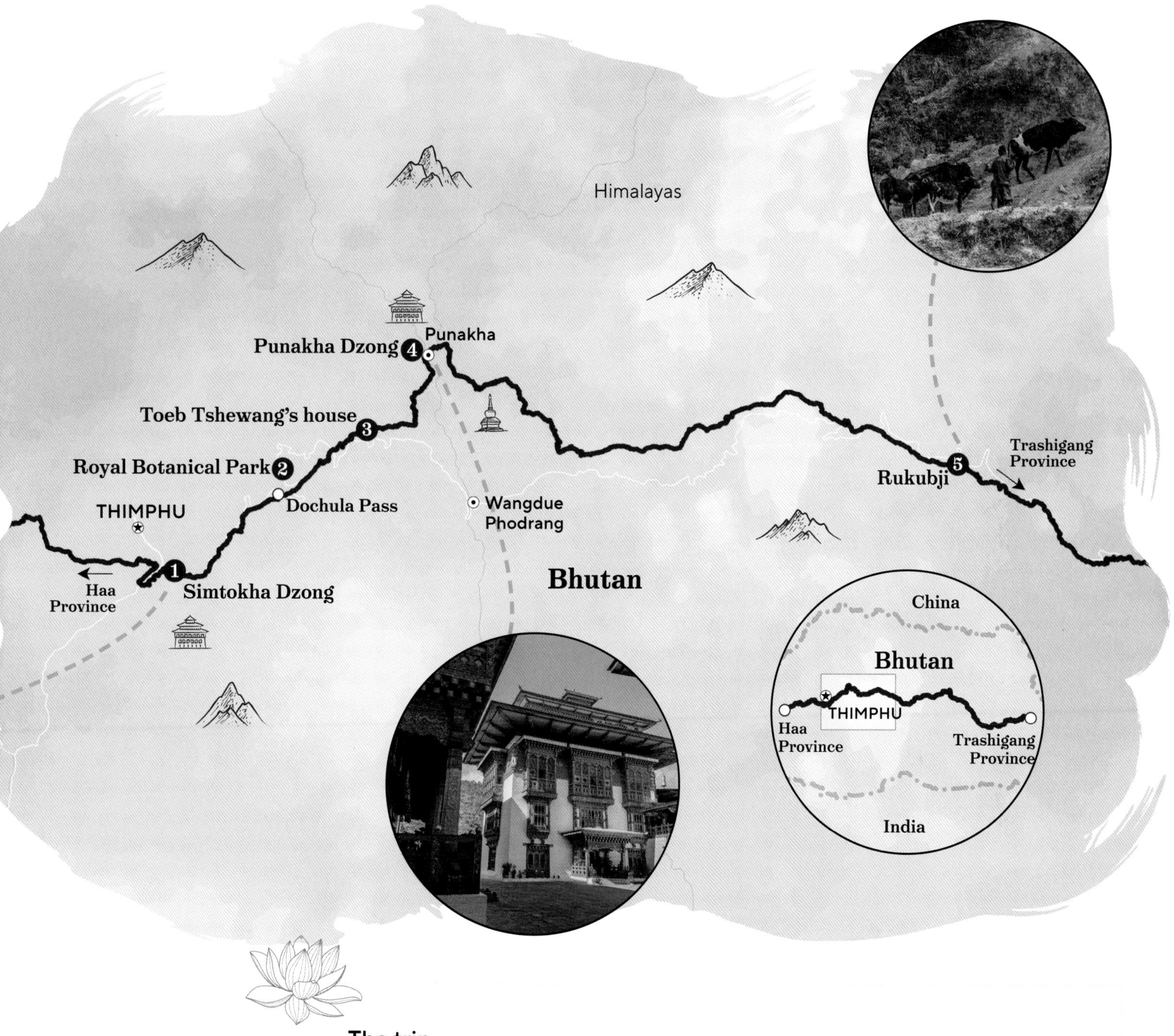

The trip

Distance: 55km (34 miles)
Mode of transport: On foot
Difficulty: Variable

1 Simtokha Dzong Nationally important hillside fortress outside Thimphu, with a unique 12-sided *utse* (tower)

2 Royal Botanical Park Prehistoric forests where old man's beard drips from trees alive with birds and insects

3 Toeb Tshewang's house A building steeped in 'Divine Madman' legends, preserved as a national monument for 15 generations

4 Punakha Dzong This beautiful fortress complex sits at the confluence of two rivers

5 Rukubji A remote wooden village with a unique language, and community run veg-growing plots

Clockwise, from left: Punakha Dzong; monks at Thimpu's Simtokha Dzong; driving cattle near Rukubji village; temple buildings at Punakha Dzong

Wangmo, Rukubji village

I was very happy when they began work on reforming [the TBT]. When I see someone pass through here, I remember the days when I used to walk it myself as a child. It makes me feel more connected to my ancestors. There's another small community just over the hill, where I used to have a farm that I would visit by walking the trail. I'm happy that they will also benefit from the development of the TBT and I would love to return to that village some day. I hope the reopening of the trail will bring good fortune again.

(34 miles) – just a fraction of the trail's full 403km (250-mile) route, which runs from Haa Province in the west to Trashigang Province in the east.

Locals believe Simtokha is built on the site where a demon was swallowed by the mountains. Constructed in 1629 by Zhabdrung Ngawang Namgyal, the Tibetan lama who unified Bhutan and is considered the founding father of the country, it's also thought to be the first *dzong* built to house both administrative and monastic residences. Head to the temple's inner sanctum to reflect on how this interplay between faith and dynastic power has shaped modern Bhutan.

National history lesson completed, hikers can walk straight out from the *dzong*'s gates onto the TBT. Heading east, it's a moderate-grade hike with steep descents and ascents, and river crossings via wooden bridges. Bhutan's environmental protection ethos dictated that the TBT reconstruction should be low-impact, and there are no concrete or solid stepping-stones to smooth the way. Along the trail, trekkers may spy the tell-tale signs of bears in the forests – claw marks cut deep into the trunks of rhododendron and oak trees – while small-scale farm plots growing flame-red chillies (Bhutan's cash crop) and swathes of apple orchards highlight the country's self-sufficiency.

Entreating forest spirits

Locals burn bundles of twigs and incense as spirit offerings at mist-swathed Dochula Pass (elevation 3100m/10,171ft). In a country that is 80% Buddhist, the Bhutanese are firm believers in the interdependence of all life forms, and this extends to the mountains and forests. Offerings are one way in which they show their respect for, and connection to, the spirits that inhabit the natural environment. Hikers can watch the offerings being made, but should be careful not to treat Bhutanese living culture as a tourist show.

© EDWIN TAN / GETTY IMAGES

> "The path is quickly swallowed by thick forest, where the ruffled canopies wrestle with penetrative shards of sunlight"

Below Dochula Pass, inside the Royal Botanical Park, the path is quickly swallowed by thick forest, where the ruffled canopies wrestle with penetrative shards of sunlight. Old man's beard clings to branches like epic spiders' webs; insects screech like harpies. The muddy descent – often accompanied by leeches – is tough, but being among the naked, native woodland is an opportunity to immerse in the conservation efforts of the world's only carbon-negative country – and contemplate how nature can thrive when governments make serious environmental commitments.

In the belly of the valley, a fern-strangled riverside *chorten* (shrine) is a reminder that the TBT has been a spiritual as well as physical journey for centuries. Every Bhutanese person is supposed to build one *chorten* during their lifetime and some of these sacred Buddhist markers, filled with human ashes and religious relics, are considered wish-fulfilling. They also played an important role in the reconstruction of the TBT – in some places prolific jungle growth and relative isolation meant they were the only remaining sign that the Bhutanese had ever walked this path.

Divine madman territory

The key subject of the TBT section through Thinleygang and Toeb Chandana is Drukpa Kunley – a Tibetan monk who revolutionised the Bhutanese relationship with Buddhism when he passed through this valley in the 15th century. Often heralded by the Bhutanese as 'the master of truth', Drukpa's unorthodox life led to his affectionate nickname, the 'Divine Madman', and this TBT section is packed with cultural and spiritual sites related to his life. At Thinleygang Lhakhang, established by Drukpa Kunley's brother and painted with lotuses and jewels that symbolise purity, hikers may be invited to take a tour of the two-storey temple. On the ground floor, a

Below: the Bazam bridge at Punakha Dzong
Left: Druk Wangyal Khang Zhang shrines, Dochula Pass

Tantric Buddhism

Bhutanese life is rooted by the principles of Tantric Buddhism, which is practised across the Himalaya but forms the core religious system in only Bhutan and Tibet. For visitors to Bhutan, one of the notable aspects of Tantric Buddhism is iconography showing Buddha in sexual union as a path to wisdom, which is considered a female principle. Even those who have no religious inclinations will be bowled over by the colourful aesthetic of Bhutan's ornate temples, which represent a 3D depiction of the life and legends of Buddha and his disciples. Hand-worked *thangka* paintings and tapestries usually line every wall, presenting a rich backdrop for golden statues of deities and ornate plaster reliefs. Paper decorations of clouds and rainbow shards of light hang down from the ceilings, as a manifestation of what the Bhutanese believe the journey up to heaven would look like.

giant gold statue shows Buddha in a tantric embrace with a female deity, alongside cherished *thangkas* (paintings), including one of the Divine Madman with his signature bow and arrow and following of dogs.

Descending past hydropowered prayer wheels flushed by streams and waterfalls, the trail from Thinleygang Lhakhang passes into a secluded valley housing one of Bhutan's most sacred places – the landing site of the arrow that Drukpa Kunley is said to have fired from Tibet in the 15th century, to determine his path through Bhutan. The house where it struck has long been preserved as a national monument, and the temple beside it was established to ward off the evil energy of a demoness. Shrouded in legend, the house and temple lie between two hills said to resemble the knees of Ngawang Chogyal, the Tibetan founder of Bhutan who chose Toeb Chandana Lhakhang as a religious seat.

Unlocking Punakha

Whether you arrive on foot or by car, the former capital of Punakha is a key point along the Trans Bhutan Trail – and its huge *dzong*, sitting at the confluence of two rivers, is one of the country's most photographed sights. Visitors might be startled by the enormous beehives that hang off its ornately painted wooden entranceway – Buddhists won't remove these because of their belief that no animal should be harmed. Hikers can meditate on this principle inside the *dzong*'s incredible assembly hall for monks, which houses giant statues of Buddha, Zhabdrung and Guru Rinpoche.

Behind the *dzong*, a dirt path leads to a giant rope-bridge spanning the river. Beyond it lies one of the Trans Bhutan Trail's homestays, run by Dawa Zam out of her family home. Accommodations like this are being developed along the breadth of the reconstructed trail to ensure tourism creates economic opportunities for the communities that

© DYLAN HASKIN / SHUTTERSTOCK

"Hikers can turn the prayer wheels of Rukubji's temple for meditative musings or to amass blessings for the onward journey"

coexist with it. But the exchange works both ways: tourists benefit by getting to experience living inside one of Bhutan's uniformly beautiful houses – this one is traditionally made of rammed earth, wattle and daub, with intricate woodwork around windows, doors and roofs. Trekkers staying overnight with Dawa might get to eat village-harvested *dandin dshering* rice, and be invited for a peek inside the household's private altar room – a common feature of Bhutanese homes.

Chasing yak shadows

Tracing the curvature of a valley towards the remote village of Rukubji, the section of the hike from Pellala Pass presents a very different side to the Trans Bhutan Trail – one where horses and cows roam free on alpine meadowland in summer and yak graze in winter, when the high passes disappear inside a cocoon of snow. Wooden yak-herders' huts dot the valley, along with the occasional farmhouse where women shoulder huge loads of rotting ferns to build soft beds for their cows. The trail threads through burbling streams, brushing through patches of medicinal plants such as artemisia, which the Bhutanese collect for hot-stone baths to soothe aching limbs.

Rukubji itself is an eye-opening example of the isolated lives many Bhutanese live in a country that's topographically characterised by steep, divisive Himalayan peaks and has more than 70% forest cover (it's written into Bhutanese law that at least 60% of the country has to be covered in forest at all times). Locals here still use one of Bhutan's oldest languages, now spoken by fewer than 1000 people.

Village vegetable plots fill the crevices between wooden houses that are gaily painted with tigers, dragons, snow lions and *garudas* (mythical bird-men) – Bhutanese symbols of discipline, generosity, purity and fearlessness. At the centre of the village, hikers can turn the prayer wheels of Rukubji's small temple for meditative musings or to amass blessings for the onward journey. If you're lucky, there might even be offerings of steaming yak-butter tea and sugary biscuits from the caretaking monk.

From top: Punakha Suspension Bridge over the Pho Chhu; yak on the Trans Bhutan Trail near Rukubji village

Practicalities

Region: Bhutan
Start: Simtokha Dzong, Thimphu
Finish: Rukubji

Getting there and back: India has overland border crossings with Bhutan, but most visitors enter via a flight with national carrier Druk Air (the only airline that flies into Bhutan) from Delhi or Bangkok. The international airport is at Paro, a 1hr drive west of Thimphu. From Rukubji it's a 5hr drive back to Paro; you'll need a private car to make the journey.

When to go: The shoulder seasons are the most reliable and accessible trekking months; spring (March/April), when rhododendrons bloom, is particularly scenic.

What to take: Leech socks (gaiters), layers, a foam roller to soothe aching legs between hikes, and swimwear for the traditional Bhutanese hot-stone baths.

Where to stay: Arrange adventure camping at pop-up sites via the TBT webaite; there are also signature camps with such comforts as hot showers and hot-stone baths, a growing network of homestays, and luxury rural lodges – some of the best are run by Amankora. All stays need to be arranged in advance.

Where to eat and drink: Local tour guides arrange picnics for remote hiking days, or organise for travellers to visit roadside restaurants run by locals.

Tours: Bhutan is unique in that visitors cannot travel alone and must either hire a private local guide or book a tour. Everything can be arranged directly through the Trans Bhutan Trail website (transbhutantrail.com).

Essential things to know: Bring a reusable water bottle with a filter so you can fill up at taps; despite Bhutan's impressive environmental legislature, it falls down on single-use plastics and water bottles are part of the problem. To enter temples, you'll need clothes that cover your legs and shoulders.

A surfing loop around Sri Lanka

Circling Sri Lanka in search of the perfect wave, you can graduate from first-timer to salt-seasoned pro on some of Asia's most enjoyable surf.

If you're after a crash course in surfing, Sri Lanka is ideal, with beginner breaks where you can find your feet and brilliant barrels where you can put your newfound skills to the test. First-timers can rest easy: at easygoing hubs such as Arugam Bay, you can set up camp in a driftwood cabin by the sand and upskill at your own pace.

Some of Sri Lanka's best breaks are off the tourist map, but locals can help you find them. Swapping tips with Sri Lankan surfers over a plate of rice and curry is almost as much fun as catching a wave.

Picture the scene: a tropical morning bursts over a sweep of caster-sugar sand. A crooked palm tree swishes gently in the breeze as you grab the board propped against your beach hut and run down to the breakers. There's no wriggling into a still-damp wetsuit – you don't need one when the water temperature hovers above 27°C (81°F). Welcome to surfing, Sri Lankan-style.

Encircled by breaks to suit every level of surfer, Sri Lanka is a relative newcomer on the global surf

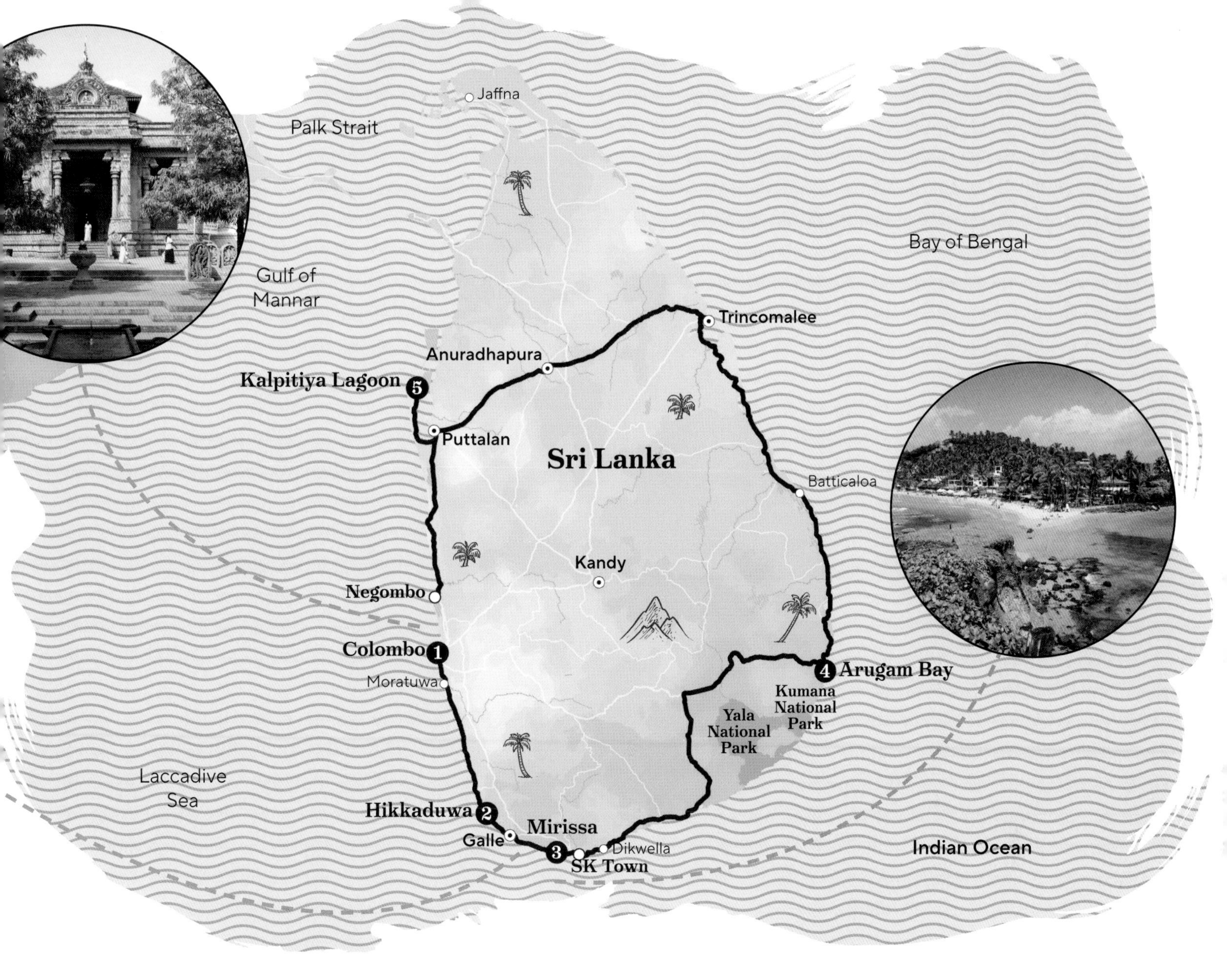

The trip

Distance: 1000km (621 miles)
Mode of transport: Bus and train
Difficulty: Easy to moderate

1 Colombo Enjoy cool coffeeshops, crab curry, colonial relics and terrific temples before hitting the surf

2 Hikkaduwa Bath-warm waters, accessible surf breaks and snorkelling right off the main beach

3 Whale-watching off Mirissa Mix rewarding surf breaks with offshore safaris in search of blue whales

4 Arugam Bay The sand-floored, sea-splashed hub of the east coast surf scene

5 Kalpitiya Lagoon Get back to basics on a kitesurfing camp on this lovely, windy lagoon

Clockwise, from left: south-coast surfing at Mirissa; Wijaya beach, near Galle; Kelaniya Raja Maha Vihara, Colombo; Mirissa beach

Fawas Lafeer, Arugam Bay surf teacher

The best thing about surfing in Sri Lanka, especially on the east coast, is that we have safe and easy waves for every level of surfer. Surfing for me is like meditating – when I go into the water, I'm not thinking about anything else in the world. What I've learned from surfing around Sri Lanka is patience, tolerance, sharing waves and being happy with people, wherever they are from – there's no age, gender, colour or language barrier.

scene, but being the new kid on the block means there's little snobbishness about beginners taking to the waves. Even if you're completely wet behind the ears, you can turn up and learn on the water, without having to worry about snooty pros hogging the best breaks.

And if you're a newbie with a few waves under your belt, this idyllic tropical island is the perfect place to hone your skills, graduating from gentle beach breaks to challenging A-frame reef breaks and zippy barrels. Best of all, with Sri Lanka's manageable size, it's easy to loop right around the island, sampling all the best breaks on one trip.

Catch your first Sri Lankan wave

Sri Lankans are used to surfers turning up at train or bus stations with a bulky board-bag, so hopping from break to break is rarely a hassle. Kick off the surf circle by boarding a local train running south from Colombo, on the British-built rail line that hugs the west coast.

One top tip for Sri Lankan train travel: stake out a spot by one of the open doorways, so you can make the most of the sea breezes and watch the tropical coastline go swishing by. Snag a space on the right side of the train and you'll see the first breakers bursting on the beach before you've even left the city limits.

The first stop on this surf safari is Hikkaduwa, an unpretentious west-coast hangout with memorable beaches, offshore coral reefs for snorkelling, plenty of budget beach resorts and places to eat, and perfect

> "You can turn up and learn on the water, without having to worry about snooty pros hogging the best breaks"

waves for newbies. Newcomers should start with the gentle breaks on the main beach at Narigama – the patient teachers from the Sri Lankan-run Reef End Surf School will get you standing up on a board on your first day on the waves.

More skilled surfers can try their luck on intermediate reef breaks such as shallow and popular North Jetty and consistent left-hander Bennys – both deliver rewarding overheads in good conditions. Pop your board on top of a three-wheeler (autorickshaw) to reach calmer Owakanda, an in-the-know spot to the south where you won't face such a wait for a ride.

Take it up a notch on the south coast

After finding your feet at Hikkaduwa, take things up a gear at Weligama, a short bus ride around the coast (on the far side of charming, Dutch-built Galle). As at Hikkaduwa, the benign breaks are ideal for new surfers – surf shacks stud the beachfront renting out boards, and local schools offer lessons for novices and improvers.

Weligama translates as 'sandy village' and the name fits – there's plenty of sugary shores to bask on between surf sessions, and friendly hostels behind the palms ensure plenty of après-surf camaraderie. Spend your days riding the chest-high, rock-free main break and your nights enjoying chilled Lion lagers and fresh-off-the-boat seafood in the cafes behind the beach.

With a few waves under your belt, flag down a three-wheeler to reach more challenging breaks at Mirissa – also a popular hub for whale-watching – or the expanding surf hubs at Ahangama and Midigama. The scene on this stretch of coast is boisterous and international, drawing surfers in sun-bronzed droves. It's a little bit Bali, and you may encounter smatterings of localism, but spend a few days sharing the breaks with local characters and you'll get some invaluable tips about the best places to catch a wave, including insider waves such as the left-hand point break at Hiriketiya near Dikwella, or the fun beach break at SK Town.

Below: Arugam Bay sunset
Left: a humpback whale off the coast of Mirissa

Rice & curry: the fuel of Sri Lanka

A slow, carb-crammed Sri Lankan lunch followed by a doze in the shade to digest is the best way to escape the oven-like midday heat and build up your energy levels for an afternoon on the waves. Enter Sri Lanka's national dish – rice and curry – or, more accurately, rice and half a dozen curries, blending meat, fish, vegetables and coconut, with an obligatory side-splodge of *sambol* (coconut pounded with chilli), sometimes a portion of hoppers (rice and lentil-flour pancakes) and a nourishing bowl of dal (lentils). It's an essential introduction to homestyle Sri Lankan cooking. Seek out authentic rice and curry in local roadhouses, usually found behind the beach on the main road through town. Expect a safari of flavours – dishes are often at the incendiary end of the chilli scale, so keep a sweet drink handy to take the edge off the heat.

Let it all hang out in the east

For all the action on the south coast, you may find yourself craving a less showy scene. If so, make a beeline for the calmer east coast, where Arugam Bay serves up cosy wooden beach huts, a gentle buzz after dark and enough surf to keep everyone happy, from fresh-faced beginners to grizzled old-timers.

The point breaks around Arugam perform best from April to October, when double overheads are not unheard of, but getting here involves a detour inland, skirting the Yala and Kumana national parks – both worth a stop for reliable leopard and elephant encounters. The trip inland will take you into the cooler Central Highlands, with several changes of bus along on the way – it's a bit of a chore with a board in tow, but worth the effort to enjoy one of the most relaxed spots on the Sri Lankan coast.

Novices kick off at Surf Point and Baby Point, two long and lovely but busy beginner breaks off the rounded headland in the heart of Arugam. If you're ready for more of a challenge, local schools can help you over the hump to intermediate level – try the respected Safa Surf School, run by enthusiastic Fawas Lafeer.

Then, put your talents to the test on the quieter intermediate breaks to the north, where rocky points tuck in between shallow, sandy coves. Pottuvil Point, Whiskey Point and Lighthouse Point serve up slow right-handers that perform brilliantly in the right conditions, with much less bustle both on and off the water.

> "Arugam Bay serves up cosy wooden beach huts, a gentle buzz after dark and enough surf to keep everyone happy"

Closing the loop

The surf potential of the Tamil-dominated far north is weakened by the breakwater effect of the Indian coastline, but there are a couple more surf stops to consider before you close the loop in Negombo. The sheltered point break in temple-crammed Trincomalee is certainly worth a moment of your time, but the serious action is over on the west coast on the Kalpitiya Peninsula – one of the top kitesurfing destinations in South Asia.

Follow the sandy tracks snaking down to the broad Kalpitaya Lagoon and you'll find plenty of places upskilling kitesurfers for the steady winds that whicker across the peninsula from May to September. It's a great place to learn from experienced pros. A little of the barefoot magic that first drew surfers to Sri Lanka lives on in Kalpitiya, where days start with kite prep and end with a cold beer sipped by moonlight behind the whispering dunes. It's the perfect spot to say a slow goodbye to Sri Lanka.

© ELKE MEITZEL / GETTY IMAGES

© SHANTI HESSE / SHUTTERSTOCK

From top: lagoon and ocean at Arugam Bay; boards on the beach, Arugam

Practicalities

Region: Sri Lanka
Start: Colombo
Finish: Negombo

Getting there and back: Sri Lanka's main international airport is 33km (20 miles) north of Colombo near Negombo. To reach the capital, take a taxi or walk out of the airport and flag down a three-wheeler to the bus station in Katunayake, then board a local bus.

When to go: From April to September, the southwest monsoon brings rain to the west and south, while the east and north stay mostly dry. From November to March, the northeast monsoon soaks the east coast. To catch good surfing weather on both coasts, surf the southwest in April and the east coast in May.

What to take: There's no need to bring a board if you're sticking to the main breaks; just rent one from a local surf shack. When travelling to more remote breaks, stow your board on the roof of a three-wheeler (autorickshaw); drivers are used to accommodating this.

Where to stay: Hotels and beach resorts hug the Sri Lankan coast, backed up by rustic beach huts and laidback surf camps – often with a barefoot, Robinson Crusoe vibe. Book surf camps ahead; most are only open during the surf season.

Where to eat and drink: Most places to stay offer meals, and restaurants and beach bars line every resort strip. Seek out local roadhouses for the best rice and curry.

Tours: Plenty of agencies offer surfing holidays to Sri Lanka, including surf-camp packages for beginners – surfholidays.com is a good place to start. Links to the surf schools in this journey are: Reef End Surf School (reefendsurfschool.com); and Safa Surf School (safaarugambay.com).

Essential things to know: Political unrest is a periodic problem in Sri Lanka; monitor local news channels and avoid potential flashpoints such as political rallies and demos.

Trek Nepal's Langtang Valley

Prepare to be inspired by the resilience of Nepal's hill communities and humbled by the awesome power of the Himalaya on one of Nepal's classic treks.

Remind yourself of the magnificent scale and sublime beauty of these mountains, and embrace the liberating acceptance of knowing that there are powers here beyond your control.

The Langtang Valley trail is lined with lodges whose owners have overcome tragedy to rebuild their lives and businesses. Hear their inspiring stories of loss and resilience.

Test yourself on the physical challenge of a week-long Himalayan trek, hiking above 4000m (13,123ft) and impressing yourself with the knowledge of just how far you can go, one step at a time.

Nestled in a mountain valley bordering Tibet, China, Nepal's Langtang Valley promises one of the best hikes in a country blessed with an abundance of incredible walking trails. The route climbs through bamboo groves and epiphyte-draped forests to reach lush yak pastures surrounded by snowcapped peaks in a classic U-shaped glacial valley – few Himalayan treks offer such a dramatic range of scenery in return for such a short investment of time. The pastoral scenery and villages decorated with prayer flags and stupas seem idyllic, but keep your eyes and ears open and you soon realise that this trekkers' paradise has been through tough times.

The 2015 earthquake

On 25 April 2015, Nepal's worst earthquake for a generation rocked the central part of the country. Centuries-old temples collapsed in Kathmandu, a

© MEZZOTINT / SHUTTERSTOCK

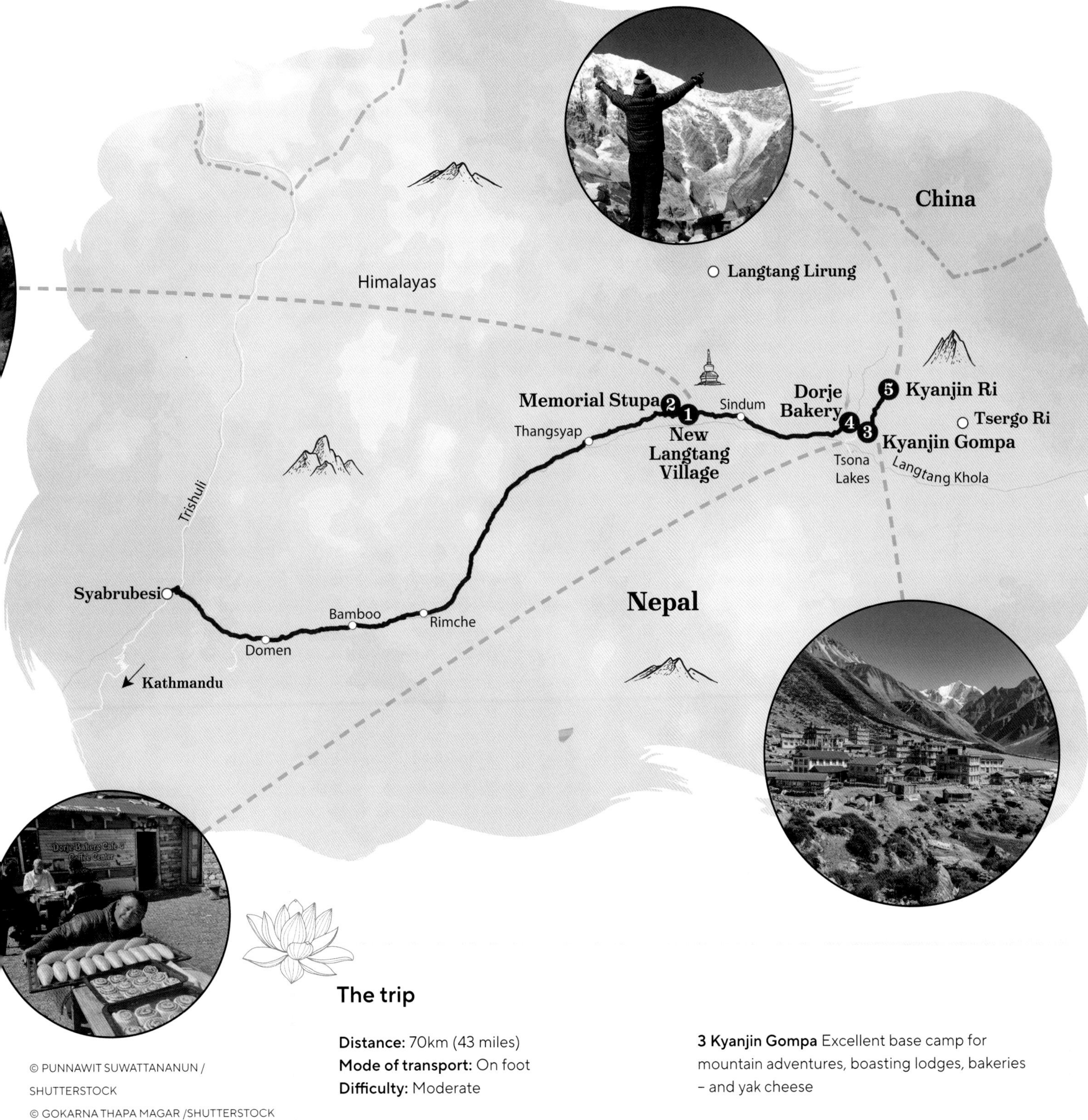

The trip

Distance: 70km (43 miles)
Mode of transport: On foot
Difficulty: Moderate

1 New Langtang Village Collection of lodges and restaurants, largely built since the 2015 earthquake

2 Memorial Stupa This New Langtang monument is inscribed with the names of those who died during the earthquake

3 Kyanjin Gompa Excellent base camp for mountain adventures, boasting lodges, bakeries – and yak cheese

4 Dorje Bakery Excellent coffee, cake and information from local baker Lhakpa

5 Kyanjin Ri Wonderful day-hike from Kyanjin Gompa, with soaring views of 7234m (23,773ft) Langtang Lirung peak

Clockwise, from left: Tsergo Ri, Langtang Valley; New Langtang Village; prayer flags atop Kyanjin Ri; the village at Kyanjin Gompa; Dorje Bakery

Bradley Mayhew, Langtang Valley trekker

Walking the Langtang Valley trail after the 2015 earthquake, I was moved to see trailside memorial plaques to Israeli and Spanish trekkers who had died during the earthquake. In Kyanjin Gompa I stayed with a Nepali widow whose lodge was definitely not the most comfortable, but whose eyes revealed a sorrow that deserved my support. It is moments like these that push this trek far beyond the carefree pleasures of a simple walk in the hills.

deadly avalanche swept through Everest Base Camp, and in the Langtang Valley a 1km-wide (0.6-mile) wall of rock, snow and ice detached from Langtang Lirung peak and dropped unseen down near-vertical slopes to bury the village of Langtang, the valley's main settlement and one of the most popular stops for trekkers. Around 155 villagers were killed instantly, along with some 40 foreign trekkers and many Nepali porters and guides. Most trekkers were settling down to lunch at the time, while many locals from up and down the valley were attending a funeral in the village. Only a single house in Langtang village survived, nestling at the base of an overhang that sheltered it from the avalanche.

Walking across the Langtang landslide today is an eerie experience. Not a single piece of wood or trace of the original village remains. It takes about 20 minutes to cross the tomb-like rubble that sits atop the former village, until you finally get to a collection of prayer flags and a memorial stupa that records the names of the dead. Read the rapidly fading lettering and you can make out the names of villagers from nearby Langtang, Gumba and Sindum; seven trekkers from Spain; four French walkers from the same family.

Rebuilding & rebirth

Within two years of that fateful day, Langtang's teahouses had been rebuilt and trekkers from around the world were once again hiking the valley's trails. In the months after the earthquake, families rebuilt their homes by hand and lodge owners reconstructed, shouldering risky bank loans rather than waiting for help from Nepal's hopelessly bureaucratic government. Tourists who had visited the stunningly beautiful valley sent donations or returned to offer help, repaying kindnesses that they had received years before as younger trekkers. Trails were rebuilt by locals, who kept faith that trekkers would one day return.

© NITI THANOMSRI / SHUTTERSTOCK

> **"It's hard not to be moved by the beauty of the valley, while also being reminded of the fragility and impermanence of life here"**

The trek today

Make your way up the Langtang Valley today, pausing for cups of tea and conversation at teahouses en route, and the traces of lingering loss are clear to see. But also visible in abundance is stoicism and resilience, hope and determination. Almost all the locals and lodge owners you meet will have lost someone in the 2015 quake – a parent, brother, husband or wife; sometimes an entire branch of a family. Many of the lodge owners are widows, who have had to reinvent their lives and livelihoods, alone, and have taken the brave decision not just to rebuild but expand and upgrade their lodges. It's hard not to feel admiration for the self-reliance, fortitude and grace on display throughout the valley. Even more than other treks in Nepal, this is one trip that will likely leave you feeling inspired and grateful, in equal measure.

Today, the scenic highpoints of the out-and-back Langtang trek are still the mountain views around Kyanjin Gompa, a community of lodges built around a monastery, a day's walk up valley from Langtang. Climbing to the stunning mountain viewpoints outside the village – such as Kyanjin Ri – and looking out over crumbling moraine and cracking glaciers, as avalanches echo around the snowcapped peaks, it becomes apparent that the mountain landscape that may seem so immutable is in fact in constant flux. It's hard not to be moved by the beauty of the valley, while also being reminded of the fragility and impermanence of life here. Rarely do mountains seem so terrible in their magnificence.

Yet it is the human stories that linger the longest. This is a journey that challenges and rewards in ways you might not have imagined. And with every night you spend in a lodge and every *daal bhaat* (rice, lentils and vegetables) you order on your walk, you know that you are directly contributing to a community that has managed to pull itself through its darkest days.

© THEPLACEWEARE / SHUTTERSTOCK

From top: looking down to Kyanjin Gompa; trekking the Langtang Valley

Practicalities

Region: Bagmati Province, Nepal
Start/Finish: Syabrubesi

Getting there and back: Buses run every day from the Nepali capital Kathmandu, but take about 9hr due to the poor quality of the mountain roads. Book a seat a day in advance. A more comfortable option is to hire a jeep for the journey.

When to go: Autumn (October to November) and spring (March to April) offer the best trekking. Autumn brings the clearest views and most stable weather, but spring sees rhododendron and other blooms. Avoid the monsoon months (June to August).

What to take: You'll need a comfortable pair of walking shoes, a three-season sleeping bag, a down jacket and good sunglasses, plus a system for water purification.

Where to stay and eat: Lodges (also called teahouses) line the trek route, offering a wide range of hot drinks and food as well as simple accommodation, so you'll never be more than an hour from a hot cup of tea. Popular overnight stops like New Langtang and Kyanjin Ri also boast cafe-bakeries, and the latter even has its own yak-cheese factory.

Trekking agencies: Kathmandu has dozens of trekking agencies that can arrange anything from just a porter to carry your bag to a whole organised trek, with a guide and teahouse accommodation and food included. It's possible to do the trek independently without a porter or guide, but you should never trek alone.

Essential things to know: Allow yourself two or three full days at Kyanjin Gompa to acclimatise to the altitude of 3870m (12,697ft), and to make the excellent day-hikes out to the viewpoints of Kyanjin Ri or Tsergo Ri and the five lakes of Tsona.

Ride in the hoofprints of Kyrgyz nomads

Ride through Kyrgyzstan's celestial Tiān Shān Mountains on a long-distance loop, taking in gorges, forests and the lakeside summer paradise of Song-Köl.

A long-distance horse-trek builds a powerful and long-lasting connection of trust, confidence and affinity between rider and mount.

Being welcomed into nomads' yurts offers a new perspective on offering hospitality to strangers. You'll likely come down from the mountains with a deeper understanding of how the harsh natural environment birthed these values, and of their importance to the survival of nomadic Kyrgyz tribes.

Even for experienced riders, rustic Kyrgyzstan creates an opportunity to learn new methods of horsemanship. Plus, spending so much time at high altitude ensures riders' red blood cells will be bursting with energy upon their return.

The history of the Kyrgyz highlands, and the stories of those who live here, would be incomplete without one central thread: the horse, a creature woven as tightly with the region's past as the fabrics once transported here along the Silk Road. The winds and currents of the centuries have changed the Earth, but human and horse still live here in a delicate harmony. Herds of hardy horses remain the pride of the Kyrgyz people.

The trip

Distance: 400km (249 miles)
Mode of transport: Horse
Difficulty: Difficult

1 Konorchek Canyons Explore fascinating dry gorges, peaks and red-rock columns

2 Kerkebes Pass On clear days, the views from this high-altitude pass stretch as far as the waters of Issyk-Köl

3 Song-Köl Rest in a lakeside yurt on the wild north side of Song-Köl

4 Shamsi Gorges A vast and verdant reserve of pine and fir forests

5 Chüy Valley Admire views of endless wheatfields as you descend to Rot-Front

Clockwise, from left: horse-trekking the Kyrgyz steppes; weaving through the mountains; wild camping in the Tiān Shān; traditional Kyrgyz yurt

Ashley Parsons, horse-trekker

Travelling unguided across Kyrgyzstan with three horses was the most difficult and yet most important thing I've ever done. Opening a constant flow of communication with my horses helped me trust my instincts and learn to listen. The Kyrgyz nomads I met along the way taught me about graciously hosting, living in the moment, and honour. I try each day to keep these lessons close to my heart.

Riding is one of the more comfortable ways to get around mountainous Kyrgyzstan, where the often hazardous roads make for slow progress by car. It's also the fastest route to immersion in the hospitable culture of nomadic Kyrgyz shepherds – and the not-so-hospitable weather of the capricious Tiān Shān Mountains. You may choose to ride the whole 400km (249-mile) loop out of Rot-Front, or just a section of it, but whether for three days or for 30, a horse-trekking journey through Kyrgyzstan is a ticket to follow in the footsteps – hoofprints – of old Silk Road traders while exploring a land where the steppe, mountains and sky meet.

Start on the right hoof

The proverb 'For Want of a Nail', in which a kingdom is lost due to a poorly shod horse, is worth taking to heart before setting out on a horseback adventure here. The Kyrgyz mountains and steppe are wild and remote: a thorough preparation phase is key. If you've rented horses, you'll want to check that they're suited to the journey, in good weight and good health. It's also essential to make sure your equipment is in good order: you'll be riding for between four and six hours each day, and saddle sores can stop a trip in its tracks. Note, too, that inexperienced riders should ride with a guide. Certainly, this changes things, but often for the better: guides speak the local language and can bring a deeper understanding of Kyrgyz culture.

© QUENTIN BOEHM

"The concept of time might flit out of your grasp, as mind and body adjust to the rhythm of the outdoors and the horse"

Riding onto a new normal

Once the saddlebags are packed, it's time to say goodbye to sedentary comforts and depart for the highlands. Many horse-trekkers load up their smartphones with a collection of digitalised Soviet maps – but these old maps are rarely up to date. Often, asking for directions is needed. The knowledge of the nomadic shepherds who drive their flocks through these mountains will quickly become essential: these are the same paths their ancestors have walked for centuries to reach the *jailoo*, or high pastures.

Approaching the dry microclimate of Konorchek Canyons, a routine is forged between horse and rider: wake up, breakfast, break down the camp, tack up the horses; ride, eat lunch; ride, set up the camp, eat, sleep. The first days are difficult. And even though the average speed on horseback is only 4kph (2.5mph), you'll be surprised at how quickly the landscape changes. As the days slip by, the concept of how time passes might flit out of your grasp, as your mind and body adjust to the rhythm of the outdoors and the horse. Each day is full of challenges, from keeping to the trail and finding grass for the horses to graze overnight, to creating a sense of cooperation and teamwork between mount and rider.

Enter the central Tiān Shān

The Tiān Shān is one of more than 88 major mountain ranges in Kyrgyzstan, and the rocky paths on this route are tough. Riders must dig deep into their mental and physical reserves to reach the high passes en route to Song-Köl, but there's ever-present inspiration in the strength and endurance of your horse.

Until the sun browns the grass in August, the mountain flanks are green with spring growth, though their peaks are still deeply packed with snow – the nuances of the colours are magnificent. Kerkebes Pass, at around 3600m (11,811ft), reveals a 360-degree view of the summits, hills and valleys where, in any village, riders can ask for lodging at a farm and enjoy a scrub in a Russian *banya* – something between a steam room and a bathhouse.

Below: the route ahead through remote Tiān Shān mountains and valleys
Left: Kyrgyz shepherds

Social steppes

Hospitality is a pillar of steppe culture; travellers must always be invited into the yurts of Kyrgyz nomads, even if the family can only afford to offer them simple vitals like bread, *caj* (tea) or *kumis*, the national drink, made with fermented mare's milk and ready to serve from mid-summer onwards. 'Caj?' or 'Kumis?' are usually the first questions that nomadic shepherds will ask their guests upon entering their yurt. *Caj* is served any time of day, often with lumps of sugar or jam stirred in to sweeten the flavour. *Kumis* may be a bit of an unknown to Western stomachs, though it's said to have medicinal properties. It's quite sour and, frankly, mare-y, but a well-made *kumis* has a pleasant fizzy taste, and can be very refreshing when served cold. If it's not to your taste, sip slowly – politeness calls for the host to refill an empty glass.

Climbing the trail towards Song-Köl, you'll wind through valleys bursting with flowers and populated by red marmots, easy to spot as they scurry from burrow to burrow; above, eagles and other birds of prey swoop on the updrafts.

Lakeside at Song-Köl

By the time Song-Köl comes into view, the trip is halfway over. By now, you'll likely have become attuned to your horse, and learned how to interpret its unspoken words, remaining mindful that its needs are priority number one. On the banks of the sparkling lake, with rich pastures stretching as far as the eye can see, rider and horse can take a few days' rest. Dotted around the lakeside are spread-out neighbourhoods of nomads, where shepherds and their families spend the summer. More than three yurts typically signify a tourism business and can easily accommodate guests. Only one or two yurts? You can still ask to stay, but it's unlikely you'll have private quarters. But the camps are very social. When chores are finished, shepherds visit each other's yurts. Horse-trekkers are often invited in, too, for tea or even a meal; don't hesitate to accept. These social visits are a cornerstone of steppe hospitality. Visitors can also lend a hand with daily chores, like milking the cows and mares or preparing *plov*, a popular Central Asian dish of rice cooked with vegetables, meat (often lamb) and whole heads of garlic. In cooler weather, nomadic shepherds may play games like *kok boru*, a type of horseback keep-away – it's not for the faint of heart, even when spectating: the ball is the headless carcass of a dead sheep.

From top: weeks spent in the saddle create a strong bond between horse and rider; lakeside yurt camp at Song-Köl

Behind the reins to Rot-Front

Song-Köl may tempt you to stay forever, but a return to Rotfront is in order – and after a few days staying with a family, you may well relish the liberty of being back on the trail. It becomes easy to enjoy simple pleasures: watching the horses napping in the morning sun as you sip hot coffee, or noticing that each horse has its own grazing tastes.

Riding into the forests of the Shamsi Gorges reserve inspires reflection and appreciation of the journey. The world seen from in between the ears of a horse has a different flavour, and you may begin to understand why the Kyrgyz have so long prized their relationship with these animals. An understanding of reliance and communication between horse and rider has forged their bond. As you descend from the last pass toward Rot-Front, drinking in spectacular views of the Chüy Valley, you may well find yourself promising to return each year. As the saying goes: 'Time spent in the saddle is never wasted.'

Practicalities

Region: Tiān Shān Mountains, Kyrgyzstan
Start/Finish: Rot-Front

Getting there and back: It's possible to reach Rot-Front from Kyrgyzstan's capital, Bishkek, via a local *marshrutka* bus; these depart several times daily from the city's Eastern Bus Station.

When to go: June, July and August are the best months to horse-trek in Kyrgyzstan. Before June, the mountain passes may still be snowed over; after August the pastures will be degraded, with little grazing for horses.

Where to stay: You'll likely mix camping with overnights in yurts owned by Kyrgyz families. There's no need to arrange yurt stays in advance – just ask if you can be accommodated. Expect to pay a modest fee.

What to take: A tent, and gear for temperatures as low as -5°C (23°F). Daytime temperatures can reach 40°C (104°F) on the valley floors, but snow is possible even in July; pack accordingly. You'll also need to bring lightweight food that cooks quickly but doesn't perish easily: pasta, lentils, rice and so forth. You can buy some fresh vegetables in villages en route.

Tours: One of the longest-running equestrian-tour outfits in Kyrgyzstan, Asiarando (asiarando.com) offers guided tours, ranging from 11 to 21 days and including your horse, meals and accommodation. It can also provide horse rental and training for would-be solo riders; the latter is good option if you've never travelled alone on horseback, covering itinerary planning and horse-trekking skills. See formationvoyageacheval.wordpress.com for more details.

Essential things to know: A horse trek should not be undertaken lightly, as there are real risks involved for both horse and rider. Inexperienced riders should go with a guide or prepare extensively via training with Asiarando. Finally, do not expect daily mobile phone coverage – it's rare to have signal in the mountains.

Bangkok to Singapore by rail

Discover what makes Southeast Asia tick on this epic train trip, a tale of three thrilling countries told over 1900km (1181 miles) of track.

A two-hour flight is hardly enough time to ponder life's complexities. With three countries to cover and some of the most fascinating stops in Asia, this epic train trip offers time to breathe deeply and contemplate how you really feel about upcoming stops on the journey of life.

This journey is as much about the people you meet as the scenery flashing by outside the windows. On the multiday ride from Bangkok to Singapore, you'll get cultural insights from Confucian traders, Buddhist monks, Hindu families, Muslim pilgrims and legions of students who can't wait to share their perspective on life in Southeast Asia.

The Eastern & Oriental Express – Southeast Asia's most luxurious train – whisks passengers from Singapore to Bangkok in four lavish days, but you'll need a Gordon Gekko-sized budget to board it. Fortunately, there is also a low-cost way to zip along the Malay Peninsula by rail, and it delivers a deep dive into the region's rich cultural melting pot as part of the package.

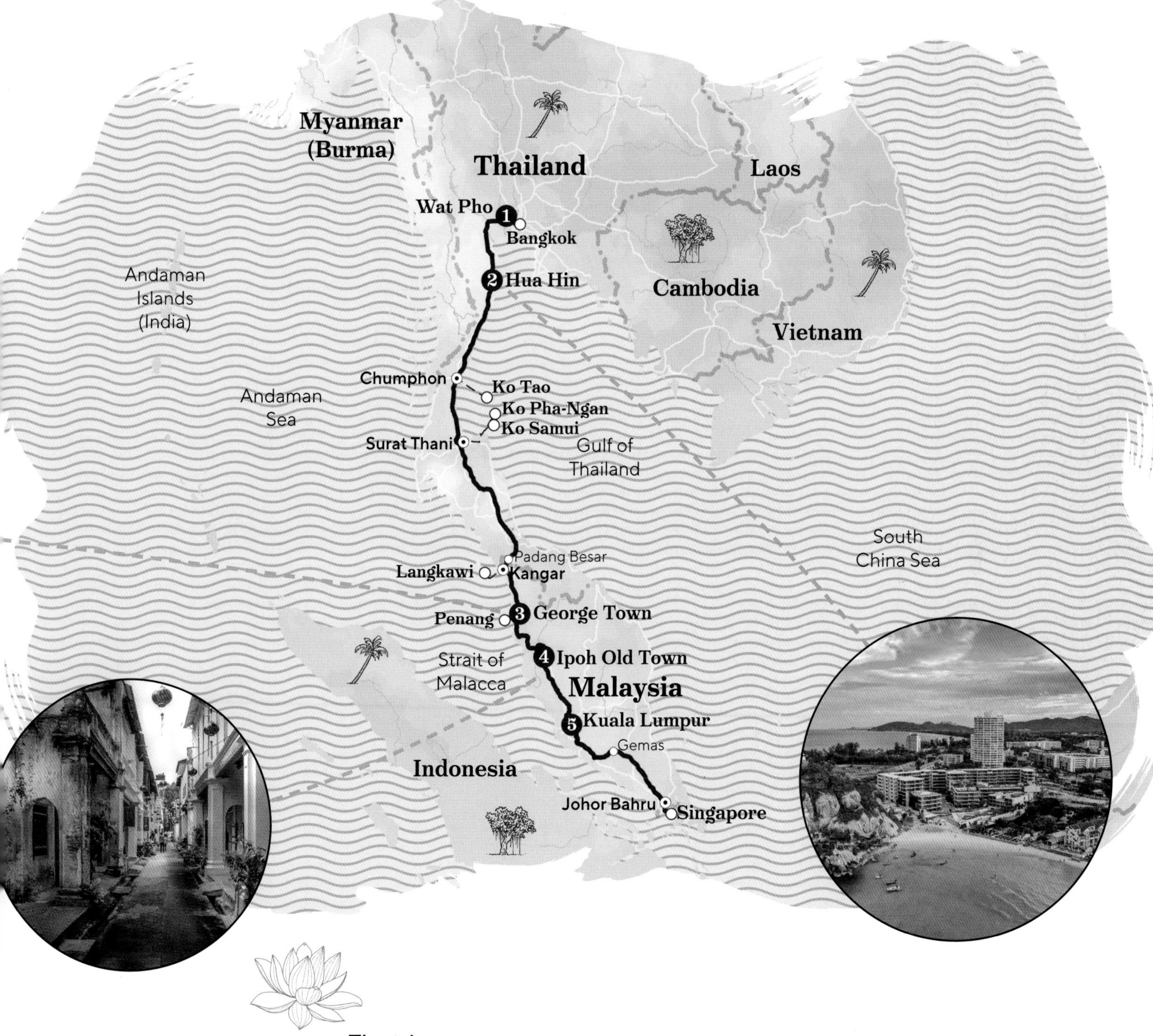

The trip

Distance: 1900km (1181 miles)
Mode of transport: Train
Difficulty: Moderate

1 Wat Pho, Bangkok Say goodbye to Bangkok with a relaxing rub-down at the spiritual home of Thai massage

2 Hua Hin, Thailand Pause the journey south for sea, sand and seafood in the country's oldest resort

3 George Town, Penang An almost mandatory stop for fabulous Malaysian food, terrific temples and time-travel street scenes

4 Ipoh Old Town Gorge on history (and beansprout chicken) in this foodie former mining town in northwest Malaysia

5 Kuala Lumpur Globetrotting menus and skyscraping towers define Malaysia's futuristic capital

Clockwise, from left: Hua Hin station, Thailand; George Town on Penang Island, Malaysia; Hua Hin beach resort; Ipoh Old Town, Malaysia

Joe Bindloss, train travel enthusiast

I first took the train from Bangkok to Singapore to break up a long-haul flight from London to Australia, and I got so wrapped up in the journey that I changed my onward travel plans twice. It was an intense cultural education. Travelling in 3rd-class carriages was an awakening – from that moment, I wanted every trip to involve the sights, sounds and smells of Asia blowing in through the open windows.

Travelling on ordinary passenger trains between Bangkok and Singapore, you'll mingle with people from every walk of life – monks heading to monasteries, students on college breaks, city families travelling for up-country reunions – while villages zip by in a blur of temple spires and minarets. It's a crash course in the cultures and customs of Thailand, Malaysia and Singapore.

OK, luxuries are limited compared to the Eastern & Oriental, but you'll appreciate every magnificent mile. In place of champagne and silver service, you'll be unwrapping banana-leaf packages of *nasi lemak* (Malay coconut rice) and drinking fizzy Kickapoo Joy Juice as a rainbow-coloured tableau of Southeast Asian life unfolds outside the windows. Expect priceless memories, for a ticket price of under US$100.

Let the journey be the destination

If you make the trip without stopping, it's possible to travel from Bangkok to Singapore in just 48 hours, but it's much more fun to break the trip repeatedly along the way, in the style of a 19th-century adventurer, filling your scrapbook with photos and ticket stubs and the pages of your journal with first-time travel experiences.

Connecting trains on the Thai and Malay rail networks provide easy access to a string of historic cities as well as junction towns for ferries to offshore islands. Spread the journey over a week, and you'll have time to bask on tropical beaches and feast in some of Asia's most famous foodie hubs, ticking off three fascinating capitals in the process.

© MATT MUNRO / LONELY PLANET

> "Travelling on trains is a crash course in the cultures and customs of Thailand, Malaysia and Singapore"

The easiest start point for the journey is boisterous Bangkok, where Buddhist monks meditate in gilded cloisters just yards from the 24-hour street party that is the Khao San Rd. Hit the nearest 7-Eleven for some travel essentials – drinking water, anti-bac hand gel, perhaps some Kopiko coffee candies and larb-flavoured Pocky pretzels – then it's all aboard for adventure.

Know the peninsula through its people

With 17 hours of travel time between Bangkok and the Malaysian border, it's tempting to cocoon yourself in air-conditioned comfort in 1st or 2nd class, but that means cutting yourself off from the country you're travelling through. Split the journey into daytime stages in fan-cooled 3rd class and you'll feel Thai culture swirl around you like a wave.

Sure, it's crowded, but 3rd-class travel comes with perks such as windows that open for easy photo opportunities and through-the-windows access to the delectable street food hawked on station platforms. You'll also be a novelty to your fellow passengers – something highly conducive to conversation.

You may find yourself chatting away the hours with hijab-wearing Malay university students heading home for the holidays, or debating world politics with gregarious Bangkok workers visiting relatives in the steamy south. In the process, you'll start to piece together the complex cultural geography of the peninsula.

As the train trundles south, there's a tangible shift as Thai Buddhism gives way to Malay Islam. Headscarves become more common, and mosques replace the gilded Buddhist monasteries in the villages beside the tracks. At the same time, a subtle Indian and Chinese influence creeps into the cuisine and culture – look out for Hindu temples, Sikh *gurdwaras* and Chinese clan houses flashing past your window.

Muse away the miles to the border

Deciding where to break the journey south is a matter of personal taste. You could jump out at Hua Hin after

Below: Ko Tao sunset, Thailand
Left: Khao San Rd, Bangkok

Kuala Lumpur skyline, Malaysia

Tastes of the peninsula

The Malay Peninsula is one of the world's great culinary melting pots, but this festival of flavours didn't come about overnight. Early traders from India, China and the Middle East brought in most of the spices used in Thai and Malay cooking, but the Portuguese provided the humble chilli, pushing every dish sharply up the Scoville scale. Before European colonialism, the peppercorn was king – imported from India along with the word 'curry', a corruption of the Tamil word *kari*, meaning 'spiced sauce'. The British had a more circuitous impact. To serve the empire's insatiable appetite for trade, labourers, tea pickers and mine workers were brought in from India and southern China, adding new layers of complexity to the local cuisine and cementing *teh tarik* ('pulled' tea) as the national drink. Dining out on the peninsula today is a sampling menu of flavours from across Asia. Bring an appetite!

four and a half hours for a seafood supper and a day on the sand. You could pause at Chumphon for a scuba-diving detour to Ko Tao. You could leap out at Surat Thani and kick back on the party beaches of Ko Samui and Ko Pha-Ngan. Alternatively, you could push straight on through to Padang Besar and settle in for some serious thinking time. With 1900km (1181 miles) to cover, you'll have ample opportunity to mull over where you are on the map, where you came from, and where you are going on life's winding path.

Put down the smartphone and tune into your inner monologue. As the palms and rice paddies slide by, you may find yourself wondering where life is heading, and how you feel about the ultimate destination. If you're at one of life's junctions, don't be surprised if you end up changing your onward flight at the end of the trip – this kind of thoughtful travel can be addictively moreish.

Follow your tastebuds through Malaysia

At Padang Besar, you'll complete customs formalities and change over to the Malaysian rail network, starting with the Komuter train to Butterworth. Over the next few hours, there'll be several chances to hop off the train and buzz over to the gorgeous, sand-circled island of Langkawi.

Alternatively, save the island-hopping for Penang. Fast passenger ferries zip from Butterworth to George Town, where you can gorge on bowls of *laksa* (spicy noodle soup), penny-priced *roti canai* (buttery flatbread with spicy dipping sauces) and *kari kapitan* – Malaysia's signature Indian-influenced curry.

© NG HOCK HOW / GETTY IMAGES

"Don't be surprised if you end up changing your onward flight – this kind of thoughtful travel can be addictively moreish"

If you can tear yourself away from Penang's treats and temples, more culinary education awaits two hours south in languid Ipoh, a former hub for tin miners and 18th-century gangsters, with a historic Old Town. Here, Chinese flavours dominate – head to the junction of Jalan Dato Tahwil Azar and Jalan Yau Tet Shin to sample the town's best *tauge ayam*, steamed chicken and rice with seasoned beansprouts.

Electrified ETS trains roll on to the historic train station in Malaysia's capital, Kuala Lumpur, where you can chow down on claypot chicken in Chinatown, graze on Malay curries with a view of the skyscrapers in Kampung Baru, or take a tasting tour around the Indian subcontinent in Brickfields. By the end of your trip, your mental map of Malaysia will be marked with main courses rather than monuments.

The final leg to Singapore is more complicated than it should be. First comes a seven-hour train ride to the border town of Johor Bahru, with an inconvenient change in Gemas. Then you can finally cross the causeway to Singapore – a convoluted process by train, but an easy hop by taxi or bus via the Woodlands Mass Rapid Transit station.

Phew! After a journey of ample thinking time, cultural immersion and fabulous food, bookend the trip with a slap-up dinner at Michelin-starred Hill Street Tai Hwa Pork Noodle in Kallang, or feast on treats from across the peninsula in the teeming Chinatown Complex. If you didn't know the Malay Peninsula before you started, you certainly do now.

© JIMMYFAM / GETTY IMAGES

From top: claypot cooking, Kuala Lumpur; Chinatown, Singapore

Practicalities

Region: Thailand/Malaysia/Singapore
Start: Bangkok, Thailand
Finish: Singapore

Getting there and back: Many airlines linking Europe and Australia fly to Bangkok and Singapore, so you can book an open-jaw ticket and cover the distance between these two megacities by rail. For a greener alternative, trains and buses connect Bangkok to cities across Southeast Asia; ferries fan out from Singapore to Indonesia's islands.

When to go: The optimum time to travel varies, depending on where you plan to break the journey. For stops on the Andaman Coast, the driest months are from November to February. For destinations on the Gulf Coast, the dry months are from February to April.

What to take: As for any long-distance rail journey, you'll benefit from a travel pillow, ear plugs and, for instant privacy, a music player with earphones.

Where to stay: If you're riding straight through from Bangkok to Padang Besar, consider a sleeper berth (you'll get less cultural immersion but more comfort). Otherwise, ride the rails by day and sleep in hostels, beach cabins or hotels wherever you stop for the night. Bookings are recommended during the busier dry months.

Where to eat and drink: Thai and Malaysian trains have dining cars for sleeper-class passengers; in 3rd class, hawkers wander the carriages and station platforms selling portable meals such as fried chicken with sticky rice.

Essential things to know: Book train tickets through 12GoAsia (12go.asia), State Railway of Thailand (dticket.railway.co.th) or Malaysian operator KTM (ktmb.com.my). Reserve ahead for sleepers; for short 3rd-class trips, buy tickets at the station on the day. Keep your passport handy for border immigration checks at Padang Besar and on the Singapore causeway.

Walk like a Tibetan – Lhasa's pilgrim circuits

Experience Lhasa's spiritual sites shoulder-to-shoulder with Tibetan pilgrims, retracing the city's sacred geography in a walk of shared solidarity.

Almost all visitors to Tibet leave with a great empathy for the Tibetan people, admiring their openness, self-confidence and charm. Finding ways to incorporate some of this into one's own behaviour can be a lasting legacy of a trip here.

Visiting Tibet is often a profound experience on many levels. See for yourself the political and religious situation for Tibetans under Chinese Communist rule.

Merge the physical and the spiritual as you walk side-by-side with pilgrims; you may well find yourself relaxing, your thoughts emptying and your emotions calming in a form of walking meditation.

Walking one of Lhasa's pilgrim circuits is like jumping straight into the pages of Chaucer's *Canterbury Tales*. Rough-and-ready pilgrims wearing yak-fleeced robes and cowboy hats walk side-by-side with elegant women from Amdo adorned with elaborate necklaces of coral and turquoise, and striking men from eastern Tibet sporting distinctive red khampa braids in their hair. Most pilgrims ceaselessly twirl hand-sized personal prayer wheels; a few carry larger versions supported by a harness. Some even carry out full-body prostrations around the circuit, dressed in protective leather aprons and wearing wooden blocks on their hands. All walk in a single direction, towards an invisible common goal.

Inside the temples that line the circuit, the air is thick with the heady smells of yak butter and juniper, as devotees spoon fragrant herbs into incense burners

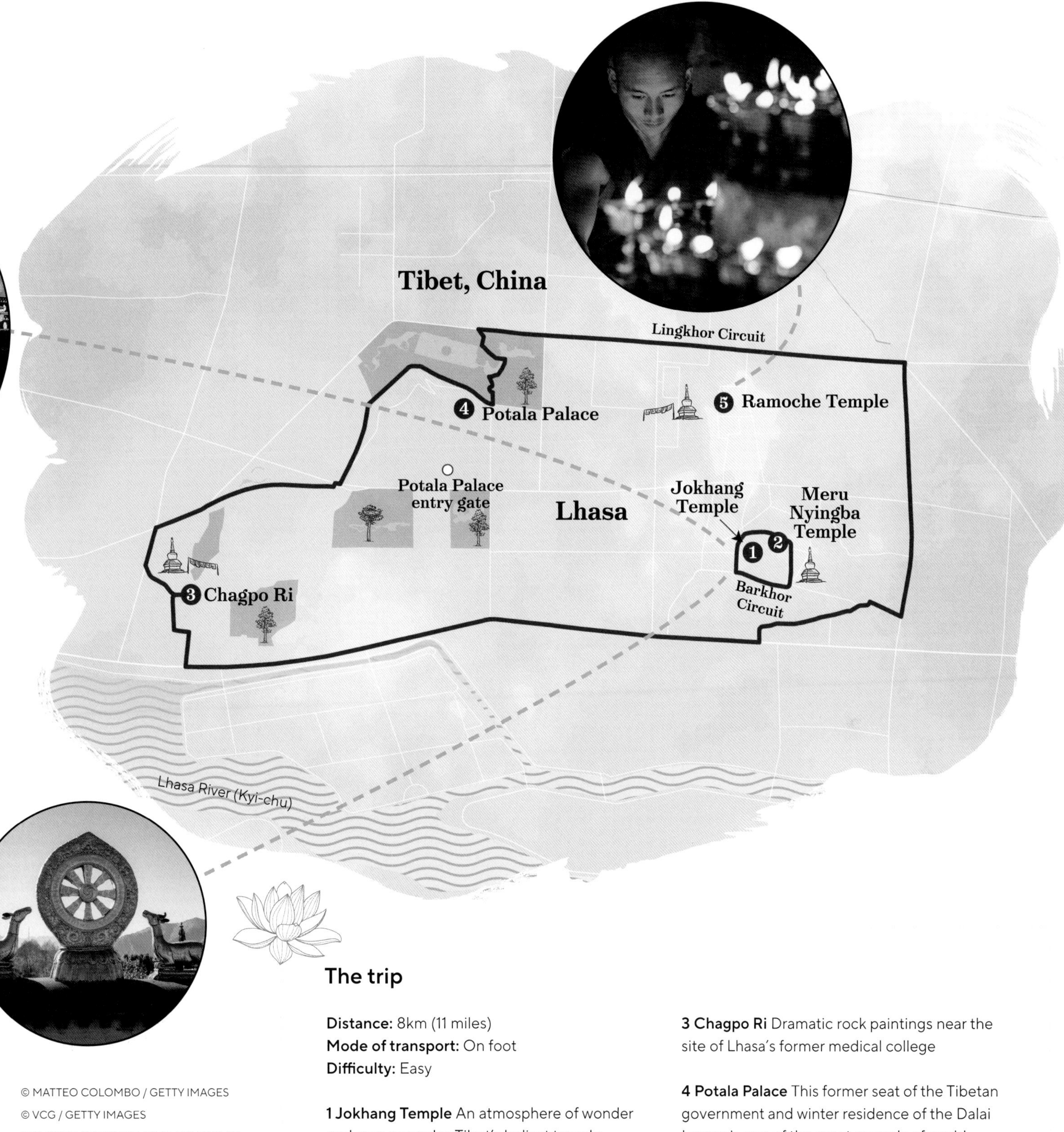

The trip

Distance: 8km (11 miles)
Mode of transport: On foot
Difficulty: Easy

1 Jokhang Temple An atmosphere of wonder and awe pervades Tibet's holiest temple

2 Meru Nyingba Temple Don't miss this incredibly atmospheric temple as you walk the Barkhor Circuit

3 Chagpo Ri Dramatic rock paintings near the site of Lhasa's former medical college

4 Potala Palace This former seat of the Tibetan government and winter residence of the Dalai Lamas is one of the great marvels of world architecture

5 Ramoche Temple Venerated 8th-century temple housing a fabulously ornate image of Shakyamuni Buddha

Clockwise, from left: Potala Palace; Jokhang Temple; lighting yak-butter lamps at Ramoche Temple; *dharma* wheel atop Jokhang Temple

Bradley Mayhew, Tibet traveller

A circuit of the Barkhor is always the very first and the very last thing I do on any trip to Tibet. It's a strange feeling, but the moment I lose myself in the moving crowd I feel a deep sense of belonging, almost of coming home. Whether at a monastery, a holy mountain or a lake, I always ask if there's a *kora* path to walk. For me, the sense of purpose and inclusion it offers brings profound joy and peace.

or top up lamps from personal flasks of molten butter. Monks bless visitors with a splash of holy water or a tap from an ancient relic, as Tantric drumming booms from deep in the building like some primeval heartbeat. It is a scene that has scarcely changed in centuries, either in appearance or in the intensity of religious devotion. This is the spiritual heart of Tibet.

The Barkhor Circuit

Tibet's capital, Lhasa, boasts the region's most important urban *koras*, or pilgrim circuits, and Tibetans travel here from across the plateau to gain merit by walking them. The city's holiest shrine is the Jokhang Temple, founded in the 7th century around a particularly sacred statue of Buddha aged 12, which pilgrims bow their head to as they file past in a hushed, shuffling line. Forming a belt around the Jokhang's exterior lies the Barkhor Circuit, a 20-minute stroll that leads pilgrims past backstreet temples, historic buildings and giant room-sized prayer wheels. At any time of the day, hundreds of Tibetans walk the circuit in a tidal flow, muttering mantras and exchanging gossip with friends and neighbours. Lining the path are fabulously atmospheric monastic buildings like the Meru Nyingba Temple, as well as religious stalls and shopping malls selling Buddhist paraphernalia, precious stones and monastic accessories in a seamless blend of high religion and hard-nosed commerce. It's a jovial and fascinating mix of the sacred and profane, infused in classic Tibetan fashion with equal measures of piety and bawdy humour.

The Lingkhor Circuit

Lhasa's second major pilgrim path is the Lingkhor, a longer half-day circuit that encircles the traditional

© HUNG CHUNG CHIH / SHUTTERSTOCK

"There's a secret hidden here – one that lies at the root of the Tibetan people's resilience, compassion and equanimity"

Old Town before venturing into more modern parts of the city. Morning is the best time to walk it, as families, friends and Tibetan grannies perform their daily religious workout. In its southwestern sweep, the route passes Chagpo Ri, a dramatic cliff face where pilgrims prostrate themselves in front of thousands of colourful rock paintings. A short distance away is one of Lhasa's former *ling* (royal temples), where pilgrims rub their backs against a sacred rock to ward off arthritis and lumbar pain. The path then swings around the back of the Potala Palace, the towering former home of the Dalai Lamas, passing teahouses, rock shrines and tiny temples filled with chanting nuns, as well as the city's best yak-milk yoghurt stalls.

Walking the walk

The wonderful thing about Tibet's pilgrim circuits is how inclusive they are. Follow a few simple rules – keep stupas and shrines on your right, spin prayer wheels clockwise, be respectful in temples – and you'll be welcomed with open arms and a warm smile. As you lose yourself in the centrifugal force of the crowd, you find yourself merging into it as the line between tourist and pilgrim dissolves.

Several things stand out as you do a *kora*: the joy in the eyes of most pilgrims; or the wide-eyed awe in the faces of those visiting the candlelit inner sanctums of temples like the Ramoche or Jokhang. Above all, it's the faith and devotion that underpins these walks that is so humbling and inspiring to most visitors. Sixty years of Chinese Communist rule has failed to dampen the joy, faith and good humour of most Tibetans. There's a secret hidden here – one that lies at the root of the Tibetan people's resilience, compassion and equanimity.

After completing your *kora*, celebrate your improved karma by following your fellow pilgrims into a traditional Tibetan teahouse for a flask of *cha ngamo*: hot, sweet milky tea. The Tibetan pilgrim experience is not austere denial, rather a joyous form of cultural expression. Rolling up your sleeves and joining in is the absolute best way to get straight to the soul of Tibet.

From top: Jokhang Temple, on the Barkhor Circuit; circumnavigating Potala Palace on the Lingkhor Circuit

Practicalities

Region: Tibet, China
Start/Finish: Lhasa

Getting there and back: High-altitude trains run to Lhasa from Beijing, Xian, Xining and other Chinese cities; you can also make the adventurous week-long overland trip from Nepal. Flights run from Kathmandu (Nepal) and from many Chinese cities.

When to go: Lhasa can be visited almost any time of year (except March), though most people arrive between April and November. Try to avoid the first week of May and of October, as both are busy Chinese holiday periods. The Saga Dawa festival in May or June sees thousands of pilgrims walking the Lingkhor Circuit, making it a great time to visit.

What to take: Bring layers for big swings in daily temperatures. A sun hat and sunscreen are essential to protect against the high-altitude sunlight.

Where to stay: Lhasa offers a wide range of accommodation. The most interesting place to stay is in or near the traditional Old Town, where several boutique Tibetan-style guesthouses lie hidden in the whitewashed backstreets.

Where to eat and drink: Lhasa has lots of Sichuanese restaurants, as well as Hui Muslim and Tibetan places to eat, and even a couple of Western-style cafes.

Tours: Foreigners need to book a guide for their entire trip in Tibet, and transportation if travelling outside Lhasa. Only then can a tour agency process the permits you need to board transportation to Tibet. Recommended Tibetan agencies include Tibet Highland Tours (tibethighlandtours.com) and Explore Tibet (exploretibet.com).

Essential things to know: Lhasa is at an altitude of 3656m (11,995ft), so be sure to take it easy for the first couple of days to allow your body to acclimatise.

Gunung Mulu to Limbang on the Headhunter's Trail

Strenuous effort, nerve-testing challenges, deep cultural immersion and plenty of rainforest know-how: every day is a classroom on this classic Borneo trek.

Trekking with guides from the Iban community is like walking with an encyclopedia of rainforest knowledge. As you hike, you'll learn the ropes of rainforest life from the descendants of the people who gave the Headhunter's Trail its name.

You'll have to dig deep to endure the heat and humidity and the challenging climb on the way up to the Pinnacles, but you'll likely emerge feeling stronger than when you set out. You'll also sweat gallons: consider it a free detox en route to a river swim before bedtime, or a well-earned shower at the end of the trail.

On some journeys, you look at the scenery. On the Headhunter's Trail in Borneo's Gulung Mulu National Park, the scenery – or at least, those parts of the scenery with fur, feathers, scales and eyes – looks at you, warily, as if you were a visitor from another world.

Most travellers to this legendary reserve fly in and out, missing the best part of the experience. On the old overland trail between the park headquarters and Limbang on the Brunei border, you'll make your way through dense, dripping vegetation, half-deafened by the noises of the jungle. After long, draining days of hiking, meandering boat trips on jungle rivers and overnight stops in tribal longhouses, you'll emerge with a newfound respect for the rainforest and the people and creatures who call it home.

© ROBAS / GETTY IMAGES

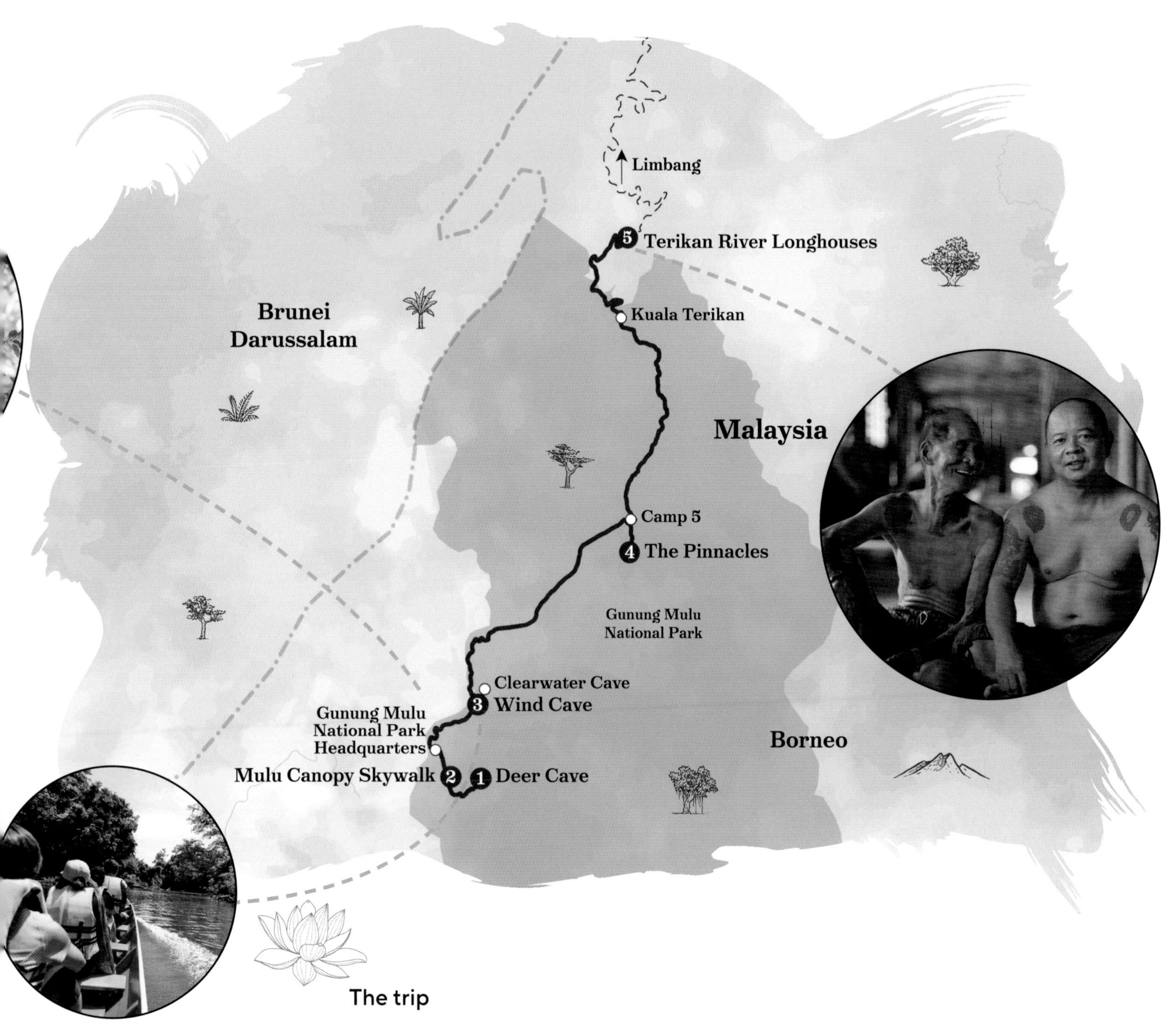

The trip

Distance: 22.5km (14 miles)
Mode of transport: On foot, boat and bus
Difficulty: Moderate

1 Deer Cave One of the largest caves on earth – a cathedral in eroded limestone

2 Mulu Canopy Skywalk Climb into the canopy for a hornbill's-eye view of the rainforest

3 Wind Cave Travel by river from the park HQ to reach this stalagmite-filled monster

4 The Pinnacles A tough hike delivers big views at these rocky spires above Camp 5

5 Terikan River longhouses Cultural immersion comes as standard during a stay at an Iban longhouse

Clockwise, from left: river-crossing in Gulung Mulu National Park; rhinoceros hornbills along the trail; Iban longhouse life; by river to the Wind Cave

Brandon Presser, Headhunter's Trail hiker

Moving from community to community, exchanging sugar, rice and packets of ramen for lodging, gave me invaluable insights into Iban traditions and the tales told on the wide verandas of the longhouses. Physical journeys undertaken by young Iban were commemorated with ritualistic tattoos – motifs, like passport stamps, that acknowledge the transformative power of travel, and how each step taken on a trail is also a voyage deeper into the mind.

Stepping into the rainforest

While the trek can be as sweaty as a wrestler's armpit, the trail is mostly flat, a boon in the energy-sapping heat. On the way, you'll tap into the insider knowledge of the Iban people. The trails through Gunung Mulu were hacked through the forest by the ancestors of the current park guides, so pay attention – they'll point out things you'd easily miss, like the venomous pit vipers lurking perfectly camouflaged in the greenery.

Most hikers start from the park headquarters, after a few days exploring the supersized limestone caverns nearby. Kick off with the out-and-back hike from the park HQ to Deer Cave – a whopping 2km (1.2 miles) long and 174m (571ft) high – and take a vertiginous climb into the treetops on the Mulu Canopy Skywalk.

The Headhunter's trek proper begins with an hour-long boat ride on a gushing jungle river, with more limestone cavern action en route at Wind Cave and Clearwater Cave. It's an atmospheric trip; the canopy parts to reveal glimpses of jungle birdlife – hornbills, if you're lucky – and karst outcrops dribble stalactites over the water's edge.

Next comes a sweat-drenched 9km (5.6-mile) hike, snaking between the buttress roots of rainforest trees and wobbling over rope bridges to reach Camp 5, the first overnight stop before you hit the Pinnacles trail. Check your body for leeches – these slippery customers can wriggle through the tiniest gaps.

Climbing the perilous Pinnacles

There's no way to sweeten the pill – the Pinnacles climb is tough. You'll start the 2.4km (1.5-mile) slog at dawn, scrambling over rocky outcrops using tree roots and tethered aluminium ladders for balance, and return to camp seven long hours later. The reward at

> "You'll emerge with a newfound respect for the rainforest and the people and creatures who call it home"

the top is a cinematic view over the treetops, perched above a surreal landscape of eroded, knife-like limestone crags. The hike isn't any easier on the way down, but expect to feel a buzz of achievement as you roll back into Camp 5 and collapse onto your mattress in the communal dorms. Most Pinnacles climbers turn back here to the park headquarters; those who leave Camp 5 on the 11km (7-mile) trail to Kuala Terikan earn the right to call themselves proper explorers.

© FADHLI ADNAN / SHUTTERSTOCK

Life among the Iban

As you tramp through the leaf litter, grill your guide for tips on the local wildlife. Most animals make a home high in the canopy, but guides are experts at pointing out birds, reptiles and amphibians hiding at ground level, as well as the plants used by Iban people to make blowpipes and shelters. You'll discover more about the Iban on your next overnight stop, after a boat ride on a stretch of river frequented by kingfishers and other darting rainforest birds. A stay at an Iban longhouse typically begins with donations of rice, noodles and other kitchen staples to your hosts, as compensation for bringing extra mouths to the table.

The Iban can trace their origins back to the trail's original headhunters, but longhouse life today comes with creature comforts such as electric lights and piped water. Staying in one of these stilt-propped homes is still a lesson in communal living – mealtimes are the highlight of the day, as multiple generations come together to prepare food, feast and get giggly on *tuak* rice wine.

It is theoretically possible to hike all the way to the coast of Sabah, but local people long ago worked out that jungle rivers make the best thoroughfares. Your trip back to the relative comforts of Limbang will involve more hours on the water, then a minibus journey on a smooth, tarmacked road. You likely won't be ready to survive alone in the jungle by the end of the trek, but you'll definitely feel a newfound respect for Borneo's threatened natural environment, perhaps inspired to spread the conservation message and help preserve this natural wonder for future generations.

© ANDREA ZANCHI / GETTY IMAGES

From top: the Pinnacles, high point of the Headhunter's Trail; Iban longhouse

Practicalities

Region: Malaysian Borneo
Start: Gunung Mulu National Park Headquarters
Finish: Limbang

Getting there and back: Unless you fancy walking in both directions, the main route to Gunung Mulu is by air, via the MASwings turboprop flights from Miri, Kuching or Kota Kinabalu. At the far end of the trail, Limbang has regular buses to Kota Kinabalu in Sabah and Bandar Seri Begawan in Brunei, where you can pick up transport to Sarawak.

When to go: Borneo doesn't really have a 'dry' season – it's hot and humid year-round and rain showers are pretty much a given. For Gunung Mulu, June to September is the travel peak – prices for accommodation and flights soar, and you'll need to book ahead for tours and activities, particularly trips to Deer Cave and Wind Cave and the Mulu Canopy Skywalk.

What to take: The rainforest is unforgiving. Bring sturdy footwear, socks (to deter leeches), breathable cotton clothes, a sleeping bag (or sleeping-bag liner), and a mosquito net and strong mosquito repellent.

Where to eat and drink: Your guides will arrange meals along the trail, usually breakfast and dinner at overnight stops and a simple packed lunch. If you like to snack while you trek, stock up before coming to Gunung Mulu. Bring gifts of food and drinks for longhouse stays – remember, for jungle dwellers, the nearest shop could be days away.

Tours: Making your own arrangements is tricky: organised trips are the way to go, covering guides, meals and pre-arranged accommodation at Camp 5 and in longhouse homestays. Operators include Borneo Eco Tours (borneoecotours.com) or Sticky Rice Travel (stickyricetravel.com) in Kota Kinabalu, and Borneo Touch (facebook.com/BorneoTouchEcotour) in Limbang.

Walking the Jeju Olle Trail

The Unesco-listed South Korean island of Jeju-do is the location for a meandering long-distance walking path traversing stunning natural landscapes.

Hiking the Jeju Olle is slow travel at its best – a chance to disconnect from day-to-day pressures and routines and nurture your inner soul as you absorb the beautiful natural surroundings and connect with fellow walkers and locals. The mindful rhythm of walking can also open mental space for personal reflection, clearing and refreshing the mind.

The trail offers many opportunities to learn about Jeju-do's history, unique culture, landscapes and traditions.

Most people opt to hike just a few sections of the Jeju Olle Trail, but tackling all 437km (272 miles) is a physical challenge that will take a month to complete.

In 2006, Suh Myung Sook was so burnt-out by her career in journalism that she resigned her job and set off to walk Spain's Camino de Santiago. It was while making this spiritual pilgrimage that she decided to create a similar long-distance walking trail around her birthplace, the island of Jeju-do. Formed by volcanic eruptions some 2 million years ago, and dotted with 368 small parasitic volcanoes known as *oreum*, Jeju-do is South Korea's largest island, and a popular holiday destination for Koreans; hikers are drawn here to summit the nation's highest peak, Halla-san, a 1947m-high (6388ft) dormant volcano.

Creating the trails

Returning to Jeju-do, Suh began searching out walking paths and drumming up interest in her plan. Local officials initially didn't have much faith in the project. Undaunted, Suh launched the first Jeju Olle

© E5CAN / SHUTTERSTOCK

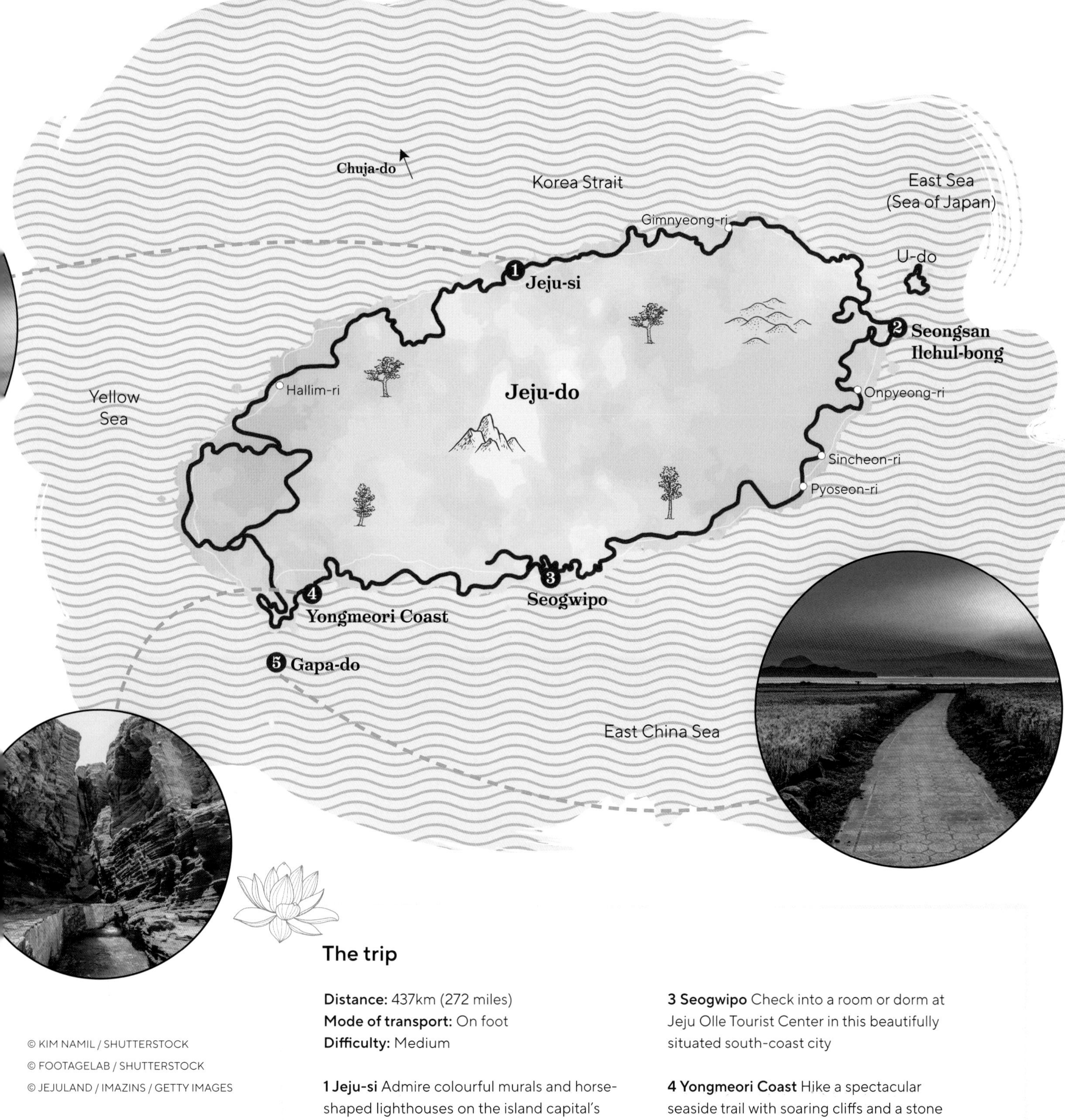

The trip

Distance: 437km (272 miles)
Mode of transport: On foot
Difficulty: Medium

1 Jeju-si Admire colourful murals and horse-shaped lighthouses on the island capital's seafront promenade

2 Seongsan Ilchul-bong Marvel at this punchbowl-shaped tuff volcano, one of Jeju-do's most impressive sights

3 Seogwipo Check into a room or dorm at Jeju Olle Tourist Center in this beautifully situated south-coast city

4 Yongmeori Coast Hike a spectacular seaside trail with soaring cliffs and a stone Buddha in a cave

5 Gapa-do Sail out to pancake-flat Gapa-do island for simple rural pleasures, sea views and contemporary architecture

Clockwise, from left: hiking the Jeju Olle Trail; lighthouse at Iho Tewoo Beach, Jeju-si; barley fields, Gapa-do; on the trail along the Yongmeori Coast, Jeju-do

Dr Anne Hilty, Jeju Olle hiker

For me, solo-trekking the entire Jeju Olle Trail system – twice, second time in reverse order – was nothing short of profound. I began each trek at sunrise, not only to enjoy the waking nature and stunning scenery, but to ensure solitude. This walking meditation, with its slow pace, sea air, small mountains, volcanic forests, historical and cultural sites, shrines and temples, wildlife and village life, served as a healing balm – and I was reborn.

Trail in 2007, the same year parts of Jeju-do were awarded World Natural Heritage status by Unesco.

Olle in the Jeju dialect describes a narrow path leading from a home to the main street. It's a word that, for Suh, resonated with one of her project's aims: a route to connect the world to Jeju-do's unique culture and scenery. 'Jeju's nature is not too big, not too wide, not too vast, yet still very beautiful and lyrical,' says Suh. 'Standing in front of vast and magnificent forms of nature, people are not only in awe, but also daunted and intimidated, reminded of human insignificance. However, Jeju's soft *oreums* and wide ocean nurse humans and nurture their minds. That's why Jeju Olle is a healing trail.'

A non-profit foundation was established and gradually the trail expanded to circumnavigate the island. Today, hundreds of thousands of people head to Jeju-do every year, specifically to tread lightly along its routes – marked by orange and blue arrows, and similar-coloured ribbons dangling from trees. You'll also encounter the trail's logo, the 'Ganse', a simplified form of a Jeju pony inspired by the word *gansedari*, Jeju dialect for a person who plods along like a horse. The Ganse logo encapsulates the Olle's philosophy of slow, meandering travel – an antidote to South Korea's stress-inducing *pali pali* (hurry, hurry) culture.

The Jeju Olle is not fixed in stone: routes are constantly being added to or adapted. There are currently 27 sections: 24 around Jeju-do, and three on the nearby smaller islands of Chuja-do, Gapa-do and U-do; the routes are between 4.2km (2.6 miles) and 21km (13 miles) long. Some hikers tackle the entire Jeju Olle, but most visitors opt for walking one or two sections. Jeju-do's excellent public bus network means it's easy to return to where you started your walk.

© KIM WONKOOK / SHUTTERSTOCK

"Jeju's soft *oreums* and wide ocean nurse humans and nurture their minds. That's why Jeju Olle is a healing trail"

Jeju Olle jewels

The first route – from Siheung Elementary School, on the east side of the island, down to the black and white sands of rocky Gwangchigi Beach – remains a fine introduction to the trail. The landscape is classic Jeju-do: verdant fields, grazed by cows and fringed by black basalt-stone walls; quiet fishing villages beside azure seas; salt flats and the ruins of Japanese WWII army fortifications. The section's highlight is getting an up-close view of magnificent Seongsan Ilchul-bong, a 182m-high (597ft) tuff volcano that rises out of the sea like a giant moss-coated punchbowl. Many find witnessing the sun rising over Seongsan Ilchul-bong to be a life-enhancing experience.

The Jeju Olle routes also provide fascinating glimpses of Jeju's people and their culture. Between Onpyeong-ri and Pyoseon-ri on Route 3, you'll pass through the village of Sincheon-ri, where there are around 100 murals painted on the sides of houses and walls. One of these depicts Jeju's famed female freedivers – the *haenyeo*. For centuries, *haenyeo* have been diving to collect seaweed and seafood, working as cooperatives and sharing their catch. Until recently they didn't wear wetsuits, despite diving for long hours in all weathers. It's not uncommon to spot these feisty women – most in their 60s, if not older – on the seashore preparing to dive, or bobbing in the ocean.

Island of wind & waves

Sections of the Jeju Olle have been designed to be wheelchair accessible – including the shortest trail, a 4.2km (2.6-mile) route around the compact island of Gapa-do, just southwest of Jeju-do. Even at its highest point, Gapa-do is barely 20m (66ft) above sea level, and the route slices through the barley fields the island is famous for. You'll also notice the wind turbines that make the island self-sufficient in energy, and the smattering of contemporary architecture that is part of a community project to help stem the tide of Gapa-do's population decline. For anyone looking for a breather from the stresses of modern life, hiking the Jeju Olle on Gapa-do is the tranquil answer.

© ERIC HEVESY / GETTY IMAGES

From top: Seongsan Ilchul-bong, highlight of the Jeju Olle's Section 1; *haenyeo* diver

Practicalities

Region: Jeju-do, South Korea
Start/Finish: Jeju-si

Getting there and back: Jeju-si's airport has many daily flights from mainland South Korea, and direct flights to nearby Asian countries. Ferries sail between Jeju-si and South Korea's Wando, Yeosu, Mokpo, Busan and Nokdong ports.

When to go: The Jeju Olle trails can be hiked year-round. April and May are good months to enjoy cherry blossom and wildflowers as they come into bloom across the island. Late October and early November are pleasant times to experience the autumn colours; this is also when the annual Jeju Olle Walking Festival is held over three days.

What to take: Trekking shoes and appropriate all-weather gear depending on the season – note that Jeju-do can experience heavy rains between July and September. Also be sure to pick up the free Jeju Olle guidebook from the many information points around the island, and a passport in which to collect the individual stamps once you've completed walking a section; the passport also provides discounts at various businesses along the trails.

Where to stay: Jeju-do offers a wide range of accommodation, from luxury resorts to simple homestays. The Jeju Olle Tourist Center in Seogwipo has a cafe, souvenir shop and the Olle Stay, with both private rooms and dorms. If you're planning on camping, the trail reports on the travel blog Going the Whole Hogg (goingthewholehogg.com) are a great resource.

Tours: Jeju Olle Trail Official English Guide (jejuolletrailguide.net) lists the schedule for the free guided walks. These are conducted in Korean but some of the guides do speak English.

Essential things to know: Check the Jeju Olle website (jejuolle.org) for up-to-date information on the trails.

Circuiting a sacred mountain: the Kailash kora

Journey through one of Tibet's most beautiful corners to walk around Asia's holiest peak, sacred to more than a billion people.

What you get from the Kailash trek depends on what you bring to it, but there is something in the transcendent high-altitude landscape here that inspires introspection, change and renewal.

Walking with pilgrims from across Tibet and seeing how they respond to the mountain and its sacred sites is one of the most affecting aspects of the walk.

Making the remote overland journey from Lhasa to Mt Kailash, and completing the trek over a 5630m (18,471ft) pass, is a physical achievement that will stick in the memory for a lifetime.

Pilgrims and trekkers who reach Shiva-Tsal, high on the second day of the Mt Kailash trek, know that they have reached a turning point. Tibetans and foreigners alike leave behind a lock of hair, an item of clothing or a photo of themselves and undergo a symbolic death here, leaving this life and its attachments behind before continuing through a netherworld (referred to by Tibetans as the *bardo*) to be reborn atop the windswept pass of the Drolma-la. At 5630m (18,471ft), the pass is the physical and spiritual highpoint of the three-day walk around Mt Kailash. It is a profound moment, in a place of redemption and renewal, as a lifetime of sins are absolved against a backdrop of prayer flags fluttering in the cerulean sky.

The sacred mountain

As befits a semi-mythical mountain, Kailash is a long

© YONGYUT KUMSRI / SHUTTERSTOCK

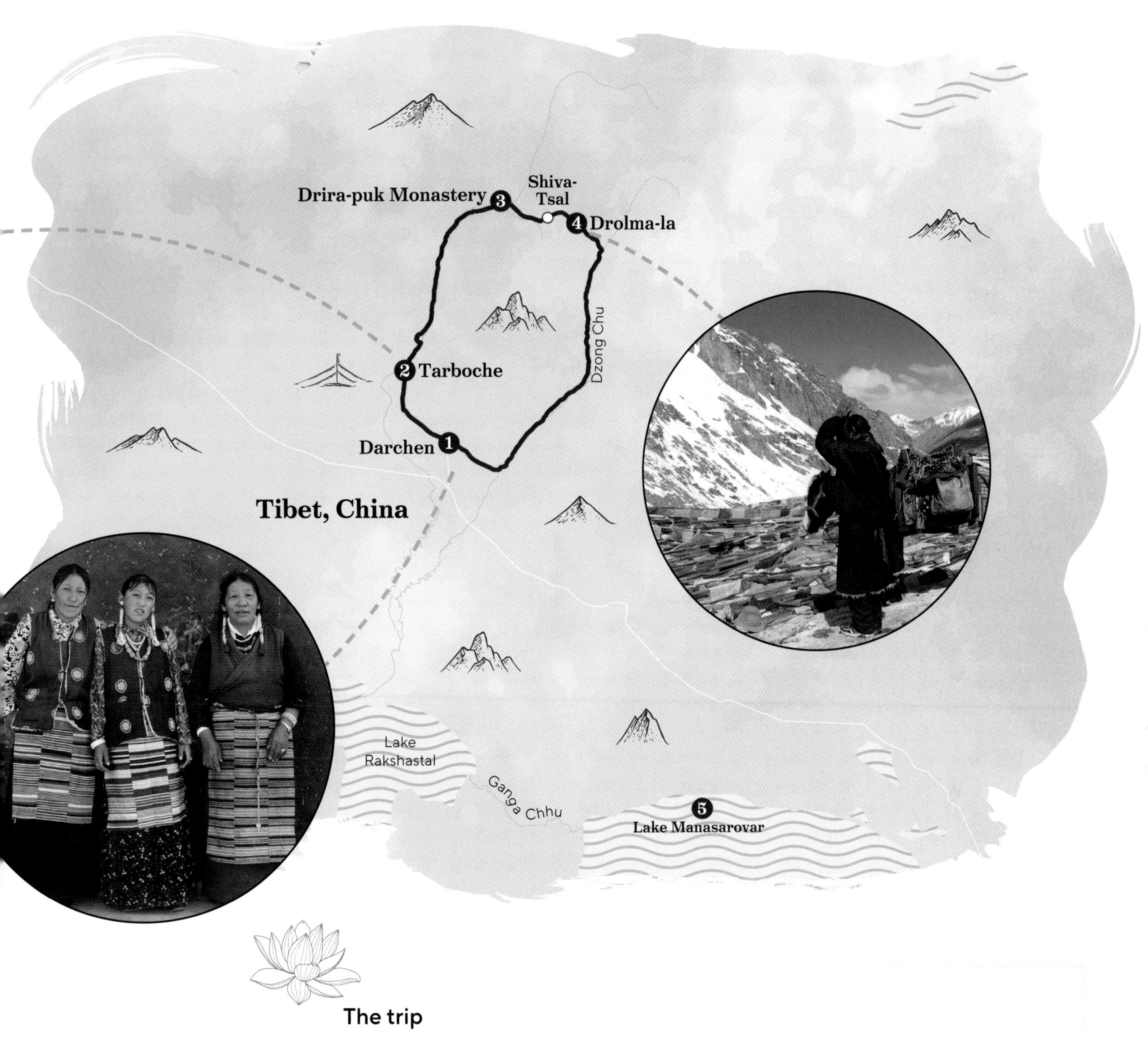

The trip

Distance: 52km (33 miles)
Mode of transport: On foot
Difficulty: Moderate

1 Darchen Staging post for the Kailash trek, with hotels, restaurants and hot water

2 Tarboche An important *darchen* (prayer pole) marks the southwestern corner of the circuit

3 Drira-puk Monastery and guesthouse with astonishing views of Mt Kailash's north face

4 Drolma-la This snowy pass marks the high point of the trek

5 Lake Manasarovar Magnificent turquoise lake, with its own pilgrim circuit

Clockwise, from left: Buddhist pilgrims walking the Kailash kora; the Tarboche prayer pole and Mt Kailash; a moment of prayer at Drolma-la Pass; Darchen village residents

Bradley Mayhew, Mt Kailash trekker

I've done the Kailash kora three times now, for my sins, and have been inspired every time. I remember being amazed by the devotion of my Tibetan guide who completed a second, overnight *kora*, straight after finishing the one he did with me. My favourite encounter, though, was with the jolly elderly Tibetan pilgrim whose only visible equipment for the walk was a half-opened can of Lhasa Beer. Respect. There's a lesson in there somewhere.

way from anywhere, hidden beyond the Himalaya in the far western corner of Tibet. Sacred to a billion Buddhists, Jains and Bönpos, it is Asia's holiest mountain. Tibetans call it Kang Rinpoche, or 'Precious Jewel of Snow', and believe it to be Mt Meru, the 'navel of the world', and the site of a series of epic spiritual battles between Buddhist saint Milarepa and rival Bön master Naro Bönchung (Bön is the pre-Buddhist religion of Tibet). Followers of Bön know Kailash as the 'Nine Stack Swastika Mountain', where their founder Shenrab ascended to heaven. For Hindus it is the earthly abode of Shiva.

Marooned to the north of the main Himalayan range, Kailash is a lone sentinel, sitting unrivalled (and unclimbed) at around 6714m (22,028ft) in the Kangri Tise range. Tradition says that four mythical rivers flow off its four cardinal faces, and the reality is hardly less incredible: four of Asia's greatest rivers – the Karnali, Sutlej, Indus and Yarlung Tsangpo (Bhramaputra) – all have their source within 100km (62 miles) of the peak. Lying at the foot of the mountain is sacred Maphum Yum-tso, or Lake Manasarovar, a liquid remnant from the Tethys Sea that once separated Asia and India. It's no wonder that pilgrims have been drawn to the region's sacred geography for two millennia.

Walking the kora

As you might expect, getting to Mt Kailash takes some serious effort. A four-day, 1200km (746-mile) drive from Lhasa gets you to the base at Darchen, from where the three-day kora, or circumambulation, of the peak begins. The walk itself follows wide glacial valleys, with the mountain always on your right, and only really gains significant altitude on the second day as it ascends past Shiva-Tsal to cross the oxygen-depleted

© MEIQIANBAO / SHUTTERSTOCK

> "There is something in the transcendent high-altitude landscape here that inspires introspection, change and renewal"

pass of Drolma-la. Porters or yaks are on hand to carry your bags if you wish, and simple guesthouses provide accommodation at key points, with snack tents offering sweet, milky Tibetan-style tea to flagging walkers.

You can tell a lot about the pilgrims from what they carry: foreign trekkers have their expensive Gore-Tex-lined gear toted by yaks or porters; Tibetans saunter along the trail with little more than a bag of *tsampa* (roasted barley powder); Hindu pilgrims from India slump on ponies, shivering in down jackets and balaclavas and often suffering from the altitude and cold. Almost everyone walks in the same direction: clockwise. Only occasionally will you come across a Tibetan walking against the tide, a clear giveaway that they are a Bönpo (follower of the Bön religion).

The mountain has its own sacred geography and is peppered with monasteries, rock paintings, two sky-burial sites and four *chaktsal gang*, or prostration points. Not satisfied with simply walking around the mountain, some hard-core pilgrims prostrate themselves around the entire route in a three-week endurance test, dropping face down to the ground and up again like inchworms, crossing the rocky and snowy terrain one body length at a time in a remarkable show of devotion. This is clearly no ordinary trek.

A once-in-a-lifetime experience

For Tibetan and Indian pilgrims, a visit to Mt Kailash is a lifelong-held dream. A single circuit wipes out the sins of a lifetime, believers maintain, while 108 circuits guarantee nirvana. A circuit during the full moon in the year of the horse (every 12 years) is considered so auspicious that it opens access to the mountain's secret inner kora.

For foreign trekkers the walk can be a profound experience, a perfect blending of the physical and the metaphysical, akin to a three-day walking meditation. For others it is simply a rejuvenating adventure in what feels simultaneously like the centre of the universe and the ends of the Earth. At the very least, it is a world-class trek around one of the world's most beautiful and remarkable mountains.

© ALEXANDER MANDL / 500PX

From top: the rising sun hits Mt Kailash; prayer flags at Lake Manasarovar

Practicalities

Region: Ngari Prefecture, Tibet, China
Start/Finish: Darchen

Getting there and back: Foreigners have to arrange private transport for the four-day, 1200km (746-mile) drive from Lhasa to Darchen. Getting to Lhasa involves a flight from either Kathmandu or a dozen airports in China, or a train from Xining.

When to go: Summer (May to September) is the only feasible season to undertake the trek. The Saga Dawa festival in May draws thousands of pilgrims to the mountain, but foreigners are rarely given permits to visit at this time.

What to take: Tea and coffee (boiling water is available everywhere); a down jacket and sleeping bag; Diamox in case of mild symptoms of altitude sickness.

Where to stay: Monasteries at Drira-puk (night one) and Zutul-puk (night two) offer simple accommodation in rooms, dormitories or tents.

Where to eat and drink: Food is simple and often limited to milk-tea and instant noodles, so bring supplies of your own.

Tours: Foreigners can only visit Tibet as part of a pre-arranged tour through a Tibetan travel agency such as Tibet Highland Tours (tibethighlandtours.com) or Explore Tibet (exploretibet.com). You can specify your own itinerary, and travel in any group size (three or four people is ideal). Your agency will have to arrange multiple permits, so need several weeks' advance notice.

Essential things to know: The whole of western Tibet sits at well over 4000m (13,123ft), so be sure you're acclimatised to the altitude by spending at least three days in Lhasa before setting off. Don't rush the incredibly beautiful drive out to the west, and try to add on a three-day side trip to the amazing ruins of the Guge Kingdom while there.

Hiking in historic footsteps on the Lycian Way

The first long-distance trekking route in Turkey, this varied, dramatic trail along the country's rugged southern coastline connects hikers with rural communities and ancient civilizations.

Adaptability and openness to new experiences are assets on this hike; this can prove challenging to some, but also freeing.

Ancient Lycians, as well as Greeks, Romans, Persians and Ottomans, all left their marks on this region. The ruins of their civilizations can be seen and explored all along the trail; Andriake's Museum of Lycian Civilizations provides historical insights.

This often steep and rocky route has some segments featuring tough ascents and tricky descents of around 1000m (3280ft) in a single day. The remote high-mountain sections can be especially testing of your mettle – and of the preparations you've made for the walk.

In the mountains above Turkey's Mediterranean coast, the village of Bel is a slumberous place, consisting only of a mosque, a small market and a few dozen homes scattered among olive groves and grazing fields on a high plateau. But Fatma Pansiyon here is abuzz with activity thanks to its namesake proprietor, whose personality is as vibrant as the orange paint covering her two-storey house. With bossy charm, she directs her newly arrived visitors to keep an eye out for other travellers coming off the trail as she races out the door, calling out, 'I have to bring in the sheep and goats now!'

Like other villages along the Lycian Way trekking route, Bel is so small that there is no regular bus service. But Fatma's husband arranges for a group of hikers to ride out to the main road in the morning on the local school bus, where they can flag down an intercity bus to their next destination.

© HSBORTECIN / SHUTTERSTOCK

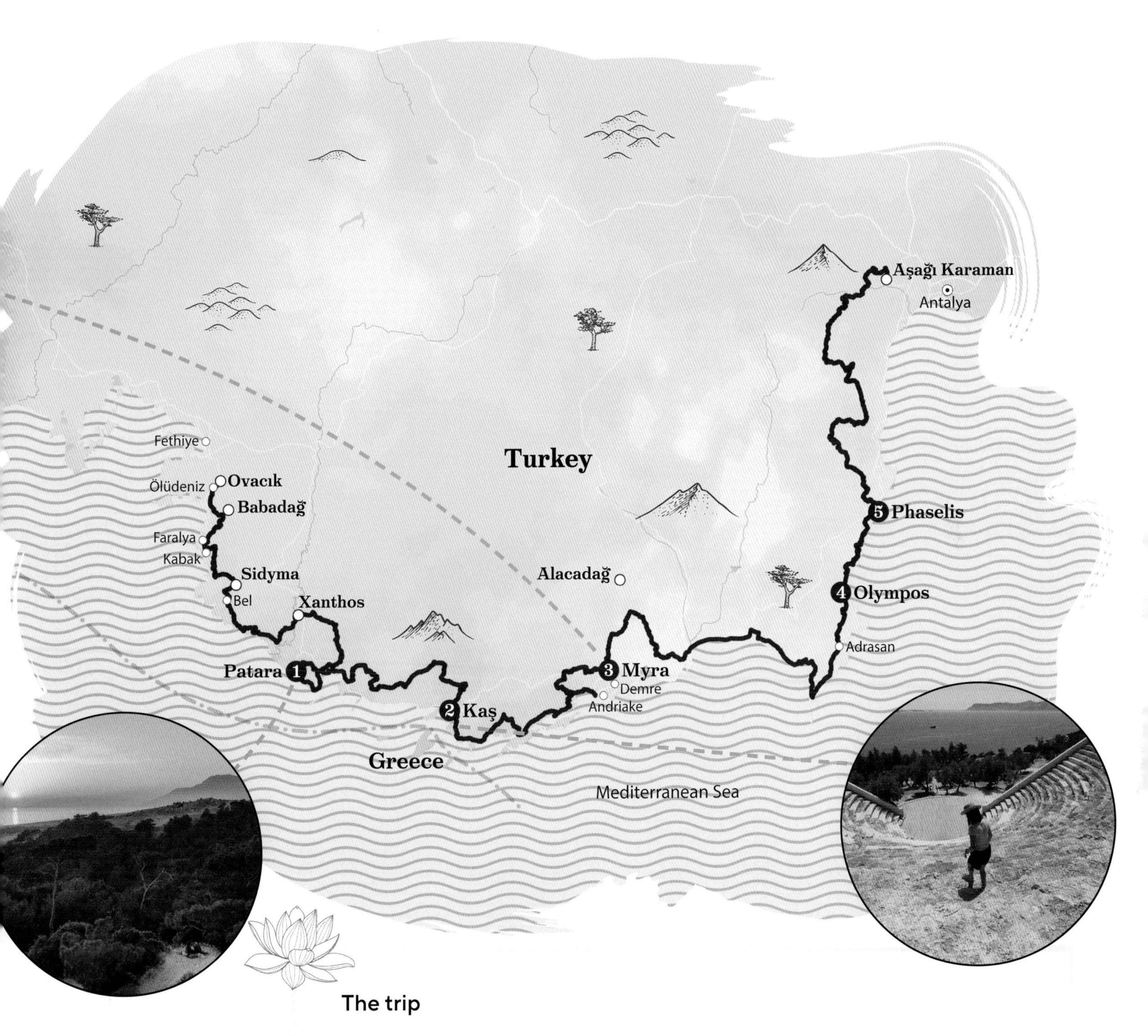

The trip

Distance: 710km (441 miles)
Mode of transport: On foot
Difficulty: Variable

1 Patara Ancient ruins near one of Turkey's longest, most beautiful sandy beaches

2 Kaş Picturesque seaside town with great restaurants – perfect for a short trail break

3 Myra Monumental rock-cut tombs towering over a well-preserved ancient theatre

4 Olympos Sprawling ancient city with a dramatic location where cliffs meet sea

5 Phaselis Ruins scattered among dense pine forest, with access to secluded beaches

Clockwise, from left: Lycian-era ruins at Phaselis; Myra's rock-cut tombs; Antiphellos Theatre, Kaş; sunset over the sands at Patara

Els Hom, Lycian Way trek developer

Coming from the Netherlands, I always wanted to know what was happening in advance, but hiking in Turkey shows you how to let go. There's always a way that can be found to solve any problem. If you show up wet in a village, someone will invite you to sit by the fire with them. I love to hear hikers say that the people they met along the way changed the course of their whole trip.

Village stays at places like Fatma's are one of the most distinctive features of the Lycian Way, which brings travellers into close contact with both present-day rural communities and the remnants of the ancient civilizations that once inhabited this region. The trail's full 710km (441-mile) waymarked length is wildly diverse, traversing alpine-like landscapes on 1500m-high (4931ft) passes, as well as dense pine forests, agricultural lands, rocky coastlines, secluded beaches and the ruins of 2000-year-old ancient cities. Small seaside resorts like Kabak, Kaş, Uçağız, Adrasan and Çıralı offer alluring places to rest along the way.

The Lycian Way route was originally mapped out in the late 1990s by British expat Kate Clow. She linked up ancient roads, goat-paths, forestry tracks and trails traditionally used to walk between villages with an aim of increasing small-scale sustainable travel opportunities in a region that remains at risk from mass tourism development.

Coastal splendour

Though the trail can be followed in either direction, the official route runs west to east, starting in Ovacık near the popular tourist resort of Ölüdeniz, where the sky is often filled with the brightly coloured sails of paragliders riding air currents down to the beach from the summit of Babadağ. Keen hikers can walk to Ovacık from the city centre of Fethiye, a one-day trek that takes in the haunting 'ghost village' of Kayaköy (Levissi), left abandoned when its Greek residents were sent across the Aegean as part of a compulsory population exchange in the wake of WWI.

© KENANOLGUN / GETTY IMAGES

"Hikers may stop to picnic under a centuries-old olive tree, or spot 15th-century water cisterns among the high grasses"

Ease of access as well as sweeping views of green capes and headlands jutting out into the glittering sea make this first segment from Ovacık one of the trail's most popular. Overnight stops are at Faralya, a small village perched on cliffs that tumble dizzyingly down to Butterfly Valley, an isolated sliver of coastline reputed to be home to some 100 colourful species; and the hippie-esque enclave of Kabak, which lures many a hiker to stay longer than planned. From the beach at Kabak, a steep path leads up through pine forest and then high, rocky fields where, en route to remote villages like Bel, hikers may stop to picnic under a centuries-old olive tree, encounter a herd of goats, or spot domed 15th-century water cisterns and old millstones among the high grasses.

Roman roads

Northeast of Bel, a newly marked section of the trail follows the ancient zigzagging road from the Roman ruins at Sidyma to the monumental Lycian pillars and sarcophagi at Xanthos, once the administrative and religious centre of the entire region. The old road was exposed recently when devastating wildfires swept through the area; its original 2000-year-old cobblestones are still in remarkably good condition. Walking – quite literally – in the footsteps of generations of travellers provides plenty of opportunities for contemplation of the passage of time and what can be learned from the past.

Between Xanthos and the bucolic beach town of Patara, hikers traverse over and alongside numerous examples of ingenious Roman water engineering, including pipes, canals and the stunning 200m-long (656ft) Delikkemer inverted-siphon aqueduct, which spans a scrub-filled valley overlooking the sea.

Ancient cities

In her 1937 book *The Lycian Shore*, the British explorer Freya Stark wrote: 'In Turkey particularly... a journey without history is like the portrait of an old face without its wrinkles. Every bay or headland of these shores, every mountain-top round whose classic name

Below: coastline near Antalya on the Lycian Way; **Left:** Butterfly Valley, Faralya

Historical riches

The inhabitants of ancient Lycia are mentioned in Hittite and Egyptian texts, as well as Homer's *Iliad*, in which they appear as allies of Troy. The region's strategic coastal location and rich timber resources were battled over by Greeks, Romans and Persians – Alexander the Great spent a winter at Phaselis during his military campaigns in the area – and the impressive archaeological heritage along the Lycian Way bears traces of all of these cultures. The ruins at Xanthos, Patara, Myra and Olympos were once important cities in the Lycian League, an early democratic confederation formed around 200 BCE. That era's historical legacy is explored in the Museum of Lycian Civilizations, in a restored Roman granary in the former harbour city of Andriake. The later Byzantine era added Christian monasteries and churches to the mix, including one in Demre devoted to St Nicholas, aka Santa Claus.

the legends and clouds are floating, carries visible or invisible signs of its past.'

One of the most striking examples of this ever-present past can be found among the sea of greenhouses that ring the modern town of Demre, in the central portion of the trek. Here, the house-like rock tombs of Myra, dating back to the 4th century BCE, were carved into the steep cliffs, forming a necropolis (the ancient Greek word for a 'city of the dead'). An optional detour into the foothills of Alacadağ mountain above Demre – a route that Kate Clow describes as 'tough but spectacular' – takes trekkers past early Byzantine monasteries.

Further east of Demre, the trail climbs up from the small beach town of Adrasan to the summit of Musa Dağı, crossing high meadows where shepherds graze their flocks. It then meanders down through fairy-tale-like forest, with dense trees twisting low across the path, before emerging at the sprawling ruins of ancient Olympos, nestled between mountains and sea. Past the beach here, the path rises up again to the Chimaera, continually burning natural-gas flames in the rocks that are thought to have inspired the monstrous fire-breathing mythological creature of the same name. Back on the coast, the ruins of Phaselis – the easternmost city of ancient Lycia – nestle in pine woods. The town's three ancient harbours are now pretty swimming beaches.

Like many segments of the Lycian Way, the going here can be tough. 'You have to be prepared for a serious hike; it's not a stroll through a park on a Sunday afternoon,' cautions Dutch archeologist Els Hom, who helps develop treks on the route with tour company Middle East Travel.

"You have to be prepared for a serious hike; it's not a stroll through a park on a Sunday afternoon"

Safeguarding the trail

Though much of the trail feels timeless, occasional examples of encroaching damage done by fires, overzealous forestry or coastal construction provide essential reminders that nature must be protected if it is to remain for future generations to enjoy. Before its terminus at Aşağı Karaman, near the booming seaside city of Antalya, one area around the original trail become heavily logged. So Clow and the team of volunteers who maintain and mark the Lycian Way, under the auspices of the non-profit Cultural Routes Society, rerouted part of the path. The new segment traverses around the back of towering Tahtalı Dağ (Mt Olympos) on a trail that Clow says is 'really wild, following some lovely canyon bed, with beautiful views from ridges, and spectacular plane trees'.

Though Clow has now been hiking in the area for nearly three decades, and has subsequently helped develop dozens more trekking routes all around Turkey, she says the Lycian Way remains special: 'Even when you think you've seen everything, there's always something else to discover.'

Left: the 4th-century rock-cut necropolis at Myra, near Demre

© ENDER BAYINDIR / SHUTTERSTOCK

Left: Lycian theatre at Phaselis

Practicalities

Region: Teke Peninsula, Turkey
Start: Ovacık
Finish: Aşağı Karaman

Getting there and back: Buses connect the route's start and finish with Fethiye and Antalya, both a short domestic flight (via Dalaman Airport for Fethiye) or overnight bus ride from international transit hub Istanbul.

When to go: Wildflowers bloom trailside in spring; the warmest sea temperatures are in autumn. The summer months are brutally hot. Lower-elevation trail sections are suitable for winter hiking if the weather is dry and mild, though area services will be reduced dramatically.

What to take: Swimming kit for cooling dips on coastal sections; and snow/ice gear if attempting high routes in winter. You'll need a tent and other gear if you plan to camp. Walking poles are helpful.

Where to stay: Most of the route passes through villages and towns where you can stay in small pensions; on some sections, you'll have to camp. Accommodation doesn't have to be booked in advance, though it's reassuring to ask each night's pension owner to call ahead to the next to reserve your stay.

Where to eat and drink: Pension stays will generally include breakfast, with an option for dinner and/or a simple packed lunch. Villages often have small shops for basic provisions. Carrying plenty of water is essential.

Tours: Guided and self-guided treks along part or all of the route can be arranged by Middle Earth Travel, Mithra Travel and other tour operators recommended by the Culture Routes Society (cultureroutesinturkey.com/the-lycian-way).

Essential things to know: Wild camping in forested areas is discouraged due to high fire risk in the region. Check ahead for possible closures of the trail during fire season. Note, too, that the route of the Lycian Way lies west of the areas affected by the devastating 2023 earthquake.

Looping through Ladakh on two wheels

The rugged highway to Ladakh is a journey through the soul as well as the Himalayan landscape, best attempted on an Enfield Bullet motorbike.

Being alone in the high-altitude deserts of Ladakh is a lesson in self-reliance. In a landscape almost devoid of vegetation, with just the growl of the engine for company, you'll test your limits, brave solitude and privations, and perhaps discover how much you are capable of.

Even the most experienced riders can upskill a notch or two by tackling some of the most challenging mountain roads on the planet. Flat tyres and mechanical misadventures come with the territory, along with breathtaking beauty and life lessons from the people who live year-round in this desolate, humbling terrain.

Think you know mountains? Try tackling the road that corkscrews like a python with indigestion through the high-altitude valleys of Ladakh. The motorbike trip from Himachal Pradesh to Ladakh and Kashmir is one of the world's most thrilling road journeys: a life-affirming test of resolve and endurance with a side serving of jaw-dropping natural beauty.

The trip

Distance: 894km (555 miles)
Mode of transport: Motorbike
Difficulty: Difficult

1 Rohtang La The first major pass between Manali and Leh – often snowcapped and upliftingly empty

2 Thiksey Gompa The definitive Ladakhi *gompa*, crowning a perfect hill to the south of Leh

3 Leh Palace The dust-coloured palace of the kings of Ladakh, soaring over Leh's medieval heart

4 Lamayuru Gompa Wind-sculpted badlands form the backdrop to this splendidly remote monastery

5 Dal Lake Srinagar's crowning glory – stay in a houseboat among the swirling lake mists

Clockwise, from left: prayer flags above Lamayuru Gompa; Thiksey Gompa; Leh Palace; Rohtang La

Joe Bindloss, overland traveller

I probably wasn't fully prepared for my first journey across the Himalaya by motorcycle – both in terms of riding experience and technical know-how – but by the end, I felt like a seasoned pro. The trip from Manali to Srinagar was life-affirming; by the end, I was brimming with confidence that I could go almost anywhere, with just the clothes on my back and my own resilience, and come out unscathed.

Like a mountaineering ascent, this journey of self-reliance will force you to look inside yourself as well as out at the scenery. Between Manali and Srinagar lie 894km (555 miles) of road through wind-scoured wilds, sawtooth ridges and parched Himalayan valleys, with just the odd living thing to remind you that this isn't, in fact, the surface of the moon.

The crossing from Manali to Leh takes two long days, and at least 18 bone-shaking hours in the saddle; and you'll face a similar journey on the far side to reach Srinagar, capital of beautiful, troubled Kashmir. That's plenty of thinking time – expect an emotional journey to rival the endless miles you cover on the dusty tarmac.

Essential tips for Himalayan riding

The definitive vehicle for this mettle-testing journey is the Enfield Bullet 500cc, a handsome, heavy machine built to colonial-era specifications. But the Bullet is – to quote one Ladakhi mechanic – a 'temperamental beast'. You'll need at least a smattering of technical know-how to deal with roadside oil leaks and other mechanical mishaps.

But whether you choose a classic Enfield or a more modern Japanese or European machine, it's vital to be confident in the saddle. The highways of Ladakh are corrugated, covered in loose gravel and pockmarked with potholes, and you'll be sharing the carriageway with jeeps, buses and behemoth Tata trucks, which thunder past in a barrage of loose stones and screaming horns.

© NICHOLAS BILLINGTON / SHUTTERSTOCK

"This is a life-affirming test of resolve and endurance with a side serving of jaw-dropping natural beauty"

Think of it as an intensive course in mountain riding. The chances are, you won't get more than a few miles before you have your first 'holy crap!' moment and reset to a more defensive riding style. Motorcyclists come at the bottom of the pecking order of Himalayan road-users, so ride cautiously and constantly assess the road surface for hazards.

A mission to the mountains

The highway linking Manali, Leh and Srinagar typically opens in March – when the Border Roads Organisation clears the snow from the high passes – and closes in late October or November with the first winter flurries. Overnight temperatures can touch zero even at the height of summer, so bring camping gear in case you get stuck after dark.

The opening stage from the hill station of Manali to the 3980m (13,058ft) Rohtang La is a gauntlet of fume-belching freight trucks, passenger jeeps and rattletrap buses, with the odd adventurous motorcyclist weaving in and out of the mayhem like a fighter jet escorting a bomber squadron. If you came to the Himalaya seeking solitude, you might wonder if you've picked the right journey.

But things calm down on the far side of Keylong, the hill-circled capital of Spiti and Lahaul. After a nerve-jangling climb up the python-like Gata Loops, you'll feel your pulse slow on the gently undulating More Plains. You may experience an almost meditative sense of calm on the approach to the prayer-flag-tangled Taglang La – the highest point on the route at 5328m (17,480ft).

As you descend to the Indus River, Ladakh's magnificent *gompas* (Buddhist monasteries) loom into view. Intricate murals of enlightened beings cover every inch of interior wall space inside these whitewashed wonders. If you weren't spiritually inclined before, gaze out from the time-worn gateway to Thiksey Gompa and see if that changes your mind!

Below: Thiksey Gompa
Left: on the road in Ladakh

Ladakh's gompas

Few places will get you thinking about your place in the universe quite like Ladakh. Locals have been mulling over the nature of existence since at least the 8th century CE, when monks from India and Tibet brought the light of Buddhism to these mountain-shadowed Himalayan valleys. Dotted among the wind-whipped outcrops, Ladakh's whitewashed monasteries seem to grow organically from the landscape, but their interiors explode with colour: rainbow-hued murals of *bodhisattvas* (enlightened beings) and mythical creatures, hand-carved visions of heaven, psychedelic butter sculptures, gleaming golden statues of Buddhas and *dharmapalas* (protector deities). *Gompas* mark every overnight stop on this journey: visit at dawn to attend morning prayers, complete with clanging gongs, honking horns and chanted mantras – it's the closest thing you'll find to the music of the spheres.

The inner journey begins at Leh

You'll be aching, wind-chapped and crunching grit between your teeth by the time you roll into Leh, where normal life reasserts itself in Ladakh's biggest city. After a well-earned shower, take a few days to acclimatise to being 3524m (11,562ft) above sea level. Explore the maze-like old city, climb the terraces of Leh's Royal Palace – as much an architectural icon as Lhasa's Potala in Tibet, China – or tune into the city's spiritual frequency on a yoga or meditation course.

It may be a wrench to leave the Ladakhi capital, but true enlightenment awaits on the empty highway running west to Kashmir. You'll encounter brief snatches of greenery where meltwater dribbles down from lofty glaciers, but in between lie miles of emptiness where you'll get lost in the landscape and in your own thoughts.

Over the next 420km (149 miles), the mountains slide past, distant and aloof, as the road becomes little more than a smear on the landscape. Pause and stop the engine where the Indus River meets its Zanskar tributary, and you'll feel the weight of all that emptiness bearing down like an avalanche. It's a spooky but liberating reminder of your own insignificance in the universe.

Moments of majesty

For the people who live year-round in this unforgiving terrain, life is a mixture of determined practicality and intense spirituality. Branching off the highway at Alchi, you'll encounter some of the most magnificent artistry anywhere in the Himalaya at Choskhor

> **"You'll feel the weight of all that emptiness bearing down like an avalanche. It's a liberating reminder of your own insignificance"**

Temple Complex, its ochre-hued chapels and *chortens* adorned inside with an explosion of 10th-century murals and statues – a Technicolour riposte to the monochrome Himalayan landscape.

Another essential stop is the 11th-century monastery at Lamayuru, a tiny cluster of whitewashed houses nestled into an almost Martian landscape of cave-hollowed hoodoos and crumbling badlands. Join the monks chanting mantras in the *gompa*'s mural-filled prayer hall, then stake out a vantage point on the edge of the ridge and mull over your own answers to life's great questions.

You may feel a pang of loneliness as you climb away from the Zanskar River towards the 4108m (12,478ft) Fotu La, the first of many exposed passes on the way to Srinagar. The road exits Buddhist Ladakh at Mulbekh and enters Muslim Kashmir at Kargil, where the cuisine makes a shift to spicy meatiness after the tamer flavours of the Buddhist Himalaya.

The road into the Kashmir Valley climbs between snowcapped peaks to the 3528m (11,575ft) Zoji La, passing through desolate Drass – reputedly the coldest place in India, where temperatures can drop to a marrow-chilling -60°C (-76°F). On the far side of the pass, vegetation reasserts its claim on the landscape as you drop through wind-smoothed alpine meadows and soaring pine forests to the cluster of tin-roofed hotels at Sonamarg.

The final descent into Srinagar will remind you why Kashmir was spoken of in hushed tones of respect on the hippy trail. Elegant arcades of poplars and towering chinar trees provide an honour guard as you slip into the outskirts of this ancient Muslim city. Book into a houseboat on mirror-flat Dal Lake for one last moment of serenity before you dive back into full-bore India in Srinagar's manic bazaars.

© NATTACHART JERDNAPAPUNT / GETTY IMAGES

From top: Leh Palace, Ladakh; houseboats on Dal Lake, Kashmir

Practicalities

Region: Himachal Pradesh/Jammu & Kashmir/Ladakh, India
Start: Manali
Finish: Srinagar

Getting there and back: Motorbikes can be rented in Manali, but most riders visit Ladakh as part of a longer trip from Delhi. In the capital's Karol Bagh bazaar, Lalli Singh Adventures (see Tours) is a tried-and-tested operator specialising in classic Enfields.

When to go: Ladakh and Kashmir's high passes are usually open from March to November, though exact dates depend on snow conditions. The Leh–Srinagar section is sometimes open as late as December.

What to take: Essentials include food, water, camping gear (for emergency overnights), cold-weather clothing and a well-stocked bike-repair kit, including spare inner tubes.

Where to stay: Between Manali and Leh, you can overnight at hotels in Keylong or at rustic government rest houses and crude tented camps along the highway. After a comfortable stay in a hotel or guesthouse in Leh, break the journey to Srinagar at Lamayuru or Kargil – both have simple guesthouses and homestays. Phone ahead to make reservations.

Where to eat and drink: Rustic *dhabas* (roadhouses) dot the highway, but meals can be basic, so make the most of hearty evening feasts and filling breakfasts of fire-cooked *khambir* bread and Ladakhi honey at the places you stop overnight. Always carry emergency food and water in case you have to camp out.

Tours: For supported motorbike trips, try Lalli Singh Adventures (lallisinghadventures.com) or Classic Bike Adventures (classic-bike-india.com).

Essential things to know: It's important to respect the altitude – break the journey with a few days in Leh to reduce the risk of acute mountain sickness (AMS), and wear sunscreen and lip balm or you'll be frazzled like a crisp.

Temple to temple on Japan's Shikoku Pilgrimage

Walk between Shikoku's 88 sacred temples, following the great saint Kōbō Daishi in a search for enlightenment – as pilgrims have done for 1200 years.

Walk the pilgrimage in 40 to 60 days and you're going to face some challenges along the way: language, weather, the daily issues of finding a place to stay, bathe and eat, plus physical and mental hardships. Overcoming adversity on the journey is part of the pilgrim's path as they strive to attain enlightenment.

Get more than a passing look at the unique culture and history of Japan's fourth-largest island, shaped over the centuries by the pilgrimage.

Walking 1400km (870 miles) is a tough proposition, and on completing the route, most pilgrims are in peak physical condition – fitter, stronger and lighter.

Until about 100 years ago, all Shikoku pilgrims walked the full extent of this 1400km (870-mile) route connecting the island's 88 sacred temples; nowadays, with the advent of bicycles, cars and buses, less than 1% complete the pilgrimage on foot. In the days of old, *henro* (pilgrims) overcame the physical hardships of the journey as a form of ascetic practice as they strived to attain enlightenment. They set out with

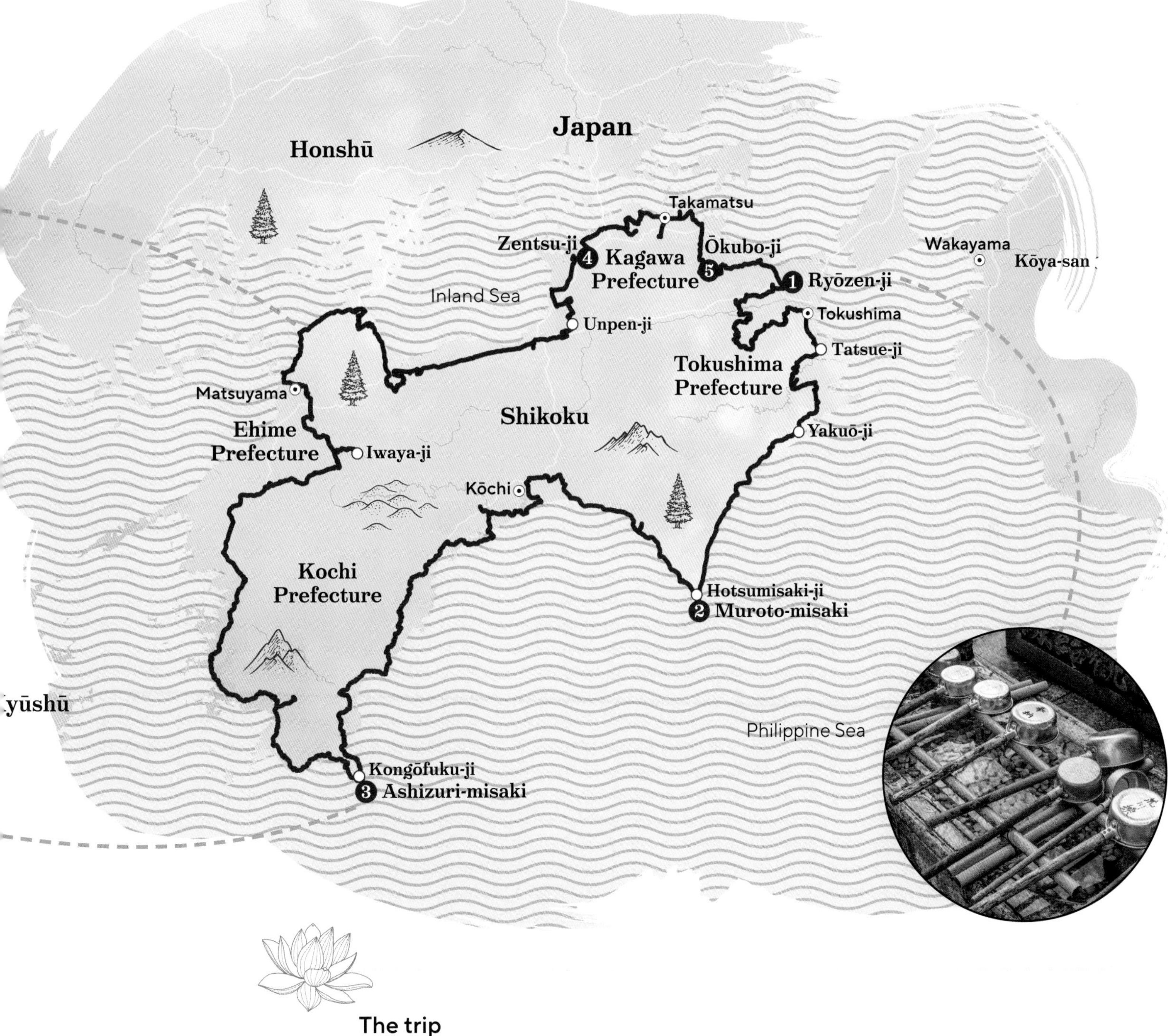

The trip

Distance: 1400km (870 miles)
Mode of transport: On foot or by bike, car or bus
Difficulty: Difficult on foot, moderate by bike, easy by car/bus

1 Ryōzen-ji Temple 1 is the start of the 1400km (870-mile) journey around Shikoku's 88 temples

2 Muroto-misaki Kōbō Daishi is said to have achieved enlightenment at this rugged cape

3 Ashizuri-misaki Temple 38, Kongōfuku-ji – the southernmost on the route – sits at this remote cape

4 Zentsu-ji Temple 75 is Kōbō Daishi's birthplace. This is the the largest of the pilgrimage temples

5 Ōkubo-ji Temple 88 is the last one on the route, but it's not the end of the journey

Clockwise, from left: Ōkubo-ji, the trail's final temple; Ashizuri-misaki's Temple 38, Kongōfuku-ji; traditional *henro* attire; *chōzubachi* washing ladles, Ryōzen-ji

Craig McLachlan, Shikoku pilgrim

I completed my first Shikoku pilgrimage on foot in 1995. It was a wondrous opportunity to slow down and contemplate life. There's a lot of time to think on the walk. I made some life-defining decisions, such as to put family ahead of career, that I'm more than happy with a quarter of a century later. I feel the pilgrimage helped make me a much better person than I might have been.

only a vague idea of where they were going. There were no maps, no guidebooks (though the first one was published in 1685), no weather forecasts, no internet and no mobile phones with which to call home when they got lonely. Some simply disappeared or died along the trail. These days, though, aboard an air-conditioned bus, most pilgrims have few hardships to face.

But the walking pilgrim is still the essence of the pilgrimage. In recent years, there has been a resurgence of walkers – both Japanese and foreign – who, disenchanted with the pace of modern life and looking for meaning and self-realisation, strike out on foot. They can be seen all along the trail.

Kōbō Daishi's Shingon is the only mainstream Japanese Buddhist sect that believes enlightenment can be achieved in this lifetime, and many set out with this in mind. *Henro* come to Shikoku with varying religious beliefs, many with none. All are welcome. In the words of Kōbō Daishi, 'Do not just walk in the footsteps of the men of old, seek what they sought.'

All the gear that a walking *henro* requires is available at Temple 1, Ryōzen-ji, including an English-language guidebook. It's not essential, but most walking pilgrims wear a white shirt emblazoned with the characters Dōgyō Ninin (meaning 'we go together') on the back; *henro* are not alone, as Kōbō Daishi's spirit is always with them. A wide pilgrim's straw hat, the kind you see farmers wearing in rice paddies, is perfect to shield the head from sun and rain, while a *tsue* (staff) helps on the multiday trail, especially in the mountains. Pilgrim's also purchase a *nōkyō-chō*, a book for the signatures of all 88 temples, with beautiful black calligraphy over bright vermillion stamps.

© AMEHIME / SHUTTERSTOCK

Below: cherry blossom at Temple 19, Tatsue-ji
Right: Temple 37, Iwamoto-ji

Starting with intent in Tokushima

Shikoku means 'four regions', these days the prefectures of Tokushima, Kōchi, Ehime and Kagawa. Ryōzen-ji, Temple 1, is in Tokushima, as are the first 23 temples. Tokushima is known as Hosshin-no-dōjō, 'The Place to Determine to Achieve Enlightenment', as this is where pilgrims realise the magnitude of their task and start to make the effort required to complete their goal.

The pilgrimage starts out in a gentle fashion, with the first 10 temples spanning barely 40km (25 miles) among farming villages north of the Yoshino River. It soon turns wild, though, in the mountains to the south, with the climb from Temple 11 to Temple 12, on a steep, forested trail, considered one of the hardest on the journey. The route then drops into a group of temples around Tokushima city, the prefectural capital.

Temple 19, Tatsue-ji, is a 'barrier temple', one that only those who are pure of intention may pass. The 'unworthy pilgrim' should go back to Temple 1 and start again!

Temple 23, Yakuō-ji, in the coastal town of Hiwasa, is a *yakuyoke-dera*, a temple to ward off bad luck for unlucky ages, the worst being 42 for men and 33 for women. While praying for luck, male pilgrims put a coin on each of the 42 steps on the men's side, while women do the same on the 33 steps on their side.

The testing grounds of Kōchi

Known as Shugyō-no-dōjō, 'The Place of Practice', Kōchi Prefecture is the pilgrims' testing ground. The trail through Kōchi makes up more than a third of the pilgrimage's total distance, but has only 16 of its

Following a spiritual path

Tradition says that the first thing *henro* should do is to visit Kōya-san in Wakayama, the mountaintop monastery that Kōbō Daishi founded in the 9th century; it's still the headquarters of his Shingon Buddhist sect. There, *henro* pray before the saint's mausoleum, asking for his support on their upcoming pilgrimage before travelling on to Shikoku. It's much easier these days, with modern bridge links, but through the centuries, all pilgrims made the crossing from Honshū by boat. The Ryōzen-ji is Temple 1 as it was the first temple that pilgrims came to on their arrival on Shikoku. *Henro* visit all 88 temples, but the pilgrimage isn't over on reaching Temple 88, Ōkubo-ji. Traditionally, *henro* must then walk the 40km (25 miles) back to Temple 1 to complete their journey – the full circle of Shikoku – and then return to Kōya-san to thank Kōbō Daishi for his support on their pilgrimage.

temples. To get to its first temple, pilgrims must walk 85km (53 miles) down the coast, to a crescendo of pounding waves, from the last temple in Tokushima.

Kōchi faces the none-too-aptly-named Pacific Ocean and was always regarded as one of the wildest and most remote parts of Japan. Its weather was considered just as inhospitable as its populace, a tough people eking out a living in a rough land, who were known to despise and distrust wandering pilgrims. While Tokushima nurtured puppet theatre through the centuries, in Kōchi they raised fighting dogs. These days though, the people of Kōchi are friendly and encouraging to *henro*.

The area around the great cape Muroto-misaki, home to Temple 24, Hotsumisaki-ji, is where Kōbō Daishi is said to have achieved enlightenment, making it one of the most important temples on the pilgrimage. Beyond Muroto-misaki, the route travels in a huge arc around the island's southern coast. Along the way, *henro* visit a cluster of temples around Kōchi city, then walk the longest distance between temples on the pilgrimage, covering 92km (57 miles) from Temple 37, Iwamoto-ji, to Temple 38, Kongōfuku-ji, which sits at the second of the island's great southern capes, Ashizuri-misaki.

Henro breathe a sigh of relief on getting through Kōchi Prefecture, for the hardest part of the pilgrimage has been completed.

The excitement of Ehime

Bodai-no-dōjō, 'The Place of Attainment of Wisdom', Ehime Prefecture is home to 27 of the 88 temples. Its southern parts are thought of as wild and remote, much like Kōchi. Popular down here, especially around Uwajima, is *tōgyū*, a form of bullfighting that pits bull against bull.

© MOKOKOMO / SHUTTERSTOCK

Below: Muroto-misaki
Right: the path up to Temple 45, Iwaya-ji

> "Overcoming adversity on the journey is part of the pilgrim's path as they strive to attain enlightenment"

Heading north, Temple 45, Iwaya-ji, is particularly atmospheric, hanging on a cliffside high above the valley floor in a dramatic location deep in the mountains. Eventually *henro* arrive in Matsuyama, Shikoku's largest city, a refined and cultured place that's home to the legendary hot springs at Dōgo Onsen and eight of the 88 temples. This is where many pilgrims get a boost of energy, knowing that another tough stage of the route is behind them. The trail is relatively genteel as it heads east, before climbing to Unpen-ji, Temple 66; this 'temple in the clouds', at 900m (2952ft) above sea level, is the highest on the pilgrimage route.

A sense of accomplishment in Kagawa

The smallest of Japan's 47 prefectures, Kagawa is Nehan-no-dōjō, 'The Place of Completion'. Pilgrims positively stride down onto the northern plain to visit the last 22 of the temples. Protected by Shikoku's central mountains, Kagawa's weather is warm and welcoming, as are its residents.

Kōbō Daishi was born at Zentsu-ji, Temple 75. It's huge: most of the other temples could fit into its car park. The pilgrimage then meanders through Takamatsu, Shikoku's second-largest city, then *henro* find themselves climbing back into the mountains to Temple 88, Ōkubo-ji. Most feel a real sense of accomplishment, and some relief, on reaching Ōkubo-ji, but there is also the understanding that it's not over. The pilgrim must walk back to Temple 1, Ryōzen-ji, to complete the circle of Shikoku – and all *henro* should complete the circle, for a circle is never-ending, just like the search for enlightenment.

Practicalities

Region: Shikoku, Japan
Start/Finish: Ryōzen-ji (Temple 1)

Getting there and back: Ryōzen-ji is a 10min walk from Bandō station, reachable via a 20min journey from Tokushima on the JR Kōtoku line.

When to go: The best seasons are spring and autumn. Avoid *tsuyu*, the rainy season, in June; summer is hot and humid, while winter can get very cold.

What to take: Walking *henro* should pack according to their budget. If you'll be overnighting in lodgings, carry the minimum in a daypack. If your budget is thin, consider taking a small tent and a sleeping bag.

Where to stay: Shikoku has catered to pilgrims for 1200 years and is set up to meet the needs of *henro*. Options include *shukubō* (temple lodgings), *minshuku* (family-run B&Bs), *ryokan* (Japanese inns) and small hotels. Budget ¥10,000 per night for dinner, bed and breakfast. Pre-booking is tricky, as walkers don't know how far they'll get each day; lodgings will generally call the next place down the line for you. *Nojuku* (sleeping out) is not frowned upon. Kōbō Daishi is said to have once spent the night under a bridge, so *henro* never tap their staffs on a bridge, in case they wake him up.

Where to eat and drink: There are plenty of places along the way. *Shukubō* (temple lodgings) generally offer only *shōjin-ryōri* (monk's vegetarian meals).

Essential things to know: Most pilgrims travel in vehicles these days. Walkers can avoid roads in many places by searching out *henro-michi* (pilgrim trails), generally marked by little white signs with a *henro* painted on them. Shikoku 88 (88shikokuhenro.jp/en) has info on the route and all the temples; *Japanese Pilgrimage* by Oliver Statler is essential pre-trip reading.

Gangotri to Gaumukh: hiking the headwaters of the Ganges

Traffic noise fades into an endless thrum of river: the end of the road is the beginning of this extraordinary high-altitude journey through India's Garhwal Himalayas.

The Gangotri to Gaumukh route is one of the Chota Char Dham, a four-site Hindu pilgrimage circuit in Uttarakhand. Devotees travel by foot or mule, prayer and conversation fill the air, and ephemeral shrines dot the riverbank. The hike can be both a spiritual pilgrimage into the heart of Hinduism and a secular journey into wild nature.

The intensely spiritual connection to landscape of Hindu pilgrims may overwhelm visitors who bring different perspectives, but a lack of common language often matters less than a gestured invitation to look at the view and share a cup of tea – communication by chai can offer a refreshing way of making relationships.

From the temple at the edge of tiny Gangotri township, the view is stunning. Saw-toothed, snowcapped mountains interrupt the skyline above rust-coloured slopes; clouds pillow the valleys. It's also daunting, as you gaze at the walking trail hugging a steady, narrow incline of shale as it parallels the river gorge. This is the Bhagirathi River, hurtling over rocks and carrying grey-green snowmelt from its headwaters 18km (11 miles) upstream at Gaumukh (Cow's Mouth) Cave, a glacial cavern that looks exactly like its name. After roaring downstream for some 200km (124 miles), this water feeds the beginning of the River Ganges at Devprayag.

The landscape here reveals itself in subtle shades of sky and earth and water. There's little greenery, except for low shrubs and an occasional stand of pine or birch trees. In these, birdsong competes with the all-pervasive sound of the river: the drill-burst of

© XAVIER GALIANA / GETTY IMAGES

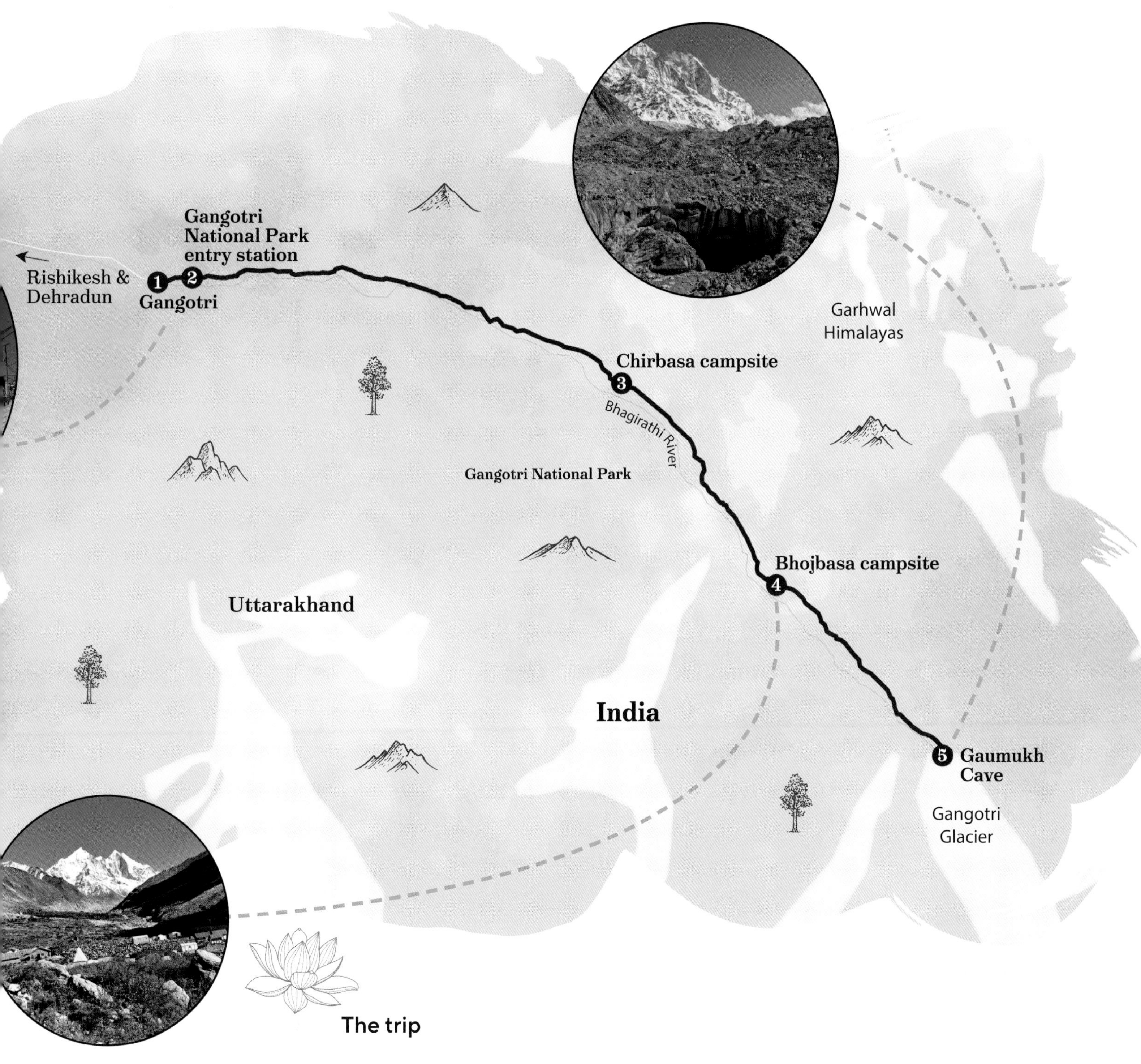

The trip

Distance: 36km (22 miles)
Mode of transport: On foot and/or mule
Difficulty: Moderate to difficult

1 Gangotri Temple Join worshippers beside (or in) the bathing ghat's icy waters

2 Gangotri National Park entry station Absorb the cheerful – if somewhat perplexing – mix of information and administration

3 Chirbasa campsite Walk and wonder at elemental views in all directions

4 Bhojbasa campsite The camp's namesake birch trees make for good birdwatching

5 Gaumukh Cave Watch the headwaters of the mighty River Ganges emerging from a glacier

Clockwise, from left: on the trail to Gaumukh; Gangotri Temple; the Ganges headwaters flowing from Gaumukh Cave; Bhojbasa campsite

Afterwords from a group of Indian high-school hikers

On time off social media: 'I felt very anxious to start; felt so free by the end.'

On changing perceptions: 'They're one of the cool kids. We hadn't spoken before and now we're friends. We've lots in common.'

On working together: 'I was happy to walk slowly and keep them company, talking or being silent together.'

Of overcoming challenge: 'I thought I wouldn't make it. I did!'

a nuthatch on bark, the trills of the laughingthrush and shyer rosefinch, the insect-eating wagtails and flycatchers. Overhead, crows caw loudly and, higher still, eagles cruise on the thermals.

This landscape is part of Gangotri National Park, and the only people around are park workers or tour guides and porters with trekkers. Flashes of colour way ahead on the trail mark the progress of visitors dressed in high-vis, high-tech hiking gear. A shifting patch of dust identifies a mule-train carrying people and supplies; the bells on the animals' loads warn of their approach, giving time for walkers to squeeze against the upside of the path to let them pass.

Looking down

This big, dramatic landscape demands a particular sort of attention. Visitors whose usual surroundings are urban, or tamed by farms and villages, or whose lives are shaped by working with screens and machines, may find themselves wondering how on earth – literally – to engage with this unfamiliar environment.

While you can cover this route in one long day, overnighting at trailside camps can offer time and space to become fully present here. The lower campsite, Chirbasa (Place of Pines), is at around 3550m (11,647ft). It's 9km (5.5 miles) from Gangotri which, allowing for the slow pace of high-altitude walking, is about a five-hour walk. Bhojbasa (Place of Birches) is a further two hours up; Gaumukh Cave at 4023m (13,200ft) is two hours beyond that.

If all this grand scale becomes too much, looking at small details can refocus and reassure. Chirbasa is on the bank of the Bhagirathi, and a slow and attentive early morning walk might show paw-print evidence of nocturnal visitors – perhaps Himalayan bears, or

© UBARAK_KHAN / SHUTTERSTOCK

> “Body not mind sets the limits of speed, breathing, the need to drink and rest; concentration is focused on being, not doing”

jackals – in the sand between waterside boulders. An apparently random pile of stones resolves itself into a small shrine containing an offering to one of the river gods. A tumbledown hut houses a *sadhu* – a holy person – who lives here permanently, worshipping and living on food offered by pilgrims. Is this life reduced to essentials, or expanded to something more? Such stories of the landscape are all around here, waiting to be read and absorbed.

Looking in

The out-and-back Gangotri to Gaumukh hike is a physical challenge in extraordinary surroundings. The wilderness makes no concessions and offers few creature-comforts to visitors. There is little shade from the elements, the only running water is the river, and toilet facilities – where they exist – are basic at best (check the ground around the trunks of birch trees: the papery bark they shed makes a decent substitute for loo paper). Being open to the likely charms and the challenges of the whole experience may make for a richer journey and for more authentic memories.

It's the journey itself, rather than the destination, that's likely to become all-encompassing. A weary hiker has to go through, and then beyond, the altitude headache of the first few hours, and the constant shallow breathing caused by the thin air. Body not mind sets the limits of speed, breathing, the need to drink and to rest; concentration is focused on being, not doing. Staying with the moment can help. Instead of longing for a hot drink, you might find it more rewarding to take time to encourage (or even join) those who immerse in the freezing river at Gangotri Temple's bathing ghat and emerge teeth-chattering and speechless, but glowing with achievement. A thumbs-up and shared grin can turn these chance encounters between strangers into a moment of surprising connection.

On returning to the relative comforts of Gangotri, an inner voice may continue to make itself heard. What have I learned about myself? How have I changed? What will I take home with me? What have I left here?

© SUPARNA HAZRA / GETTY IMAGES

From top: the the Garhwal Himalayas' Bhagirathi massif, near Gaumukh; Gangotri ghat on the Bhagirathi River

Practicalities

Region: Uttarakhand, India
Start/Finish: Gangotri

Getting there and back: To get to Gangotri, take a train to Dehradun or Rishikesh, then a bus, shared jeep or charter taxi for the 250km (155-mile) drive on to Gangotri. This last leg takes at least 8hr, and the road and all public transport stops here.

When to go: Gangotri National Park is open from April to November, depending on seasonal snowfall. During monsoon season (July to September), much of its area is inaccessible. Nights are cool to very cold throughout the year.

What to take: Clothes for all weather (hot, cold and wet); sunscreen and a hat; water bottle and water purifying tablets.

Where to stay: Accommodation is available in Gangotri township (book in advance), but designated camping areas are the only option on the route. Bring your own gear, or hire equipment and a porter in Gangotri.

Where to eat and drink: Bring all supplies from Gangotri: there are no permanent eating places in the park, though there may be basic pop-up kiosks during peak season. Several small restaurants in the township serve a range of food styles.

Tours: There are many tour operators in Gangotri. Word-of-mouth recommendations or online travellers' forums may be the best source of current information. See uttarakhandtourism.gov.in for general information on Gangotri National Park.

Essential things to know: It's a minimum three-day/two-night walk there and back for the fit, a four-day/three-night walk for the casual hiker, or a very long day-trip on a mule with mule-driver. There's phone and internet in Gangotri township, but no mobile phone reception along the track.

Oceania

THE GHAN
NR109

The Ghan, en route across Australia from Adelaide to Darwin

Hiking the Larapinta Trail

Experience the solitude and magnificence of a two-week walk through Australia's desert centre, wandering over mountains by day and sleeping beside waterholes at night.

Deserts are almost a metaphor for solitude and contemplation, and they both come in spades on this long and demanding hike. Reflection and reconfiguring of life come naturally as you walk through this wide and open landscape.

Interactions with Indigenous culture are few along the trail, but this is still a chance to experience the power of Arrernte Country. Browse up on bush tucker and the art of Albert Namatjira to gain some affinity with the land.

Don't be surprised if you feel superhuman after 223km (139 miles) and two weeks of desert and mountain walking.

At a glance, hiking the Larapinta Trail can seem a task worthy of a religious ascetic: walking for two weeks across a desert in the hot heart of Australia. But this world-class trail, now one of the most popular long-distance hikes in Australia, is an extended oasis, furnished with gorges, swim-holes, vast mountain views and well-designed campgrounds. This a hike that's about desert delights, not deprivations.

Larapinta landscapes

Stretching from Alice Springs to distant Mt Sonder, the Larapinta Trail traverses the length of the Tjoritja / West MacDonnell National Park. It explores a curious landscape in which the rivers run through the mountains rather than away from them, meaning the range is sliced into sections by gorges and cooling pools. Such is the spacing of these gorges that many days end at camps beside the waterholes.

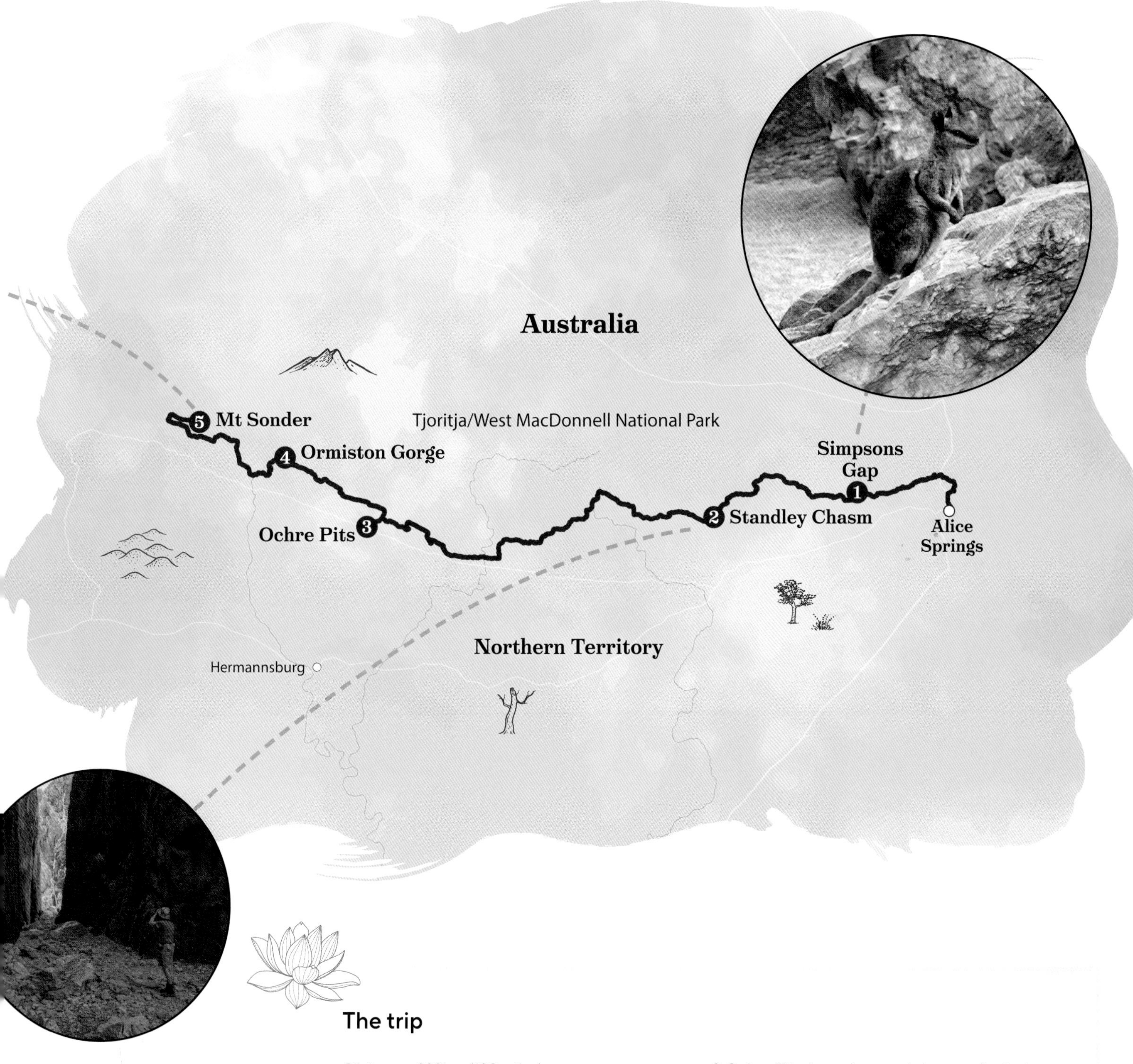

The trip

Distance: 223km (139 miles)
Mode of transport: On foot
Difficulty: Difficult

1 Simpsons Gap Scan the slopes outside this peaceful gorge to spot black-footed rock wallabies

2 Standley Chasm The trail wriggles through this narrow and spectacular slot in the range

3 Ochre Pits Arrernte people have collected ochre from this colourful outcrop for millennia

4 Ormiston Gorge A massive ring of mountains, filled with waterholes and high red cliffs

5 Mt Sonder The range's highest peak is known as the Sleeping Woman – see its profile and you'll understand why

Clockwise, from left: Ormiston Gorge; sunset over Mt Sonder; a black-footed rock wallaby at Simpsons Gap; capturing Standley Chasm

Andrew Bain, Larapinta Trail hiker

When I first hiked the Larapinta Trail in 2005, I did so with a friend going through a tough emotional time, while I was fighting the fatigue of the arrival of a new baby. The simplicity of the desert and the walking life brought perspective, each day returning us closer to our true selves until we wandered into Alice Springs rejuvenated – and almost ready to turn around and walk back the other way.

The trail can be walked in either direction. Some hikers favour having the sun at their back rather than in their face, so head west from Alice Springs. Others prefer to walk from Mt Sonder, allowing them to finish in Alice Springs whenever they choose, without the time pressure of a waiting shuttle at the hike's end.

Gorges & ridges

The Larapinta is divided into 12 distinct sections, ranging from 9km (6 miles) to 31km (19 miles). At its eastern end, the trailhead is the Alice Springs Telegraph Station, though there's a sense of rightness about beginning in the city centre and walking the hour north to the station. It's a gentle start, crossing the Stuart Hwy and the Ghan railway line and climbing onto the eastern tail of Tjoritja / West MacDonnell National Park along Euro Ridge, with its long view back over Alice Springs creating an early sense of progress.

Simpsons Gap is the first break in the range, and it's a spectacular preview of what's ahead, but it's at Standley Chasm, at the end of Section 3, that the walk really gets into its stride. Every hiker will emerge from the Larapinta with their own favourite sections, but for many it's the three or four days from Standley Chasm to Hugh Gorge that are the most wild and wonderful.

The narrow slot of Standley Chasm is a portal into one of the truest of the Larapinta's mountain stages, beginning among the tourist crowds through this fracture-like gorge and then leaving them as you climb high above. The trail then scrambles to mountaintops, balances across Razorback Ridge, and dips back to earth through Hugh Gorge, where the cliffs are as smooth as plates and as tall as city buildings. You might see a few other hikers up here, but otherwise you're alone with your thoughts and the views.

"As you rise to the summit, the sky flares with dawn colour, and not a thing seems wrong with the world"

After a long crossing of the parched Alice Valley plain, switching between the lines of mountains, it's back onto the ridges, ascending to Counts Point, one of Tjoritja / West MacDonnell National Park's finest viewpoints. The mountains in view run like ribs across the desert landscape, all seemingly pointing towards Mt Sonder, the trail's finish, even if it is still days away.

Ravishing rocks

Beyond here, it's a step into prehistory, passing the vividly coloured Ochre Pits en route to Inarlanga Pass, a dry gorge choked with cycads. Looking like a cross between a palm tree and a fern, these plants have been on the planet for 280 million years – to put that into perspective, they were once dinosaur food. Even older is the Finke River, another day of walking ahead. Though it rarely flows, the Finke is believed to be 340 million years old, making it the oldest river on Earth. The trail takes its name from the river, known as Larapinta to the Arrernte people.

The Finke's headwaters are near Ormiston Gorge, arguably the trail's most beautiful feature. At the end of Section 9, this near-circular enclosure of peaks funnels into a high and narrow gorge, dotted with waterholes and penned by high, rust-red cliffs and brilliantly white ghost gums. The Larapinta doesn't enter the gorge proper, but it's worth building in half a day to explore this impressive spot.

A Mt Sonder sunrise

One day's walk on, the Larapinta reaches the foot of Mt Sonder. Only the climb to its 1380m (4530ft) summit remains – trail tradition is to ascend in the cold desert night to reach the summit at sunrise. Guided by the narrow beam of your headtorch, it's an at-times steep ascent. As you rise to the summit, the sky flares with dawn colour – the red earth and rock glowing like neon – and not a thing seems wrong with the world. You might be tired (two weeks of walking can do that), but there's almost certainly new strength in your legs, and your mind might be as clear as the sands far below. The desert's job is done.

© VIKTOR POSNOV / GETTY IMAGES

From top: sunrise at Mt Sonder; a waterhole at Ormiston Gorge

Practicalities

Region: Northern Territory, Australia
Start: Alice Springs
Finish: Mt Sonder

Getting there and back: Alice Springs has direct flights to Sydney, Melbourne, Darwin, Brisbane and Adelaide. Larapinta Transfers (larapintatransfers.com.au) and Larapinta Trail Trek Support (treksupport.com.au) both run shuttles between Alice Springs and Mt Sonder, and points along the trail in between.

When to go: Winter (June to August) is prime time on the trail, with April, May, September and October also good options (they also have warmer evenings than the frigid winter nights). Avoid summer (December to February) when the heat is intense – often above 40°C (104°F).

Where to stay: Camping is your sole option, and the trail has more than 40 campsites neatly spaced along its length. Many of the 34 official trail sites have shelters with tent pads, allowing you to simply roll out your sleeping mat and bag under the cover of a roof. The one chance to properly sleep indoors is at Glen Helen Resort, near the trail's western end.

Where to eat and drink: The effort of hauling two weeks' worth of food on foot through the desert is mitigated by the presence of food storage facilities at Ellery Creek and Ormiston Gorge (you'll need to get a key from the Tourism Central Australia Visitor Centre in Alice Springs to access the storage). You can arrange food drops with the private operators at Standley Chasm and Glen Helen Resort, or talk to shuttle operators about including them as part of your transfer.

Tours: The Larapinta Trail has boomed in popularity across its two decades of existence, and there are now multiple companies offering guided walks. Seasoned operators running end-to-end walks include World Expeditions (worldexpeditions.com) and Trek Larapinta (treklarapinta.com.au).

Cycling the Tour Aotearoa

Ride from tip to tip of New Zealand on this two-wheeled quest through old-growth forests, past smouldering volcanoes and deserted beaches.

Get to know yourself better and how you respond emotionally to setbacks and unpredictability.

Riding the length of this nation affords the time to get to know its complex history and island culture as you pass key sites. The natural world is an ever-present companion and you'll gain an appreciation of the vulnerability of native ecosystems and species.

The physical skills you'll gain will be to do with self-sufficiency and being able to bodge basic mechanical repairs and keep you and your bike rolling. Fitness isn't a pre-requisite but the will to persevere though fatigue and discomfort might be.

It's not long after you've left Cape Reinga, having sped across hard-packed sands on the Far North Cycleway, that you'll enter the great Waipoua Kauri Forest. The road is steaming after a light shower and sounds are muffled by the giant trees around you. This is the domain of Tāne Mahuta, a 2000-year-old kauri and one of the world's largest trees. It's an enchanting stretch of the Tour Aotearoa, but Tāne Mahuta is certainly not the only awe-inspiring sight on this odyssey the length of the Land of the Long White Cloud.

Some time after the Great Financial Crisis of 2008, New Zealand's government decided to invest in building cycle trails. This was a prescient move, given the importance of sustainable travel and how well-trodden some regions were becoming. It inspired Jonathan Kennett, a celebrated New Zealand cyclist who began pioneering mountain-

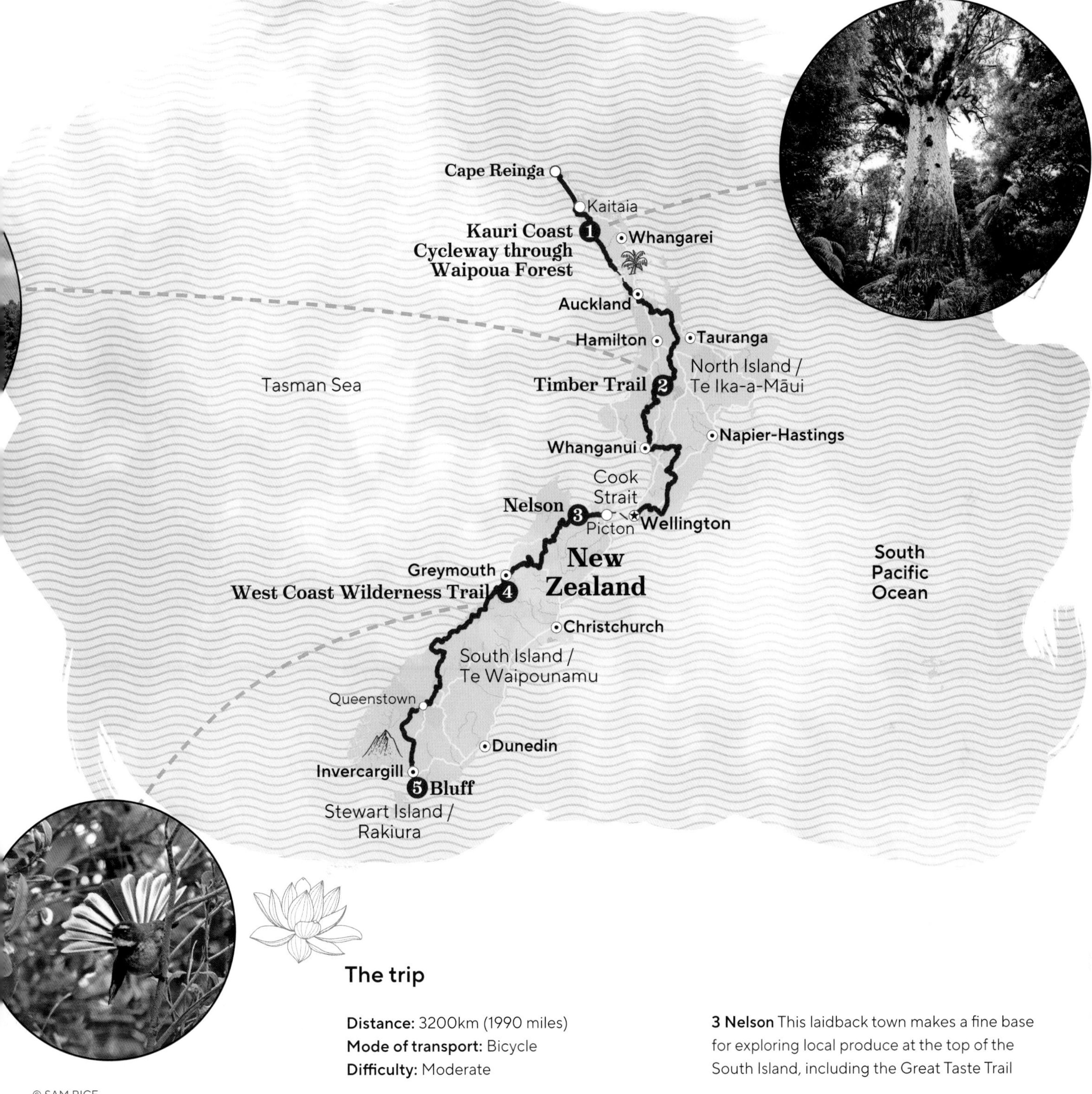

Clockwise from left: riding gravel roads south out of the kauri forest; the Maramataha suspension bridge on the Timber Trail; Tāne Mahuta in Waipoua Kauri Forest; a fantail on the west coast

The trip

Distance: 3200km (1990 miles)
Mode of transport: Bicycle
Difficulty: Moderate

1 Kauri Coast Cycleway Flitting through a forest of giant trees is a highlight of the North Island; stop to say hello to Tāne Mahuta

2 The Timber Trail This purpose-built trail combines mining history with natural regrowth in a little-visited corner of the North Island

3 Nelson This laidback town makes a fine base for exploring local produce at the top of the South Island, including the Great Taste Trail

4 West Coast Wilderness Trail Feel the awe of the South Island's wild west coast on this flat section through former gold-rush country

5 Bluff Your final stop is the tip of the South Island and the departure point for Stewart Island

Jonathan Kennett, route creator

After finishing the ride for the first time, in March 2016, I was buzzing. The journey gave me a new love for New Zealand and its people, more than any adventure I'd had before. Everyone who completes Tour Aotearoa will gain a deeper understanding of Aotearoa New Zealand – of the landscapes and the people – as long as they don't rush. I always feel sorry for people who plan every day of the trip in advance and book a flight home with no leeway. What I hope people gain is a sense of topophilia – a love for our home.

biking routes in the country in the 1980s, to attempt to piece together as many of the official new trails as possible in order to ride from the top of New Zealand to its southernmost tip at Bluff.

'Riding on the busy roads had become a horrible experience,' explains Jonathan. But his Aotearoa route, still regularly refined, takes in rail trails, river trails, forest trails, volcano trails and coastal trails. Yes, there are some stretches of highway to endure, and experienced bikepackers familiar with wilderness in places like North and South America or Central Asia won't find the challenge they expect, but for most novice or intermediate riders – the Tour Aotearoa is a big adventure, albeit one with safety nets.

The appeal of such a bike ride is that, after a bit of practice on shorter trips, you can neatly pack away everything you need on your bike for a month or more and then be as independent as you desire. Stay longer where you feel a connection. Seek out the time and space to get to know yourself. Be impressed by your newfound confidence in changing plans, dealing with the unexpected and getting through the most challenging days.

'People learn how to ride within their abilities,' says Jonathan. 'Sometimes that might mean taking a rest day when you are starting to feel tired or a storm is blowing through. But riding it with my wife on a tandem was the best holiday we've ever had.'

From the kauri forest of Northland, the route passes through Auckland and then dives into the often-overlooked centre of the North Island, using the Hauraki Rail Trail and the Waikato River Trail to reach what is Jonathan's personal highlight, the Timber Trail, an 85km (52-mile) track through old-growth forest and over several suspension bridges.

© SEBASTIAN93 / SHUTTERSTOCK

> **"There is so much beautiful forest on the Timber Trail and a long downhill that just keeps on giving."**

'Some of the small communities that Tour Aotearoa passes through survive on very little. Bikepackers who buy food and accommodation from local businesses in rural New Zealand make a big difference. In turn, those communities help bikepackers on the remote parts of their adventure.'

After the Timber Trail, you'll follow the Mountains to the Sea route and then pedal along the gravel backroads of the Wairarapa wine region before reaching Wellington, New Zealand's low-key capital. A potentially nauseous ferry transfer across the churning Cook Strait lands you on the South Island to enjoy an idyllic stretch through Marlborough. Fill up on local provisions along the Great Taste Trail and don't hurry past every winery or brewery. This is the calm before the storm; originally, the next leg shadowed the glacial mountains of the west coast, which is regularly buffeted by wind and rain. In 2022, Jonathan Kennett launched an alternative course south, the 1500km (930-mile) Sounds to Sounds route, which travels from Queen Charlotte Sound to Milford Sound via the vast sheep stations of the east, Christchurch and Lake Tekapo. Riders can pick up the Tour Aotearoa again near Queenstown if they're set on reaching the Catlins and Bluff.

Keeping to the traditional Tour Aotearoa route, riders follow the West Coast Wilderness Trail south and eventually arrive in the South Island's adrenaline-charged hub, Queenstown. 'Wilderness' is a thought-provoking concept in New Zealand. Even in a nation of just five million people (almost a third of whom live in Auckland) on a landmass larger than the UK, there are few places untouched by humans here and the truly wild spaces are very tightly curated. Wild camping is forbidden and you won't be far from civilisation for most of the ride. But being able to meet people and learn about the cultural history of Māori and colonists alike is another part of the appeal of the Tour Aotearoa.

From Queenstown, it's a few more days' riding to Bluff and the end of the road. Coming to terms with the return to your regular routine is just a part of the experience; it may inspire you to change your life.

From top: cool your feet in Queenstown's Lake Wakatipu; the Bridge to Nowhere on the Mountains to the Sea trail

© ROLF_52 / SHUTTERSTOCK

Practicalities

Start: Cape Reinga, North Island
Finish: Bluff, South Island

Getting there and back: You will likely need a shuttle from Auckland (the country's main international entry point) to Cape Reinga since there's no train service northward and the buses have limited bike-carrying capacity. A couple of companies offer this service. Research the tide times at Cape Reinga before you get there. From Invercargill (near the finish at Bluff), there's a shuttle service that takes you to Dunedin or Queenstown for onward travel and the InterCity bus service may carry dismantled bikes too. There are also flights from Invercargill to other New Zealand airports.

When to go: Jonathan Kennett recommends riding the Tour Aotearoa in late summer and early autumn (from January onwards). The winter can bring lots of rain and mud with fewer facilities open. The official brevet should be completed in fewer than 30 days (which is about 100km or 60 miles per day) but you can ride the route at your pace at any time.

What to take: A copy of Jonathan Kennett's Tour Aotearoa guide, navigation kit (GPS units are fine but a detailed map is advised), plus assorted chargers, camping and cooking kit if you're aiming for a degree of self-sufficiency. Carry the tools and spares you will need – including spokes, brake pads and chain links. You can be quite some distance from the nearest bike shop. A robust gravel bike is fine for the trip. Bring waterproofs and especially water-resistant overshoes – nothing feels better than putting on a pair of dry shoes after a storm.

Food and drink: Generally, you're not far from a bed or a place to eat but if you're planning to camp you will need to use official sites, such as the DOC's campgrounds. Wild camping is illegal.

Essential things to know: New Zealand's drivers are notoriously bad and get worse the further from urban areas you get. Ride cautiously on public roads. It is compulsory to wear a helmet when cycling in New Zealand.

Crossing a continent by rail aboard the Ghan

Embrace the art of slow travel on the legendary and luxurious Ghan, journeying across the vast emptiness of Australia's ancient red heart.

Slow travel brings plenty of time for reflection. Some find the minimalist desert landscapes liberating, a perfect backdrop for practising mindfulness and reviewing inner journeys. Others might recoil from the emptiness, seeking camaraderie with fellow travellers. There's time enough for both.

Off-train excursions offer opportunities to learn about the Outback's landscapes, history, Indigenous culture, flora and fauna, and a window into the day-to-day lives of the people who live here.

Whether it's a camel ride, a chopper flight or a wilderness trek or mountain-bike ride, the Ghan offers a rich array of bucket-list tick-offs.

The 100 billion stars of the Milky Way cast their luminescence over the desert night, myriad pin-pricks of ancient light cascading through a velvet cloak of blackness. Sparkling like jewels across an infinite sky, they offer a humbling reminder of our own universal insignificance. The Southern Cross, this hemisphere's talisman, hangs over the route already travelled, and a shooting star leaves a blazing retinal imprint: disintegrating meteorite or fried piece of space junk? Who needs a designated Dark Sky Reserve when the whole interior of the Australian continent appears entirely unlit?

Soon a faint orange glow sharpens the eastern skyline, reflected off the stainless steel carriages stretching along the tracks. The crisp, dry winter air is just above freezing, and all is quiet save for the quiet thrumming of the Ghan's two 4000HP diesel

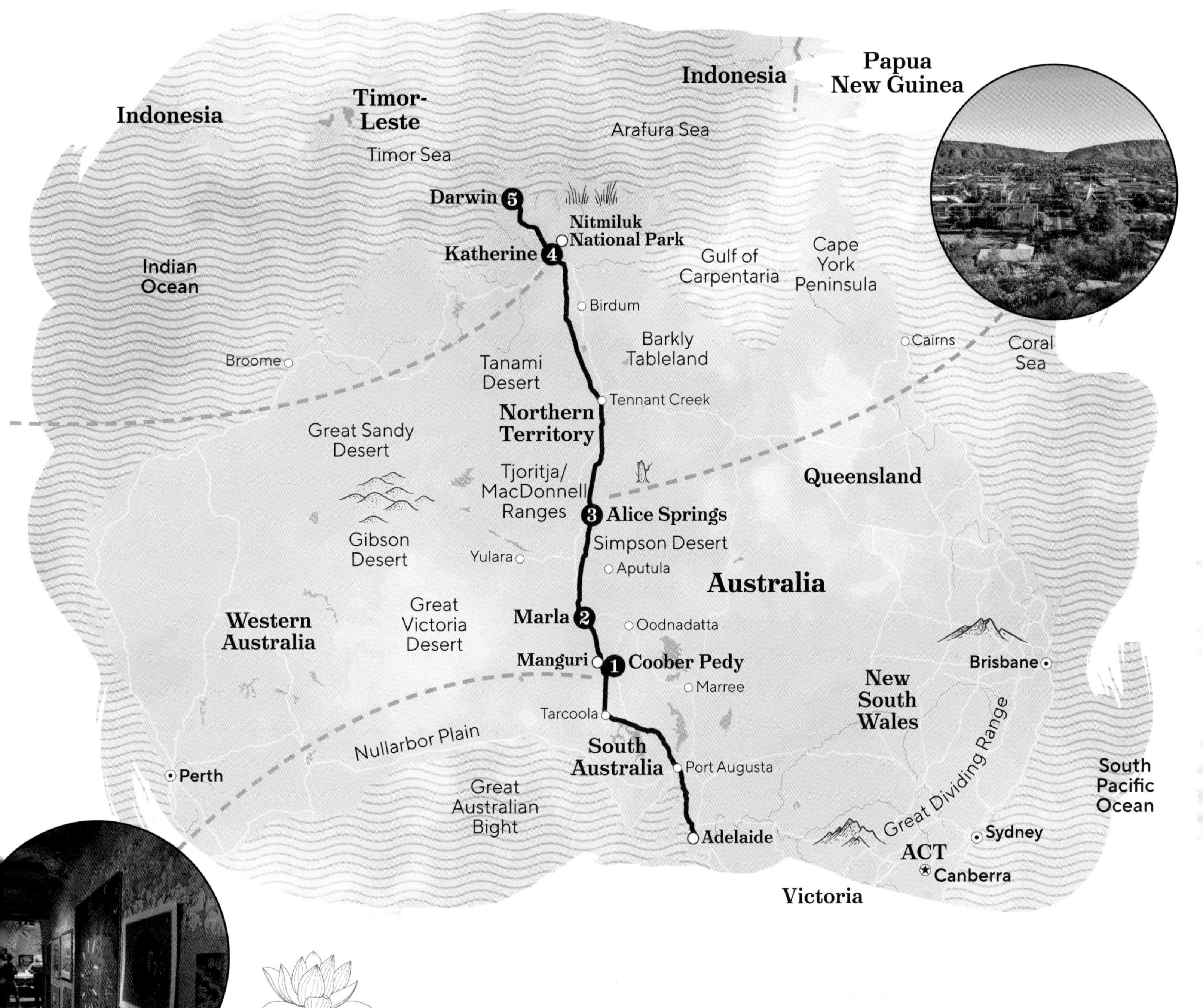

The trip

Distance: 2979km (1851 miles)
Mode of transport: Train
Difficulty: Easy

1 Coober Pedy Discover why most of this sizzling Outback town is buried underground

2 Marla Watch the Milky Way dissolve into a new day on the rails

3 Alice Springs The heart of the Red Centre offers plenty of adventure opportunities

4 Katherine Head off-train into the gorgeous gorges of Nitmiluk National Park

5 Darwin Discover the tropical delights of Australia's most laidback capital

Clockwise, from left: a memorial to Ghan cameleers at Alice Springs station; Katherine Gorge, Nitmiluk National Park; Alice Springs; an underground gallery in Coober Pedy

Steve Waters, Ghan traveller

The panorama of luminous stars on dark desert nights in Central Australia is a thing of wild beauty, unearthly in every sense of the word. Anybody witnessing the full vista from horizon to horizon cannot be unmoved by it, by our own fragility, by our own impermanence against the cosmos. It messes with your head, but it's also addictive. I always long for that sky, no matter where I am. I look up and think, 'ah, it's not as good as my desert'.

locos, far off at the front of the train. Coffee and hot breakfast rolls are served as you continue to stare at the rays escaping from the closest star of all, turning the darkness into day near Marla in the South Australian Outback.

Luxe life on the rails

Trains have always held a certain mystique among travellers, a memory of a slower, more romantic era. Pre-dating other mechanised transport by at least 50 years, railways have been responsible for the economic boom and bust of many regional areas, none more so than Australia's Red Centre. Railway tracks first reached Alice Springs in 1929 and the town grew exponentially; the only other transport had been camel trains. And while camels are no longer needed to explore the vast Australian interior, their legacy lives on in the Ghan and its luxury 'immersive' rail journey. This behemoth of a train is named for the original cameleers, who were collectively known as 'Afghans', though in fact most hailed from the deserts of what is now Pakistan, or beyond there as far as Turkey and Egypt.

Everything about the Ghan is epic. The only true north–south continental rail crossing in the world takes a minimum of 54 hours (three days, two nights) to complete the 2979km (1851-mile) journey north from Adelaide, on St Vincent Gulf, through the Red Centre to tropical Darwin by the Timor Sea. Heading south takes even longer (four days, three nights), allowing for more extensive off-train excursions (ie, the immersive bit).

© MATT MUNRO / LONELY PLANET

Below: a Ghan loco **Right:** en route to Darwin

> "It's easy to become lost in your thoughts, travelling hundreds of kilometres without turning a single page of your book"

In its full contingent of 38 carriages, including sleeping berths, lounges, restaurants, crew quarters and power cars, together stretching for almost 1km (or half a mile), the Ghan is also the world's longest scheduled passenger train. There are two classes – Gold and Platinum – and both include a sleeping berth/private day lounge, five-star restaurant meals and all the top-notch local booze you could possibly consume. Lounge cars quickly become social hubs, with generous windows that afford the best views of wide desert landscapes. There are quicker and cheaper ways to travel between Adelaide and Darwin, but none are as stylish, luxurious or with so many killer cocktails as the Ghan.

Epic landcapes

Riding the Ghan, you won't experience the sudden changes in the landscape associated with air travel. As the train rolls out of Adelaide's euphemistically named Parklands station around lunchtime, the northern suburbs melt subtly into the dry farming country of South Australia's mid-north, whilst passengers head to the restaurant car to enjoy the likes of grilled barramundi and a tasty Hunter Valley semillon.

The flatlands around Spencer Gulf slide past, perhaps over an early cocktail in the lounge car, where excited travellers congregate in the hope of finding like-minded souls to play cards with or swap anecdotes. Perhaps the self-introspection will start the next day; perhaps cocktail hour will become a daily ritual.

North of Port Augusta lie salt lakes and the stunted scrub of the Great Victoria Desert, where hours pass and the changing light brings out different textures in the landscape – a first taste of the intense isolation of the Australian Outback. Alice Springs, the next major town, is 1200km (750 miles) away.

Building a legendary line

A north-south railway crossing of Australia was first proposed in the mid-1800s; by 1884 a narrow-gauge line connected Port Augusta with Marree in South Australia, while the North Australian Railway crept south from Darwin. By 1929 the tracks from Marree reached Alice Springs, and the Ghan was born, though the line from Darwin petered out at Birdum, leaving an 800km (500-mile) gap between buffers. The original Central Australian Railway from Marree to Alice, via Oodnadatta and Aputula (then called Finke), followed the Overland Telegraph route, which in turn observed pathways that Indigenous Australians had long used linking natural water sources – reliable water being a necessity for both travellers and steam engines. But as washouts and delays were common, the track was rerouted. Completed in 1980, the new route via Tarcoola saw the Ghan reborn. In 2001, work began on tracks north of Alice; the first Ghan pulled into Darwin in 2004.

Depending on the season, most of the first section will be at night, until the sunrise stop near Marla. Southbound services halt at a siding called Manguri, where excursions head out to the underground dwellings of the insanely hot opal-mining town of Coober Pedy, and the bizarre mesas of the Kanku-Breakaways Conservation Park, followed by an atmospheric evening meal around a firepit under the celestial chandelier.

Heading north into the sunrise from Marla, the morning is filled with glorious light over a scenery of red dirt, saltbush and low scrub; staring at the featureless desert, it's easy to become lost in your thoughts, travelling hundreds of kilometres without turning a single page of your book. But the more you stare at this landscape, the more you realise that it's anything but featureless.

Off-train in Alice Springs

The red and raw MacDonnell Ranges – Tjoritja to the local Arrernte – break up the mesmerising view and herald the arrival of Alice Springs, the most renowned of Australia's 'frontier' towns, immortalised in books and film. It's here that the off-train excursions begin in earnest, a trip highlight for many travellers.

Inclusive tours pitched to various energy levels range from mountain biking or cultural hikes into nearby gorges (Standley Chasm or Simpsons Gap) to informative guided strolls to discover the regional fauna and flora of Alice Springs Desert Park, or a history jaunt around the city taking in the Old Telegraph Station and School of the Air, which broadcasts lessons by radio across the Outback. The immensity and beauty of the MacDonnell Ranges is best appreciated by air, but this will cost you extra; ditto a very slow tour on camelback. Southbound travellers also score dinner under the stars; those heading northbound perhaps a buffalo curry and a Vasse Felix cabernet sauvignon on the train.

Just north of Alice, the Ghan passes over the Tropic of Capricorn as night approaches, the enigmatic Tanami Desert no match for any number of lounge-car trivia sessions. On hazily retiring, leave the curtains open and the Milky Way will flood into your berth.

Into the Top End

By the time the train pulls into Katherine the next morning, you're travelling through plains of tropical savannah, perhaps nursing a throbbing headache after a little too much evening camaraderie. Morning excursions here revolve around Nitmiluk National Park and its stunning

gorge, where Indigenous guides share their culture on a rock-art river cruise. Other activities feature Indigenous music and stories, and a glimpse into the working life of an Outback cattle station. By afternoon it's almost a relief to get back on the train, reconnecting with its familiar rhythm as you tick off the rivers named by whitefellas for their wives, daughters or monarchs – Katherine, Edith, Adelaide, Elizabeth.

Just on sunset, the Ghan pulls into Darwin's remote Berrimah terminal, all too soon for those just getting the hang of this slowness caper. But then this is tropical Darwin – there's no need to hurry in this laidback Top End town that's closer to Jakarta than it is to Sydney or Melbourne, and where new adventures lie around every corner.

From top: fine dining at the Ghan's Queen Adelaide restaurant awaits Gold Class passengers; off-train trips include Alice Springs' Old Telegraph Station

Practicalities

Region: South Australia to Northern Territory
Start: Adelaide Parklands (Keswick) Terminal
Finish: Darwin Berrimah Terminal

Getting there and back: In typical Australian fashion, both termini are nowhere near civilisation. Parklands (Keswick) Terminal is 3km (2 miles) from central Adelaide, in an industrial wasteland with no nearby public transport. Darwin Terminal is 17km (10 miles) out in the boondocks, but city hotels offer transfers. Both Adelaide and Darwin have plentiful flights between other Australian cities, and several in Asia and the Pacific. The east-west transcontinental Indian Pacific train also passes through Parklands Terminal en route to Sydney or Perth.

When to go: Between March and November, the Ghan departs weekly from Adelaide (Sun) and Darwin (Wed); May to August is peak season. Between September and October, the wildflowers are blooming and tickets are cheaper. In the Top End, March is still wet and November humid.

What to take: Dust off that weighty tome you've always wanted to read. You'll need a warm jacket for the night-time desert stops; some folks like to dress up in bizarre fashions for the evening lounge-car trivia sessions.

Food and drink: All meals and beverages are included, with a choice of fine-dining restaurants, an informal lounge car and outdoor dining options. Platinum punters have an additional exclusive saloon.

Essential things to know: Book and get background info via the Ghan website (journeybeyondrail.com.au). Gold Class consists of single or twin (bunk) compartments that cleverly convert into daytime sitting rooms. Each cabin has an ensuite bathroom. Platinum Class has bigger everything, especially windows. Fellow travellers are likely retirees or older couples who have saved for this once-in-a-lifetime trip; children and young families are noticeably absent. Off-train excursions cater for all fitness levels, though inclement weather can affect availability.

An Antarctic awakening in the Ross Sea

Follow in the footsteps of explorers to Antarctica's Ross Sea, to marvel at dramatic snowbound landscapes and the awe-inspiring remnants of heroic expeditions.

Crack open your heart and embrace your resilience amid the staggering spectacles – calving icebergs, magnificent mountain ranges, wind-sculpted valleys – of a place that's like nowhere else on the planet.

The explorers of the Heroic Age gained a foothold here for exploration of Antarctica's interior. Study their history and learn about the crucial science still being accomplished in the region to understand where we've come from and where the Earth might be headed.

Test your mettle as you cross the great Southern Ocean and tramp across scree-covered Antarctic shores, home to Weddell seals and Adélie and emperor penguins.

No place on Earth compares to this vast white wilderness of elemental forces: snow, ice, water, rock. Antarctica is simply stunning. And as you journey through the storied but seldom-travelled vastness of the Ross Sea, you can experience Antarctica – and, indeed, yourself – to another degree. Cold and wind are magnitudes greater, tabular icebergs more abundant, wildlife scarcer.

The Ross Sea area enjoys some of Antarctica's most spectacular terrain, and as the Heroic Age explorers' gateway to the South Pole, the region has the continent's richest historic heritage: the huts built for the British Antarctic Expeditions led by Robert F Scott, Ernest Shackleton and Carsten Borchgrevink. This route also calls in at busy US and New Zealand scientific research stations and several of the wildlife-rich peri-Antarctic island groups.

Invercargill
McMurdo Sound
Ross Sea
Antarctica
Mt Erebus
Cape Royds 2
Shackleton's *Nimrod* Hut
5 Dry Valleys
Ross Island
Cape Evans
3 Scott's *Terra Nova* Hut
Transantarctic Mountains
Scott's *Discovery* Hut
4 Scott Base
1
Ross Ice Shelf
McMurdo Station

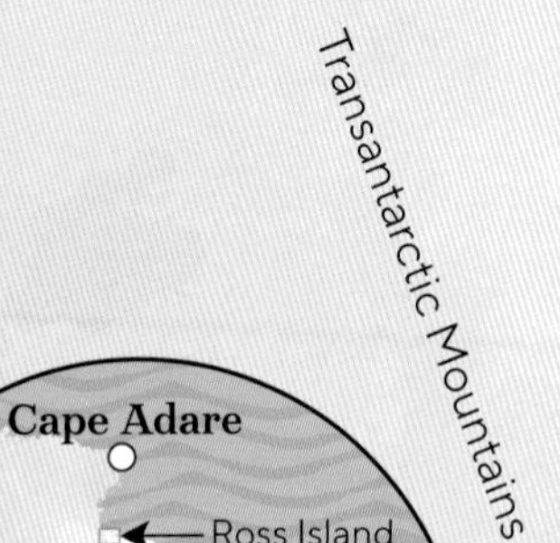

The trip

Distance: Around 3700km (2300 miles)
Mode of transport: Ship
Difficulty: Easy to moderate

1 Ross Ice Shelf Land atop this magnificent ice shelf by helicopter, or look for pods of orcas along the ice edge from the ship

2 Cape Royds Shuttle between Shackleton's *Nimrod* Hut, packed with fascinating finds, and a thrumming colony of Adélie penguins

3 Scott's *Terra Nova* Hut Feel the presence of the polar party who never returned, and marvel at their bravery

4 McMurdo Station Visit Antarctica's largest research station, huddled in the lee of enormous Mt Erebus volcano

5 Dry Valleys Puzzle over fantastical wind sculptures and soul-soaring vistas in one of the world's most unusual terrains

Clockwise, from left: Adélie penguins; Shackleton's *Nimrod* Hut, Cape Royds; the Scott Memorial at Cape Evans; a Weddell seal

Alexis Averbuck, Antarctica traveller

Antarctica possesses an unnameable quality which took me beyond writing and academia to become a painter. Call it inspiration, the need to express in a non-verbal way... or simply the feeling of being a small speck in a vast, harshly beautiful land. A land where towers of striated ice float among geometric pancake ice; literally untouched mountains rear from marine mist; and wildlife lives, year in and year out, to its own rhythms, quite apart from our human concerns.

Starting from Invercargill (though trips also depart from other ports in New Zealand or Australia), you'll spend a couple of days rolling across the Southern Ocean, getting your sea legs and watching the abundant bird life. Depending on your route (and the unpredictable weather), stop-offs on the way to the Ross Sea might include Macquarie Island, Campbell Island or the Auckland Islands, all famous for their breeding seabirds and windswept aspects.

Prolific penguins & historic huts

The northernmost headland at the entrance to the Ross Sea, Cape Adare was named for Britain's Viscount Adare, one-time MP for Glamorganshire, by his friend James Clark Ross, who discovered the cape in 1841. Sprawled across the shore is Antarctica's largest Adélie penguin rookery – home to some 250,000 nesting pairs – as well as two sets of historic huts.

If the fierce wind allows, your boat might try for a landing. In February 1899, the ten-man party of Carsten Borchgrevink's *Southern Cross* expedition took two weeks to erect a pair of prefabricated structures here, the remains of which are just back from Ridley Beach. These huts are the oldest buildings in Antarctica. What little remains of another hut – built by Victor Campbell, George Murray Levick and Raymond Priestley, the 'Northern party' members of Scott's 1910–13 *Terra Nova* expedition – lies east of Borchgrevink's huts. As you're buffeted by the powerful winds, it's no stretch to appreciate the fortitude of those who survived here.

© SOTT / HERITAGE EXPEDITIONS

Below: Campbell Island, New Zealand
Right: Ross Ice Shelf

"As you're buffeted by the powerful winds, it's no stretch to appreciate the fortitude of those who survived here"

Majestic Ross Ice Shelf

As the ship turns starboard and heads south into the Ross Sea, you'll pass Cape Washington, home to one of the largest emperor penguin colonies in the world. You're also afforded a stunning view of the floating, France-sized Ross Ice Shelf. This vast white slab has inspired many awestruck responses. In 1841, the blacksmith aboard the Ross expedition ship *Erebus* was moved to write a couplet: 'Awful and sublime, magnificent and rare; no other Earthly object with the Barrier can compare.'

In 1847, six years after discovering the shelf, Ross himself wrote: 'this extraordinary barrier of ice, of probably more than a thousand feet in thickness, crushes the undulations of the waves, and disregards their violence: it is a mighty and wonderful object, far beyond anything we could have thought or conceived.'

Feel the magnificence and listen for the sounds of icebergs calving off its glittering face. Later in the journey, some helicopter-equipped ships offer flights up onto the surface of this grand ice shelf.

On to Ross Island

The next stop for most ships is Ross Island, where 3794m-high (12,448ft) Mt Erebus volcano lords it over everything, its bubbling magma lake creating a trail of steam into the sky. If you're fortunate – and the pack ice permits – you'll hit a historic-hut trifecta here, landing at Scott's 1902 *Discovery* expedition structure at Hut Point, Shackleton's 1908 *Nimrod* Hut at Cape Royds and Scott's 1911 *Terra Nova* Hut at Cape Evans, to which Scott and his men would have returned had they not perished on the way back from the South Pole.

Journey to the Pole

If your Antarctica wanderlust isn't satisfied by a voyage to its icy fringes, you might consider an even more extreme challenge: the South Pole. First reached just over 100 years ago by Norwegian explorer Roald Amundsen, the South Pole still embodies myth, hardship and glory; today, it's topped by a high-tech research station surrounded by astrophysical observation equipment. High-end tours do provide access to the Pole – at a cost. To reach it, you'll need to cross – by skiing and ice-climbing or by air – the magnificent glaciers, crevasse fields and barren snowscapes of the Polar Plateau. This vast ice cap, thousands of metres thick, covers entire mountain ranges and subglacial lakes, which are now being explored for the first time. To the visitor, a photo op with the flapping flags and globe-topped rod of the Pole is a rare opportunity and, perhaps, the accomplishment of a lifetime – standing here, you are as far south as it's possible to go.

The huts sharply convey the hardships endured by the early explorers. The most abundantly stocked, Shackleton's *Nimrod* Hut calls for close examination. Ask your guide to point out Shackleton's signature (which may or may not be authentic) in his tiny bunk room. (It's upside down on a packing crate marked 'Not for Voyage', which he had made into a headboard.) A freeze-dried buckwheat pancake still lies in a cast-iron skillet on top of a large stove at the back of the hut, beside a kettle and a cooking pot. Coloured-glass medicine bottles line several shelves, and one of the other surviving bunks still has its fur sleeping bag laid out on top. Tins of food with unappetizing names – Irish brawn (head cheese), Aberdeen marrow fat and lunch tongue – lie on the floor. The dining table, which was lifted from the floor every night to create extra space, is gone, perhaps burned by a later party that ran out of fuel.

At Cape Evans, Scott's hut from the *Terra Nova* expedition is steeped in an incredible feeling of history. Here, dog skeletons bleach on the sand in the Antarctic sun, perhaps evoking thoughts of Scott's death march from the Pole. Stand at the head of the wardroom table to recall the famous photo of what would be Scott's final birthday, his men gathered around a huge meal with their banners hanging behind; the tremendous risks taken by these early explorers feels very immediate here.

Modern Antarctic life

Most cruises visit one of Ross Island's two current human communities. The sprawling US McMurdo Station is Antarctica's largest research station, and has the feel of a bustling frontier town (albeit one with helicopters and snowmobiles). Backed by looming Mt Erebus, it's home to more than 1100 people during summer, and hosts a multinational assortment of many more researchers in transit to field camps and the Pole. Established in 1956, McMurdo takes its name from McMurdo Sound, which James Clark Ross named in 1841 after Lieutenant Archibald McMurdo of the ship *Terror*. The settlement was large right from the start: 93 men wintered here in the first year. If you're lucky, you can tour the labs: three large aquariums stocked with Ross Sea species allow scientists study the region's marine life, including the antifreeze-like properties in the blood of Antarctic cod.

Compared to McMurdo, New Zealand's Scott Base, just 3km (2 miles) away by gravel road, looks positively pastoral. An orderly collection of lime-green buildings, named for Robert Scott, it was established in 1957 by Edmund Hillary as part of Vivian Fuchs' Commonwealth Trans-Antarctic

From top: Taylor Valley in the Dry Valleys; an aerial view of McMurdo Station, Ross Island

Expedition (TAE); three original structures are historic monuments maintained by the NZ Antarctic Heritage Trust.

The exceptional Dry Valleys

Some cruises with helicopter support offer a quick excursion to the Dry Valleys, with their ancient ventifacts (wind-sculpted rocks), and bizarre lakes and ponds that sustain life in this harshest of environments. Valley air is so lacking in moisture that there is no snow or ice here, and scientists believe that these Antarctic 'oases' are the nearest equivalent on Earth to the terrain of Mars. Visiting them is a unique experience.

In Taylor Valley, Lake Hoare is permanently covered by 5.5m-thick (18ft) ice, but dense mats of blue-green algae carpet its bottom. Wright Valley's Lake Vanda has been intensively investigated for years for its microbial ecosystems; Antarctica's longest watercourse, the 30km (19-mile) meltwater Onyx River, flows into the lake from the glacier at the end of the valley. It's one of very few rivers in the world to flow inland from the coast.

Practicalities

Region: Antarctica
Start: Invercargill, New Zealand
Finish: Ross Island, Antarctica

When to go: The Antarctic tour season is roughly November to March, with each month offering its own highlights. Cruises towards the end of the season may be less crowded, but wildlife-spotting might be more limited as animals head out to sea after breeding season. December and January are optimum, with up to 20 hours of sunlight each day: penguins hatch eggs and feed chicks, seabirds soar.

What to take: As the saying goes: there's no bad weather, only inappropriate clothing. So you'll need to pack carefully. As a general rule, take clothes you can layer, with an emphasis on wool, flannel or polar fleece (not cotton). You'll also need a windproof and waterproof jacket and trousers, plus high waterproof boots with high-traction soles. Also consider specialty photographic equipment.

Where to stay, eat and drink: Visiting Antarctica by ship as part of a group tour has the advantage of combining transportation, meals and accommodation, and means that no visitor infrastructure has to be built ashore in Antarctica's delicate environment.

Tours: It's essential to book with an ecologically responsible tour operator in order to minimise pollution on sea and land. They should be a member of IAATO (www.iaato.org), the International Association of Antarctica Tour Operators. Reputable operators include Heritage Expeditions (heritage-expeditions.com), Ponant (en.ponant.com) and Oceanwide (oceanwide-expeditions.com).

Essential things to know: To get the most from your trip, do some advance reading to gen up on Antarctic explorers and adventurers. The Australian Antarctic Division (antarctica.gov.au) is a useful general resource; the Antarctic Heritage Trust (nzaht.org) has information on historic-hut conservation; McMurdo Station (nsf.gov) has background on South Pole research.

Lurujarri Dreaming – walking Country with the Goolarabooloo

Surrender to the rhythm of an ancient songline on the Lurujarri Heritage Trail, a nine-day cultural immersion into the wild and wonderful Goolarabooloo Country of Western Australia's Dampier Peninsula.

After walking this trail, many people return home with a greater appreciation of environmental and Indigenous issues, and a wish to be closer to nature. Some return again and again. Country is calling.

Indigenous Australians sharing their culture is a wonderfully generous gesture; most hikers gain a greater understanding of, and appreciation for, the Indigenous concept of Country: the landscape and everything in it.

After nine days of walking in sunshine, bathing in saltwater and eating a fresh, wholesome diet – plus no alcohol (a Lurujarri condition) and zero screen time – you'll likely be looking trim and terrific.

The message sticks are clapping. You're wrapped in a sleeping bag, high on a sand dune, the Indian Ocean lapping gently somewhere below. Stars burn brightly above, and birds are yet to rise – yet something is nagging. The sticks...

One of the lower stars is moving. A red smudge of dawn appears, far to the east. You smell of wood smoke, there's a coral reef in your hair and the sandfly bites on your legs itch unbearably. The moving star is a head-torch, the clapping sticks signal breakfast, you're on the Lurujarri Heritage Trail in Goolarabooloo Country – and it's time to get up.

The roots of the Lurujarri

Several times each Barrgana (dry season), a motley mob of around 60 strangers head off from

© STEVE WATERS

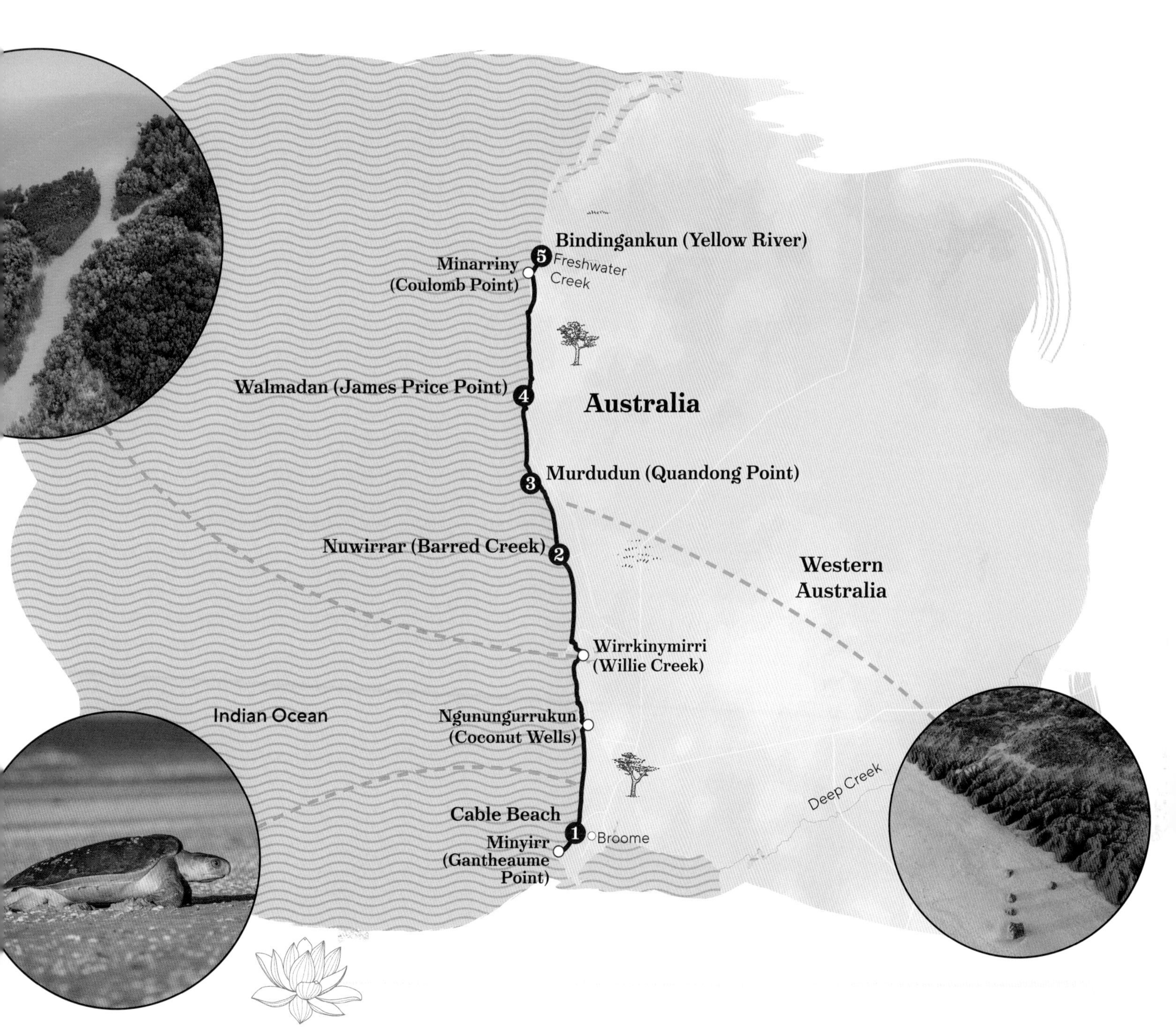

Clockwise, from left: beachwalking on the Lurujarri Trail; Wirrkinymirri (Willie Creek); pindan cliffs along the trail; a flatback turtle

The trip

Distance: 82km (51 miles)
Mode of transport: On foot
Difficulty: Moderate

1 Cable Beach Take in Broome's iconic sunset seascape on the first stretch of the trail

2 Nuwirrar (Barred Creek) Learn how to make a spear and hunt for crabs in the mangroves at this sheltered trail camp

3 Murdudun (Quandong Point) Camp under the stars, discover dinosaur prints, learn about bush tucker and scout for migrating whales

4 Walmadan (James Price Point) Watch the pindan cliffs glow crimson in the setting sun, and reflect on the headland's powerful history

5 Bindingankun (Yellow River) The final trail camp: forage for bush tucker, then witness a *corroboree* (ritual dance ceremony)

Steve Waters, Lurujarri hiker

When I returned to the city after life on the Lurujarri, I felt as though I'd stepped out of a shining, full-colour world back into a mundane black-and-white existence where everything was drabber, dimmer, more muted. On the Lurujarri I had felt whole, complete – at one with nature and the land, accepted by Country and the other beautiful souls who'd shared the journey. Such an addictive feeling – I've spent almost every Barrgana or Wilburu in the Kimberley ever since.

Minyirr (Gantheaume Point, at the southern end of Broome's iconic Cable Beach) in late afternoon, wandering en masse along these fabled sands – past the array of sunset-seeking 4WDs, their owners sipping sundowners; past the camel trains and their photogenic silhouettes; past the sinking sun and on into the starry twilight, bound for Ngunungurrukun (Coconut Wells), the first camping spot on the Lurujarri Heritage Trail. This nine-day guided walk over 82km (51 miles) stretches north into turquoise infinity, threading along the dazzling deserted beaches, crimson-red cliffs and paperbark swamps of the remote Dampier Peninsula.

Indigenous culture runs deep across the vast Kimberley region of northwestern Australia. When a Nyikina man, Paddy Roe, came to the Dampier Peninsula in the 1930s, the fading Jabirr Jabirr elders, bereft of suitable descendants, passed on custodianship of their Country (the Lore: songs, stories, customs and responsibilities, inherited from the earlier Ngumbarl and Djugun people), to the young Roe. Having founded his Goolarabooloo Community at Millibinyari on Ngunungurrukun, Roe instigated the Lurujarri Heritage Trail in 1987 as a means for his own people to reconnect with Country and their traditional heritage, and as a way of cultivating respect for Country among non-Indigenous people.

Walking a songline

The Lurujarri Trail is a section of an ancient songline (essentially a storyline, both oral history and map), which stretches from the deserts south of Broome to the Dampier Peninsula's northern tip. From Minyirr at Cable Beach, easy walking follows beaches, rocks and sand dunes north along the Indian Ocean coast to Bindingankun (Yellow River), from where the route loops back to the final pick-up point at Minarriny (Coulomb Point). Luggage is transported between campsites by truck, with walkers carrying only a daypack for water, sunscreen, swimming gear and a camera.

© STEVE WATERS

Below: hiking the Lurujarri Trail
Right: Dampier Peninsula coastline

Setting up camp after dark in a strange place, with unfamiliar equipment, can be a steep learning curve; and pre-dawn wakeups are standard in a place where the natural laws of sun and tide govern all movement. Each morning, trail walkers stagger down to a campsite kitchen area behind the dunes, where a large pot of billy tea is already boiling. Mumbled greetings are exchanged and mugs are filled. The hot, bitter black-tea tannins scald the tongue, but they have the desired effect.

On the Lurujarri Trail, you'll eat meals cooked on fires and wash in the sea. It's a chance to pack away electronic devices, forget about frivolous luxuries and unburden your mind of pressing commitments: this is life without rigorous schedules, stripped back to the bare essentials of shelter, water, food and movement across a landscape unchanged in millennia. Some find this lack of structure confronting; others discover previously hidden qualities, grasping these unaccustomed freedoms with newfound relish.

Conversations on Country

Just as a songline is more than a map, the Lurujarri Trail is more than a seaside stroll. Each ochre dig, ancient campsite, meeting ground or landmark is a line from the song, and it's these Dreaming glimpses, proudly related by Paddy Roe's descendants, that make for a rich, privileged and extremely satisfying cultural experience. Storytelling is an essential part of the trail and after each session, walkers' understanding of Country sharpens: another layer revealed, another detail or nuance exposed.

Six seasons

Indigenous groups of northern Australia, including the Goolarabooloo, commonly recognise these six seasons:

Mankala (December to March) Rainy season: hot and humid, with thunderstorms and plant growth, fruiting bush-tucker trees and flying foxes.

Marul (April) Hot but rain easing. Plants that flowered in Mankala are fruiting, and lizards fattening. Ample groundwater increases bird and animal movement.

Wirralburu (May) Temperatures cool, winds shift southeast and bloodwood trees flower. Waterholes shrink and lizards prepare to hibernate.

Barrgana (June to August) Winter and dry season: cold nights, strong southeast winds, sea fog, the Seven Sisters (Pleiades) in the sky and whales migrating north.

Wilburu (September) Transition: the heat returns, and low spring tides bring in abundant seafood along the reefs.

Larja (October to November) Build-up to the wet: hot and humid as the wind swings northwest. Turtle-mating time.

And when walking in a large group, stimulating conversation is rarely lacking: fellow hikers might be students, documentary-makers, a family with young kids or the retired parents of a camp volunteer. You can choose to engage, or set sail on your own internal journey through the shifting sands against a canvas of sun, salt and never-ending seascapes.

After you break the first night's camp, the ensuing long, hot 20km (12-mile) day includes a launch across Wirrkinymirri (Willie Creek), home to a large saltie (estuarine crocodile). Lunch is below shade-giving paperbark trees, before the beach-walking (*lurujarri* means 'coastal dunes') ends with a nervous low-tide wade across Nuwirrar (Barred Creek) to a lovely sheltered campsite and a much-needed sunset swim.

It's not just a walk

Rest days at the camps allow time for learning: making spears, jewellery and boomerangs, or fishing, crab-hunting and bush-tucker collection. Activities are laidback and unstructured: if you see a group heading off into the bush, don't wait to be invited – just join in. Nuwirrar provides the favoured timber for a morning spear-making class, which then morphs into an afternoon crab hunt, which in turn leads to an evening chilli-crab cookery lesson. Near Quandong Point, Murdudun camp, dominated by a large gubinge tree, is rich in bush tucker and the big, black 'bush pearl' seed-pods favoured for necklaces. Dinosaur footprints are found in the coastal rocks here, and whales frequently breach offshore. And from Bindingankun (Yellow River), the final camp, a day trip explores a freshwater creek in search of yams and bush honey.

Back on the trail, the route winds onward to the striking red pindan cliffs and fractal sand patterns at Walmadan, glowing crimson as the sun sets. Named for the old Jabirr Jabirr warrior who schooled Paddy Roe in the Lore, it's also known as James Price Point and, in recent times, was the scene of one of Western Australia's most bitter environmental battles. In 2013, plans to build a gas plant and port here were finally quashed by protestors, and remnants of their camp remain behind the dunes; the environmental threat has receded, but this story, too, has been woven into the songline.

Life on the trail

Continuing north, the days pass easily in a primeval cadence of pre-dawn wakeups, low-tide beach-walking, nature spotting and culture stops. Shady lunches are followed by afternoons hiking to stunning dunal campsites, sweat washed away in seas burnished by relentless, searing sunsets. Communal

dinners are eagerly anticipated, the wholesome food freshly caught and plentiful. Afterwards, billy tea is taken by the fire, where someone pulls out a ukulele or the professor next to you explains orbital mechanics with two pieces of fruit and a mussel shell. Wander wearily back to your sleeping bag, searching the sky for shooting stars... until the sticks start calling once again.

All too soon, it's almost over as you bed down after the final night's *corroboree* ceremony, perhaps reflecting on the amazing things you've seen – the incredible flocks of cockatoos, brolgas and pelicans, the solitary sea eagles, jabiru storks, sea turtles and endless whales. All the experiences, the stories, the people; the sun setting on the blood-red cliffs or boiling off west towards Africa; the smells, the smoke, the salt, the bitter scalding tea. You've been welcomed to Country. And now Country knows you.

From top: watch for migrating humpback whales from Murdudun (Quandong Point); tool-making sessions at a Lurujarri Trail camp

Practicalities

Region: Dampier Peninsula, Western Australia
Start: Minyirr (Gantheaume Point)
Finish: Minarriny (Coulomb Point)

Getting there and back: Broome is a long way from anywhere, but has flight connections to Perth, Darwin, Sydney and Melbourne. Unless you love driving thousands of empty kilometres, leave the car at home.

When to go: Lurujarri Trail walks usually set off in July during Barrgana, the dry, winter season of sunny days and cool nights.

What to take: You'll need swimming gear, a sarong or towel, sunglasses, a brimmed hat, lightweight loose clothes, a warm sweater and walking shoes. Bring a daypack to carry a water bottle, sunscreen, insect repellent and swimming gear. For camps, you'll need a torch, a light sleeping bag, a swag (bedroll) and an insect-proof sleeping shelter (a mesh tent or mosquito dome), plus a bag in which to pack it all for transportation between the camps by truck. In Broome, Kimberley Camping & Outback Supplies or Yuen Wing General Store sell camping gear. Get cheap clothes and sarongs from the Courthouse Markets or Chinatown alleyways.

Essential things to know: All Lurujarri Heritage Trail walks are fully guided, and only a very limited number run each year; book well in advance via the Goolarabooloo website (goolarabooloo.org.au), which also has useful info on the trail and the area. All food and transport is inclusive, but you bring your own sleeping gear and shelter. Most dietary requirements can be catered for. Luggage is transported to each night's campsite so you only carry your daypack on the trail.

Rafting the Franklin River

Once slated to be dammed, Tasmania's Franklin River today provides rafters with a wild ride through nature at its most raw and beautiful.

Learn the intricacies of rafting, from paddle technique to reading a river's flow. Lengthy, heavy portages through the Great Ravine can do wonders for your strength.

A tiny cave on the river's lower banks was critical to this region being granted World Heritage status in 1982: with more than 300,000 discovered artefacts, Kutikina Cave is one of the most prolific archeological sites ever found in Australia.

On a map, you're never more than 150km (95 miles) from Tasmanian capital Hobart when on the Franklin, and yet there's a solitude and remoteness that no map can express. It's a place to go deep within yourself, discovering your own resilience.

Inside the Great Ravine, midway along the Franklin River, the world echoes and grumbles. A chain of furious rapids pours through a sheer gorge, its walls up to 500m (1640ft) high, and whitewater turns to white noise as the thunderous roar of the river bounces between the cliffs. It's one of the most intimidating places to be in a raft along any river in the world. Little more than 5km (3 miles) in length, the Great Ravine can typically take more than a day to traverse, requiring punishing portages and infinite patience. If rain has fallen and the river rises, it's not unusual to be stuck here for days.

There are so many outside forces at play along this river, but it's also a place for introversion. Set out on the Franklin and it's at least a week before you emerge at the other end. You're out of communication the entire time, and such is the wild nature of the river that there's not a single home

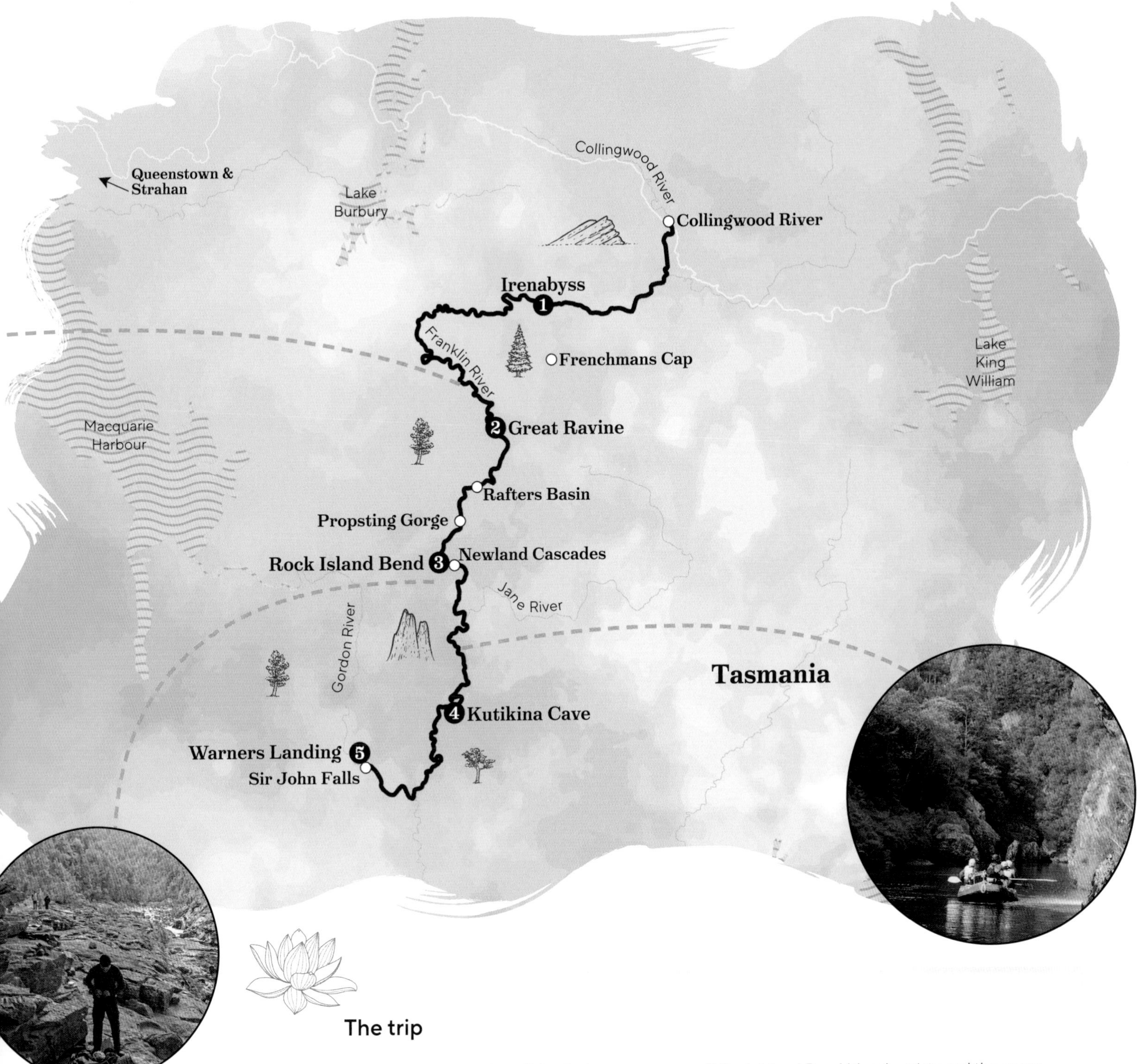

The trip

Distance: Around 100km (60 miles)
Mode of transport: Raft
Difficulty: Difficult

1 Irenabyss Moments of peace in a deep, still gorge ahead of the major rapids

2 Great Ravine The Franklin's wild heart, strung with dauntingly named rapids such as the Churn, Thunderush and the Cauldron

3 Rock Island Bend Moody, misty and the scene of Australia's most famous nature photograph

4 Kutikina Cave Small riverside cavern that tells a long story of Indigenous palawa (Tasmanian Aboriginal) history

5 Warners Landing Site of the Franklin River blockade in 1982, on the Gordon River near the trip's end

Clockwise, from left: the wild and wonderful Franklin River; a Tasmanian Devil; rafting the river; a waterside camp near Newland Cascades

Geoff Law OAM, author and activist

I was an experienced bushwalker, but when I rafted the Franklin River in 1981 I got way more than I bargained for. Eleven out of 14 days of rain, high water and frequent capsizes. But the vulnerable beauty of the gorges and forests made an indelible impression. My life-changing 'immersion' continued when I joined the hectic efforts to save the river. It propelled me into a life of activism to protect wilderness and ancient life forms.

or patch of cultivated land in its catchment. When you're handed that much time in a setting this remote and removed, there are far more reflections than just those in the river's dark waters.

It's no great stretch to suggest that the Franklin River changed Australia. In the early 1980s, plans to dam its waters were met with fierce resistance. Around 1400 people were arrested during a four-month blockade, and a new federal government was elected, promising to stop the dams. In the four decades since that environmental victory, the Franklin has flowed as deeply through the Australian psyche as it does through the remote gorges of the Tasmanian Wilderness World Heritage Area.

Picking up your paddle

Rafting trips begin beneath a remote highway bridge over the Collingwood River, a tributary that flows into the Franklin less than 5km (3 miles) downstream. River levels very much dictate the nature of the journey. When the river is high, it's a fast float into the wilderness; when it's low, it's a wet bushwalk, with rafters in and out of their boats, dragging them over rocks in a taxing start to a river run.

There are plenty of chunky rapids ahead, but things start relatively peacefully. Early in its journey, the river slows almost to a standstill inside the Irenabyss. Named by future Australian Greens leader Bob Brown, the word Irenabyss translates as Chasm of Peace in Greek, and it's hard to feel more embraced by nature than you do when drifting through this small gorge.

Below: Corkscrew rapids, Great Ravine **Right:** contemplating Rock River Bend

> "In a setting this remote and removed, there are far more reflections than just those in the river's dark waters"

Most rafting groups camp just beyond the Irenabyss' mouth, where a hiking track heads steeply up to the summit of Frenchmans Cap, one of Tasmania's most prominent and imposing mountains. The river, however, continues its journey, wriggling between mountains and heading for the Irenabyss' far rowdier companion, the Great Ravine.

Into the Great Ravine

The first European to sight the Great Ravine was so intimidated by the gorge that he described it as a 'hideous defile', but it's truly a place of awe – a Tolkienesque piece of Middle Earth in Tasmania's southwest. Cliffs rise hundreds of metres overhead, and the river brawls and battles its way through the narrow cut in between. The gorge is entered through the Churn, a rapid that sets the tone of things to come, with rafters portaging their gear and boats along narrow cliff ledges above the river.

You might camp a night inside the Great Ravine, just beyond the Churn along Serenity Reach. Here, as the name suggests, the river momentarily stills before rushing away through Coruscades rapids. As with every other night along the river, camp consists of a sleeping mat and sleeping bag on the ground, with a tarp strung overhead if there's a chance of rain. It's a return to a primal existence, a step well outside many comfort zones, but with the pretty possibility of glow-worms overhead as you sleep.

There are four more rapids requiring strenuous portages through the Great Ravine. Rafts are packed, unpacked, carried, tossed off boulders

Inspired by the Franklin

Evidence from caves along the Franklin River suggests that Aboriginal people lived here from 20,000 to 13,000 years ago, but it wasn't until 1841 that the first European settler, surveyor James Calder, sighted the river. In 1958, a group of four men made the inaugural descent of the Franklin in a canoe – it was their third attempt in six years. Eighteen years later, the first raft, holding Paul Smith and Bob Brown, followed them down the river. Brown, a young Tasmanian GP, gave names to many features along the Franklin, and returned from the trip transformed. A meeting at his home that year led to the formation of the Wilderness Society, one of Australia's leading environmental organisations; he would later lead the Franklin blockade that succeeded in stopping the river from being dammed. In 2005, he became leader of the Australian Greens political party, and remains one of Australia's highest-profile conservationists.

and unpacked and packed once again. Rinse and repeat for an entire day before the river finally slows and settles into Rafters Basin, where sleep typically comes easily and early after the rigours of the ravine. If you have no appreciation for the power of nature after this day, you're not paying attention.

The reprieve at Rafters Basin is momentary, with the river soon narrowing again into Propsting Gorge, arguably the most fun stretch for rafters. Rapids continue to churn, but they're broken up by mellow pools and the whole gorge can be run rather than portaged, with rafts tipping over drops of up to 2m (6.5ft) as they skid down the river.

Stirring scenes and a powerful past

Propsting Gorge ends with a scene so beautiful it helped sway a nation. At Rock Island Bend, the tea-coloured river curls past a swathe of rainforest and a protrusion of rock that rises from the river like the prow of a ship. A misty image of the bend, taken by legendary Tasmanian wilderness photographer Peter Dombrovskis, was used in the 1983 fight to save the Franklin from being dammed, appearing as a full-page advertisement in national newspapers beside the slogan: 'Could you vote for a party that will destroy this?' It was one of the pivotal moments in the campaign, swinging sentiment in favour of the free-flowing river.

Rock Island Bend also marks the start of the Franklin's longest raftable rapid, the 400m (1300ft) Newland Cascades. Under cliffs just beyond the rapid, rafters set up camp, rolling out mats beneath overhangs in the cliffs. If water levels are low, it's possible to wander back upstream for another audience with the natural royalty of Rock Island Bend. It's hard to sit here and not feel thankful for its preservation.

As the river flattens and widens past Newland Cascades, the current does less work and your arms and body do more, paddling hard on the slowing waterway as it flows towards the Gordon River. There are just three rapids ahead – none of them very large – but the river's tales are also more compelling at this end, as the rafts float past Kutikina Cave, first occupied by palawa (Aboriginal Tasmanians) 20,000 years ago; and Warners Landing, where a bulldozer was landed on the bank of the Gordon River in 1983 to begin work on the failed dam.

The Franklin River's greatest beauties are now all behind, but it's doubtful that you'll emerge without being changed in some small way by this waterway that both divided and united a country. It may flow out of sight to the world, but it continues to flow through so many people.

From top: sunset over the Franklin River; Pig Trough Waterfall at Rock Island Bend

Practicalities

Region: Tasmania, Australia
Start: Collingwood River
Finish: Sir John Falls

Getting there and back: The put-in point on the Collingwood River is 38km (24 miles) west of Derwent Bridge, along the Lyell Hwy. It's a 3hr drive from Hobart and a 2hr 45min drive from Launceston. Rafting trips typically end at Sir John Falls on the Gordon River; here groups are met by the *Stormbreaker* yacht, which sails them downriver and across Macquarie Harbour to Strahan.

When to go: Rafting trips run from October to April, through and around the edges of summer. Water levels fluctuate massively, depending on rainfall: keep an eye on levels through the Hydro Tasmania website (hydro.com.au) – follow the 'Water' and 'Water flow and levels' links. Rain and even snow can come at any time, and the valley can get seriously cold – come prepared. All rafting tour companies supply a full packing list.

Where to stay: There's no accommodation – even structured campsites – anywhere along the Franklin: you'll camp on the banks of the waterway and in riverside caves and overhangs.

Tours: Only the hardiest and most experienced of paddlers attempt the Franklin River independently, meaning most rafters come on organised tours, which mostly run over about nine days. There are only a few rafting operators on the river: Tasmanian Expeditions (tasmanianexpeditions.com.au), Franklin River Rafting (franklinriverrafting.com), Water by Nature Tasmania (franklinriver.com) and Wild Journeys (wildjourneys.com.au).

Get a glimpse: Want to see this storied river, but not raft it? Stop at the Franklin River Nature Trail, 13km (8 miles) before the Collingwood River bridge, for a short walk down to its banks; or from the Lyell Hwy set out a few hundred metres along the multi-day Frenchmans Cap Trail to reach a suspension bridge over the river.

Africa & the Middle East

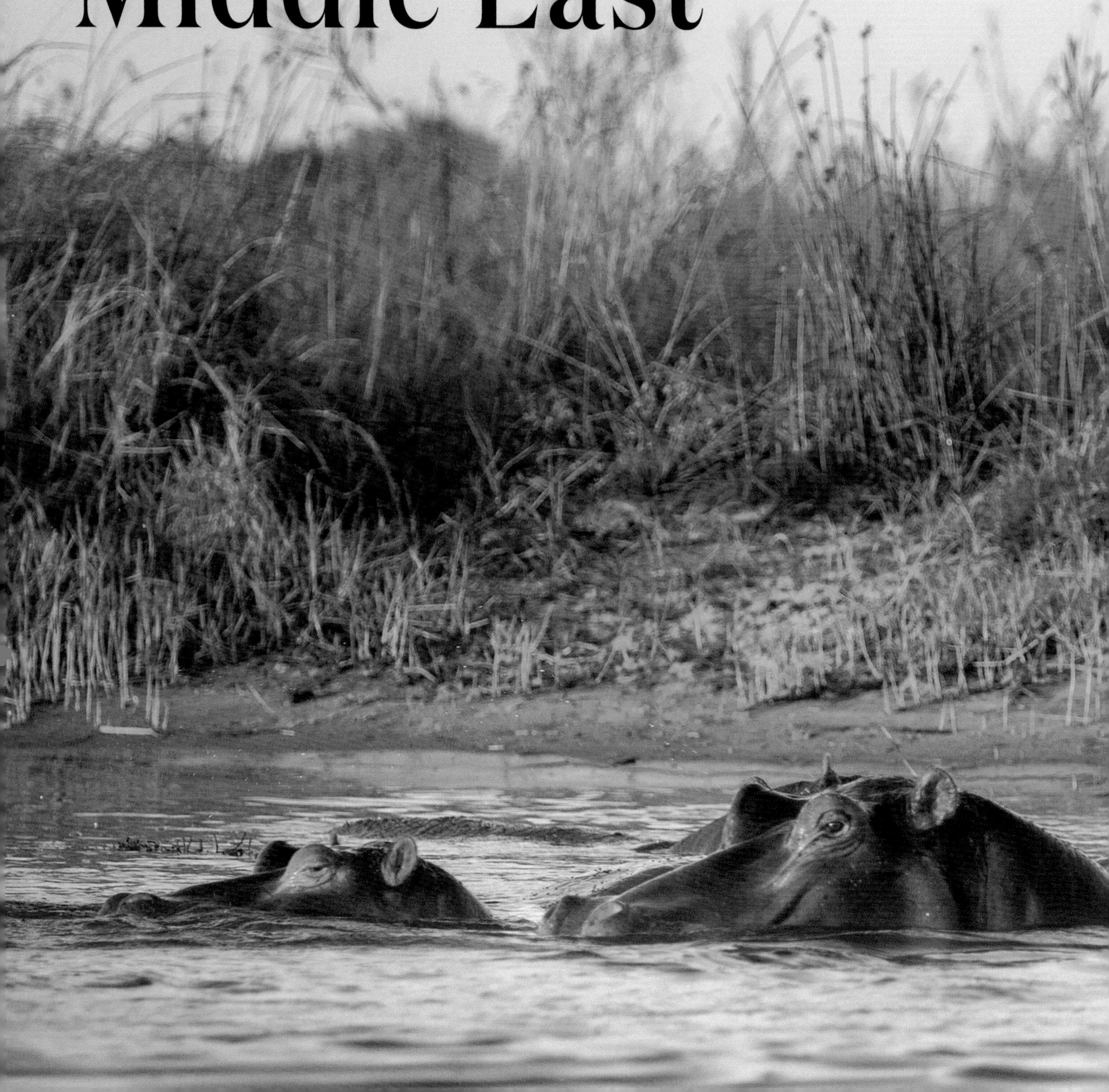

Hippos in the
Zambezi River

An overland odyssey through Africa

Taking you deep into the wilds of Eastern and Southern Africa, this overland adventure promises one adrenaline rush after another, sandwiched between moments of contemplation.

There's freedom in embracing your raw self on this adventure, because clean clothes never stay that way for long, and no one cares if you wear make-up or not.

Ubuntu, a Southern African philosophy that speaks to the oneness of humanity, is a term you may be introduced to on this trip – a journey which will likely offer many a perfect opportunity to reflect on how it can be applied to your own life.

While there isn't a huge amount of physical activity on this adventure, you may surprise yourself by how quickly and efficiently you can erect and dismantle a tent by the end of it.

Mesmerised by the immense African elephant casually crossing the dusty road in front of your safari truck, it may take you a moment to realise there's also a black rhino quietly grazing in the grasslands beyond. Never mind the family of giraffes feasting in a cluster of acacia trees nearby, their hilariously long blue tongues expertly navigating around the long thorns to reach succulent leaves.

Such is a typical moment on an overland adventure in Sub-Saharan Africa, home to the continent's most iconic game parks. It's a moment you might experience in just about any one of the region's national parks, in fact. But perhaps only via an extended trip across Eastern and Southern Africa can you begin to fully appreciate how wild and wonderful, fragile and complex this corner of the world really is.

© QUALITY MASTER / SHUTTERSTOCK

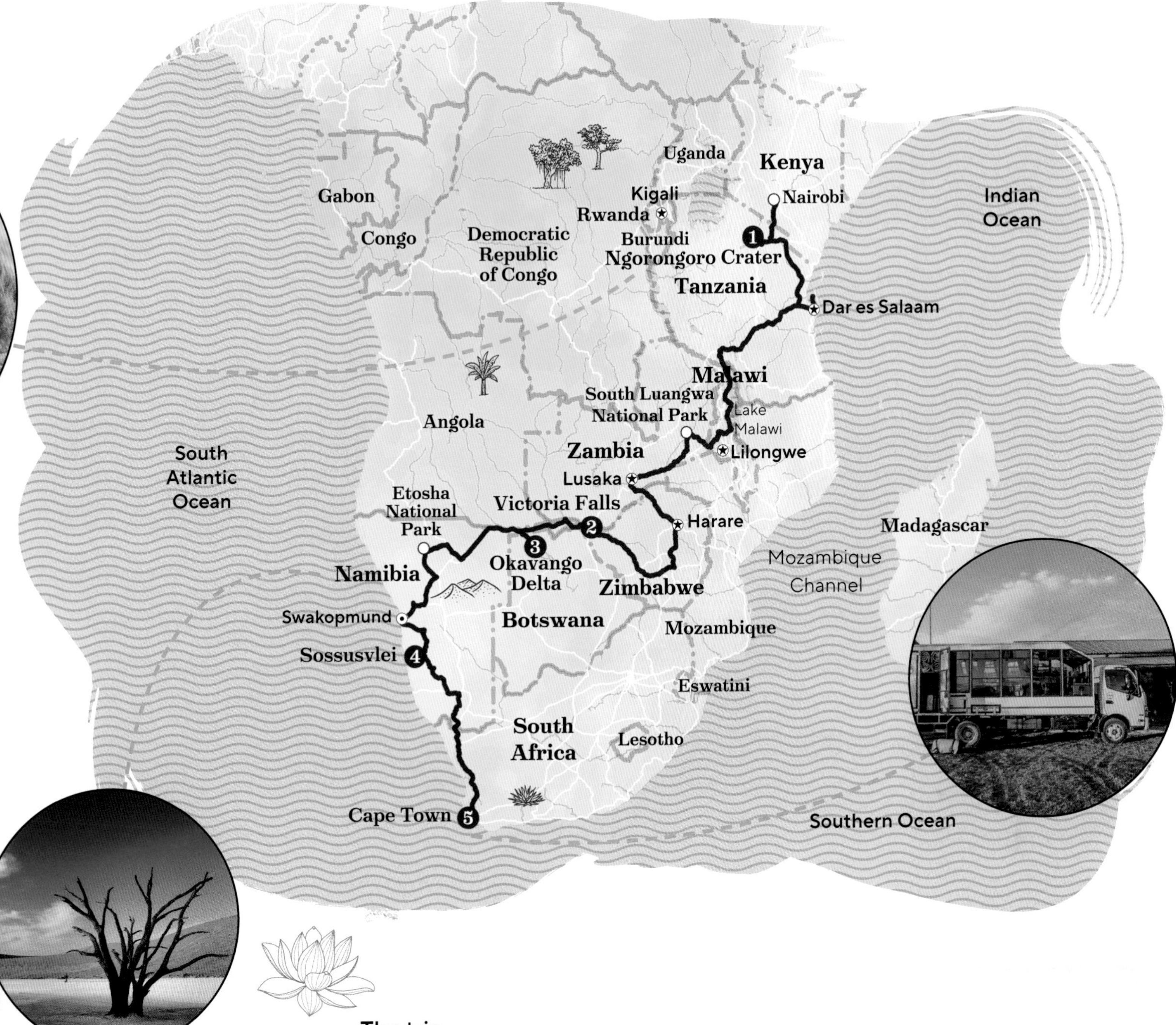

The trip

Distance: 10,000km (6210 miles)
Mode of transport: Overland vehicle
Difficulty: Easy

1 Ngorongoro Crater, Tanzania For some, this wildlife-rich caldera trumps the Serengeti

2 Victoria Falls, Zambia/Zimbabwe Feel the power of Africa's most spectacular waterfall

3 Okavango Delta, Botswana Admire elephants and other wildlife on a canoe safari

4 Sossusvlei, Namibia Clamber up towering scarlet dunes and wander ghostly salt pans

5 Cape Town, South Africa Sync to the rhythm of the Western Cape

Clockwise, from left: Cape Town, South Africa; a lion at the Ngorongoro Crater, Tanzania; a 'lando' overland truck; the red dunes of Namibia's Sossusvlei

Sarah Reid, overland traveller

Spending months living in a tent on an African overland trip reminded me that I can do hard things. But I'm not talking about sleeping just millimetres from roaming predators. As a hygiene freak, learning to accept that I would very rarely feel clean was quite liberating. And, having grown up with largely unhelpful references for understanding the continent, this trip opened my eyes – and heart – to Africa's incredible diversity, helping me to get to know the world that little bit better.

Hey Siri, play the Toto song

Group tours hosted in modified trucks known as overland vehicles or 'landos' are an immensely popular and affordable alternative to attempting this journey under your own steam; the logistics of sourcing one's own vehicle puts most people off tackling the trip independently. Among the most popular itineraries is the route from Nairobi, Kenya to Cape Town, South Africa (or vice-versa), which takes around seven weeks – though some companies offer the option of tapping in or out of the trip at Victoria Falls on the border of Zimbabwe and Zambia if time, budget or an aversion to long periods of camping prevent you from completing the entire route. But forget any concerns about a longer itinerary beginning to feel same-y, for you are guaranteed to experience something new on every single day of this adventure.

In fact, most operators make it easy for you to extend your journey if you can't get enough. Tacking on a loop of Kenya, Uganda and Rwanda before heading south from Nairobi, for example, connects you with two of the continent's most moving experiences: observing Rwanda's critically endangered mountain gorillas in their natural habitat, and visiting the capital's Kigali Genocide Memorial.

Recounting the horrors of the 100-day killing spree in 1994 that left up to 800,000 Tutsis and moderate Hutus dead, this place of remembrance and learning offers an unflinching insight into Africa's turbulent history, as well as the roles of colonial powers in shaping it. It's an experience that stays with

© 2630BEN / SHUTTERSTOCK

> "Doing everything with a truckload of strangers can test your teamwork skills and your patience. But oh, are the payoffs worth it"

you as you travel south through Kenya, Tanzania, Malawi, Zambia, Zimbabwe, Botswana, Namibia and finally South Africa, and may well be to be one of many experiences on this trip that challenge your worldview, forcing you to reckon with biases both conscious and unconscious.

Cultural connections

While this journey is foremost a nature-based experience, most overland itineraries include plenty of opportunities to interact with local characters and learn about indigenous cultures; experiencing the Unesco-listed living culture of the ethnic Sān people, typically at a 'living museum' in Namibia, is a highlight for many. National park safari guides are walking encyclopedias of wildlife behaviours, while driving days offer opportunities to learn more about the culture and life experiences of your tour leader, who will probably hail from Kenya or South Africa. While these tour leaders are usually tasked with being more of a logistics manager than a full-time guide, they will likely have more than a few fascinating tales to share about their time on the road. As will your driver and cook – if you're lucky enough to have both!

TIA (This Is Africa)

While this journey does not tend to be physically demanding, it's a trip that tests you in other ways. The days are typically long, often beginning before the morning sun creeps above the misty savannah, and sometimes not ending until the moon glows bright in the night sky. Camp meals, while usually healthy, can be basic and repetitive. Border crossings can take hours; sometimes days. And doing absolutely everything together with a truckload of strangers of various ages, nationalities and cultures – for weeks on end – can seriously test your teamwork skills, not to mention your patience.

But oh, are the payoffs worth it. From the shot of adrenaline that courses through your body when you lock eyes with an apex predator on a game drive, to quiet moments of reflection as you gaze up at the

Clockwise, from above: zebra crossing Tanzania's vast Ngorongoro Crater; canoeing the Okavango Delta, Botswana; Victoria Falls, bordered by Zambia and Zimbabwe

Wild country

If there's one thing this trip puts into perspective, it's the vulnerability of Africa's wild spaces – and the role that responsible tourism can play in protecting them. The pandemic was a disaster for Africa's wildlife, with Southern Africa seeing an increase in poaching for both subsistence consumption and wildlife trafficking. This was reportedly encouraged in part by the reduced numbers of tourists (whose presence tends to deter poachers). And as changing weather patterns transform habitats and restrict access to water, forcing the displacement of animals, climate change has emerged as a threat to Africa's wildlife. The huge privilege of travelling through the continent's incredible yet often compromised environments becomes clearer as you go – as does the importance of travelling with an operator invested in helping to conserve these spectacular places for future generations – of wildlife, locals and visitors.

seemingly endless canopy of stars twinkling above your camp at night, you may find you have never felt more alive – or more at peace. Challenging moments, too, only make you stronger, whether it's mustering the confidence to make a midnight dash to the loo as a hyena cackles in the darkness (most campsites are in or near national parks), or participating in an optional activity that takes you to the edge of your comfort zone – from navigating the unpredictable rapids of the Zambezi River near the base of Victoria Falls to tracking lions on foot in Botswana's Okavango Delta.

Just do it (responsibly)

Several free days are typically spaced throughout overland itineraries, giving you an opportunity to choose your own mini-adventure. Use them wisely. While you're unlikely to remember the day you spent doing your laundry in Swakopmund on the Namibian coast, you're bound to look back fondly on the excursion you signed up for here to uncover Africa's 'Little Five'. And while you may be reluctant to shell out extra dollars for an optional game drive if you're travelling on a tight budget, it's a rare occasion when your decision won't feel like it was worth it. If you can only choose one splurge, a night-time game drive in Zambia's South Luangwa National Park is an exhilarating opportunity to spot nocturnal predators.

If you have an opportunity to camp near – or spend extra time observing – a waterhole in Namibia's Etosha National Park in the dry season,

> "There's freedom in embracing your raw self on this adventure, because clean clothes never stay that way for long"

do it. If the conditions are good for snorkelling in Lake Malawi, strap on that mask. And if you don't need to rush home, extending your trip to explore the spectacular coastal city of Cape Town and beyond unlocks a whole new world of wonder.

You may be similarly tempted to engage in wildlife and cultural experiences along the way that are less ethical or safe. By taking the time to research responsible tourism in the region before your trip, you'll know that activities such as walking with (usually captive) lions and visiting orphanages and schools can be more harmful than helpful. And there's nothing like a Sub-Saharan Africa overland trip to motivate you to take any steps you can to safeguard this special part of the planet.

© WESTEND61 / GETTY IMAGES

From top: out on the Sossusvlei dunes in Namibia; Cape Town's Camps Bay and the Twelve Apostles mountains, South Africa

Practicalities

Region: Sub-Saharan Africa
Start: Nairobi, Kenya
Finish: Cape Town, South Africa

Getting there and back: Flying in/out of Nairobi and Cape Town is the only practical option for most travellers. Hop-on, hop-off bus services are a handy option for travelling onwards to Johannesburg.

When to go: Organised tours run year-round. The June to September dry season coincides with East Africa's Great Migration of wildebeest, but crowds can be an issue. The typically dry shoulder month of February is also a good time for spotting wildlife, with fewer tourists around.

What to take: A smartphone loaded with podcasts is ideal for driving days, especially if motion sickness prevents you from reading; power points often at a premium, so bring a rechargeable power bank. Pharmacies en route are not always well stocked, so pack all the medications you could possibly need, along with a healthy stash of US dollars in a range of denominations for visas, optional activities, accommodation upgrades, tips and meals that aren't included. A portable wash-bag makes handwashing a breeze, and a self-inflating sleeping mat typically offers more comfort than supplied mats.

Where to stay: Budget overland trips are usually all-camping, with the option to upgrade to a basic room at some stops available for a small additional charge.

Where to eat and drink: Overland itineraries typically include most meals due to the scarcity of dining options on the route. Packing healthy snacks for the long journeys is a good idea; beer and cider are often available for purchase at campsites.

Tours: Responsible operators include Intrepid Travel (intrepidtravel.com) and G Adventures (gadventures.com).

Essential things to know: Tour guests are typically expected to muck in with food prep, washing-up and cleaning the vehicle.

Journey along the Jordan Trail

Embark on an epic hike through the remote wadis and wild mountains of the Jordan Trail, ending your pilgrimage at the ancient city of Petra.

Petra's rock-hewn facades make for a magnificent finish after a long journey on foot: an overwhelming sense of achievement is guaranteed.

Fractured and fought over, the Middle East isn't known for its natural landscapes – walking through Jordan's protected wilderness may offer an illuminating counterpoint to news coverage of the region. And meeting fellow hikers, and locals including Bedouin along the way is a wonderful part of the experience.

The Jordan Trail is still in relative infancy, and walking it requires route-finding skills, stamina and discipline – especially when rationing the amount of water you drink.

Petra is a city that unveils itself with a theatrical flourish. Visitors must follow narrow, snaking canyons deep into the innards of sandstone mountains. The rock seems to close in as you tread onward, the path twisting and writhing, the shadows intensifying – until the Treasury eventually appears, like a flash of light at the end of a tunnel. Soon, the 2000-year-old facade reveals its full, heartstopping majesty. This dramatic approach has been immortalised in photographs, books, poems, paintings and, of course, in *Indiana Jones and the Last Crusade* (in which the Treasury serves as the location of the Holy Grail).

To the people who concealed their capital in the canyons – the Nabateans – it was a feat of urban planning designed to impress and intimidate, and two millennia later this magic endures. But the approach to Petra is perhaps most stirring for those

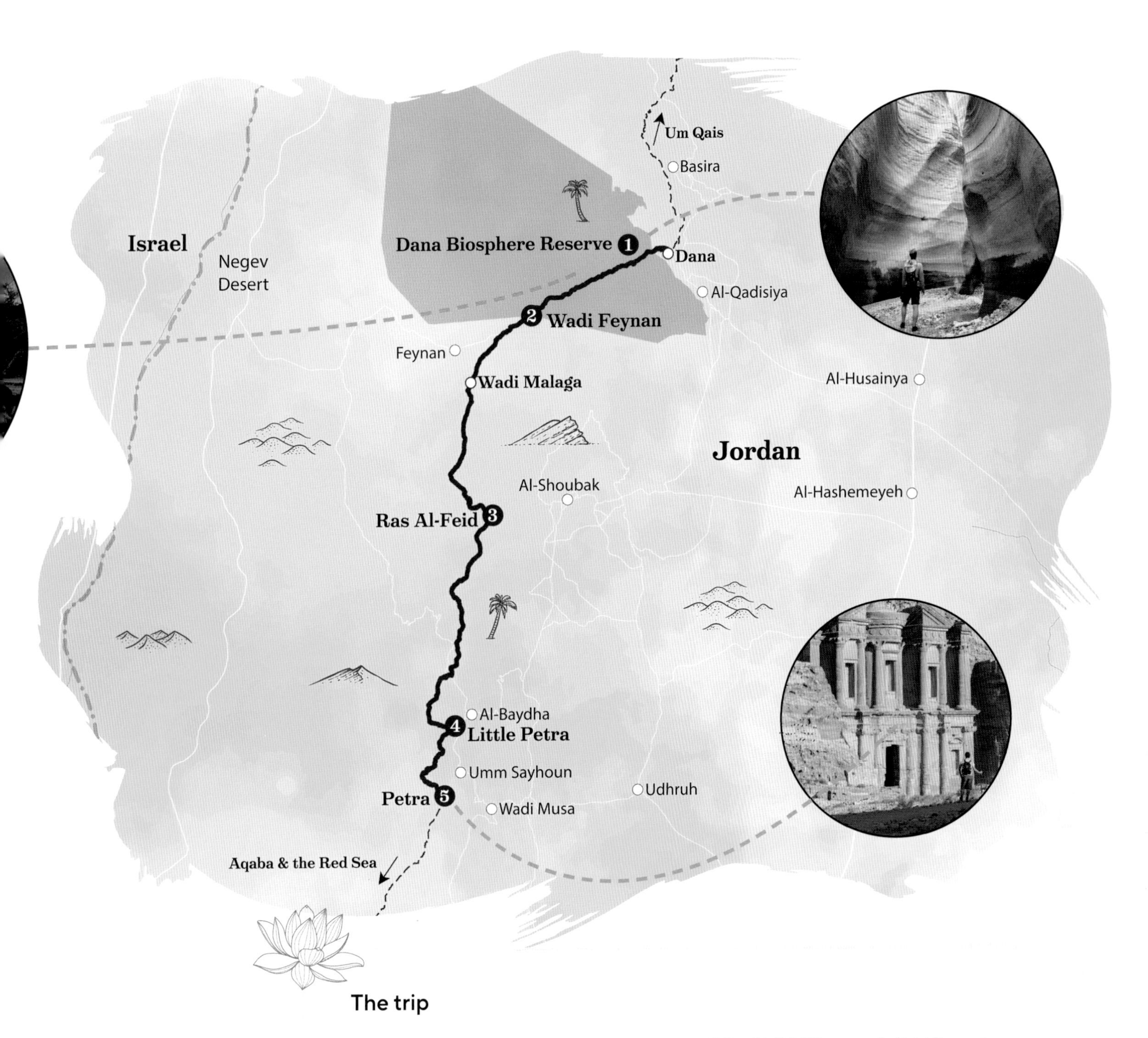

The trip

Distance: Around 80km (50 miles)
Mode of transport: On foot
Difficulty: Difficult

1 Dana Biosphere Reserve The picturesque village of Dana serves as the gateway to Jordan's largest nature reserve

2 Wadi Feynan The first day on the Jordan Trail brings you to Wadi Feynan, where you'll find the excellent Feynan Ecolodge

3 Ras Al-Feid The sound of trickling water resonates around deeply-gouged canyons at the midpoint of the trek, its most remote stretch

4 Little Petra This small outpost of the Nabatean city is a prelude to the greater riches of Petra itself

5 Petra One of the wonders of the ancient world: Petra's tombs, theatres and thoroughfares lie sprawled over a vast, mountainous area

Clockwise, from left: hilltop views of Petra; a trailside oasis at Feynan Ecolodge; through the canyons to Petra; Petra's mysterious Monastery

Oliver Smith, Jordan Trail hiker

I'm fortunate enough to have hiked widely across the world – but the Jordan Trail counts among the walks that moved me the most. The dry canyons between Dana and Petra teach you many things, but perhaps foremost is the life-giving power of water – from the springs that burst from the mountains to nourish palm and bamboo groves, to the water in your bottle that sustains you on this arid trail.

who have walked for four days to stand amongst its rose-red stones.

Laying the Jordan Trail

Established in 2015, the Jordan Trail is a hiking route that covers the entirety of this Middle Eastern nation – from the olive groves in the north to the deserts in the south – but perhaps its most glorious stretch is the central section, where southbound hikers take four days to traverse a series of remote canyons and arrive at the gates of Petra. The Jordan Trail began as a social enterprise, employing local shepherds as guides, and bringing business to local communities. In a part of the world better known for barbed wire and closed borders, it offers a rare and refreshing taste of the open road, and of freedom.

Into the Dana

The adventure begins in the Dana Biosphere Reserve – the largest nature reserve in the country, perched at the lip of a green canyon. Setting out from the higgledy-piggledy Ottoman-era village of Dana itself, the trail winds through a landscape of fig and pistachio trees – listen out for the clattering hooves of ibex in the crags. As you descend, the temperature nudges up: while the peaks above occasionally get snow drifts in winter, the desert below routinely sees temperatures simmering at around 45°C (113°F).

The first day's hike ends in the parched expanse of Wadi Feynan, where the hills are swiss-cheesed by Roman-era copper mines. Here you'll also find Feynan Ecolodge – a pioneer in sustainable accommodation run by the Royal Jordanian Society for Conservation and Nature, with candlelit rooms arranged around a central courtyard. It is a fine place to disconnect from the world. End the day joining a stargazing session from the roof terrace, peeking through a telescope at the craters of the moon and glancing down over a terrestrial landscape which – under the pallor of moonlight – also seems lunar.

© JOE WINDSOR-WILLIAMS / LONELY PLANET

From left: Petra's magnifiicent Treasury; threading past peaks on the Jordan Trail route

> "The approach to Petra is perhaps most stirring for those who have walked for four days to stand amongst its rose-red stones"

Wild canyons and ancient trails

From here, the going gets harder and the gradients steeper – hikers must dig deep to push through some of the most challenging sections of the Jordan Trail. The second day sees you scrambling up a wall of rock out of Wadi Malaga, then shimmying along a narrow ledge into the summits around Ras Al-Feid. The path here is erratic – at one moment riding high, et eye-level with the birds of prey, the next burrowing through the bamboo groves that sprout along the canyon floors. This is one of the wildest stretches of the country: the hills were supposedly the haunt of leopards just a few decades ago. Except for occasional dust clouds kicked up by shepherds and their flocks, signs of human existence are scarce.

Though the Jordan Trail itself is fairly new, the paths it follows have been carved by generations of shepherds, merchants, soldiers – and holy men carrying stories of new prophets. Perhaps some among them were heading to Petra when the city was in its prime, two millennia ago. Doubtless these wayfarers would recognise these mountains today – this is a landscape largely unaltered by the

The wider Jordan Trail

The Dana to Petra hike is a section of the full 650km-long (404-mile) Jordan Trail, which runs north to south through the entire country – but it's by no means the only highlight of this pioneering path, which crosses a tapestry of landscapes and ecosystems. The northern starting point is the Roman city of Umm Qais, where hikers set off in the shadow of tumbledown temples into wildflower-strewn hills, bound for the 12th-century castle at Ajloun. In the far south the landscape is wholly different, with wind-scoured deserts culminating in the monoliths of Wadi Rum – immortalised in the campaigns of Lawrence of Arabia (and in more recent times by Hollywood movies, where it has oft served as a stand-in for Mars). The Jordan Trail threads through the almighty cliffs and sandy tracts of Rum to reach the Red Sea – where thru-hikers often celebrate their achievement with a swim.

passage of time, unblemished by roads or permanent settlements. With no accommodation, hikers must wild-camp at Ras Al-Feid (Jordan Trail authorities can arrange for local villagers to bring simple mattresses, and barbeque a chicken on a campfire). Here you might get a small sense of a nomadic way of life: a kind of travel that feels ancient in its character. You carry only what you need, fill your days with miles and count the shooting stars from your open-air bed at night.

Little Petra

The penultimate day of the hike is the longest, the path cresting a series of sandstone escarpments crowned with acacia and juniper, and commanding godlike views westward into Wadi Araba. The dusty plains below mark the northernmost part of the Great Rift Valley, which runs thousands of kilometres southward to Lake Victoria and the foot of Tanzania's Mt Kilimanjaro. It is a view that's almost metaphysical in its majesty.

Eventually the path squeezes through a cleft in the rock to reach Little Petra – Siq al-Barid (Cold Canyon) in Arabic – where hikers get their first encounter with Nabatean culture. An Arabian tribe who grew rich controlling trade routes, the Nabateans stood at the centre of a web of commerce that extended eastward to China and westward to Europe; Little Petra is believed to have been founded to accommodate merchants travelling along these arteries of the Silk Road. As a fellow traveller arriving on foot – late in the day, after the tour buses have departed – you can imagine the ghosts of caravans here, or of traders bartering over merchandise. In truth, however, Little Petra is only an appetiser for the city that lies a half-day's march to the south.

Petra

Whichever way you approach it, Petra conjures itself out from the landscape as if by miracle. Hiking from the north, you might first spy the soaring profile of the Monastery – a mysterious landmark that was perhaps a royal tomb. You may then pass under the High Place of Sacrifice; or scramble along the narrow canyon of Wadi Al-Mudhlim to see the irrigation channels that once carried cool spring water to residents. Petra is part ancient city, part national park – it would take many days to explore all its tombs and caves, to summit all its windswept heights. Interestingly, few building blocks were laid here – it is a city largely hewn, sculpted and excavated out of the natural rock. And so, too, is it a place that leaves a deep mark on visitors – none deeper than when the first glimpse of the Treasury appears. At the end of four days of walking, it might indeed seem like a kind of holy grail.

From top: hiking the route north of Petra; a Jordan Trail guide

Practicalities

Region: Central Jordan
Start: Dana
Finish: Petra

Getting there and back: Set just outside the Jordanian capital, Amman Airport is well served by international flights. Dana lies a little way off the beaten track; get there via a JETT bus (jett.com.jo) from Amman to the town of Wadi Musa (the gateway to Petra) before taking an hour-long taxi ride north.

When to go: The Jordan Trail is best attempted in the spring (March to April) or autumn (October to November); winter can see chill winds scouring the hilltops, while the summer heat can make hiking heavy going.

What to take: Among other things, pack sturdy high ankle-high walking boots, a sun hat and plenty of sunscreen. More important than anything else is water – places to refill are few and far between, so carry a minimum of 3L to 4L per person per day.

Where to stay: In addition to Feynan Ecolodge, the Royal Jordanian Society for Conservation and Nature has another wonderful property along the Jordan Trail: rooms at Dana Guest House have swooping stone arches and spacious balconies cantilevered over Wadi Dana.

Tours: The excellent Jordan Trail website (jordantrail.org) is a one-stop shop for information on the route, with a directory of official tour operators and details of local people who can bring food, water and bedding to more remote stages of the trail on a more informal basis.

Essential things to know: The Dana to Petra hike is challenging, with daily distances of up to 24km (15 miles) per day, and elevation gains up to 1000m (3280ft). Though many tackle it on an organised tour, it's possible to walk it independently – but you'll need to be very experienced hiker with a good level of fitness, and be extremely confident reading maps. Most the trail isn't marked by signposts – GPX files can be downloaded from the Jordan Trail website.

Adventuring in the Empty Quarter's endless dunes

The Empty Quarter counts among the biggest blank spaces anywhere on Earth – making a journey into this sea of sand can be a surprising and sublime experience.

Nowhere offers a stage for self-reflection quite like the Empty Quarter – its dunes are the perfect place to order your thoughts. Though ostensibly peaceful, the sands can also be treacherous: learning to remain cool and resolute when a vehicle gets stuck in deep sand is essential.

The Empty Quarter offers glimpses of a nomadic way of life that has all but vanished. It also initiates visitors into the wildlife of a place which, to the untrained observer, may appear lifeless.

Everywhere in the Empty Quarter is furnace-hot, and shade is nonexistent: spending time here will teach you how best to acclimatise and manage extreme heat.

The Empty Quarter is not the biggest desert in the world; nor is it the driest, or even the hottest. But if you imagine a desert, it is invariably the Empty Quarter that you'll conjure up: a world of windblown sand and rolling dunes, gilded dawns and bloody sunsets, loping camels and lonely wanderers. Its featurelessness puts a heroic focus on all those who come here: anyone who marks its pristine dunes with their footprints, or casts a rare shadow over its shadeless tracts. It is known as a place of enchantment – but it is also a deadly serious place to go exploring.

Known as the Rub' al Khali in Arabic, the Empty Quarter covers some 600,000 sq km (225,000 sq miles) across Oman, Yemen, Saudi Arabia and the UAE. It's a vast a swathe of territory with no roads or settlements, few people and little mobile phone reception – a corner of the planet left on default settings, for it has yielded very little to sustain the life of *homo sapiens*. For hundreds of years, only Bedouin caravans ranged between its remote wells; in the past century, Western explorers sought fame and glory by crossing this alien terrain. Today a few tourists set out on tours to nibble at the edges of the Empty Quarter,

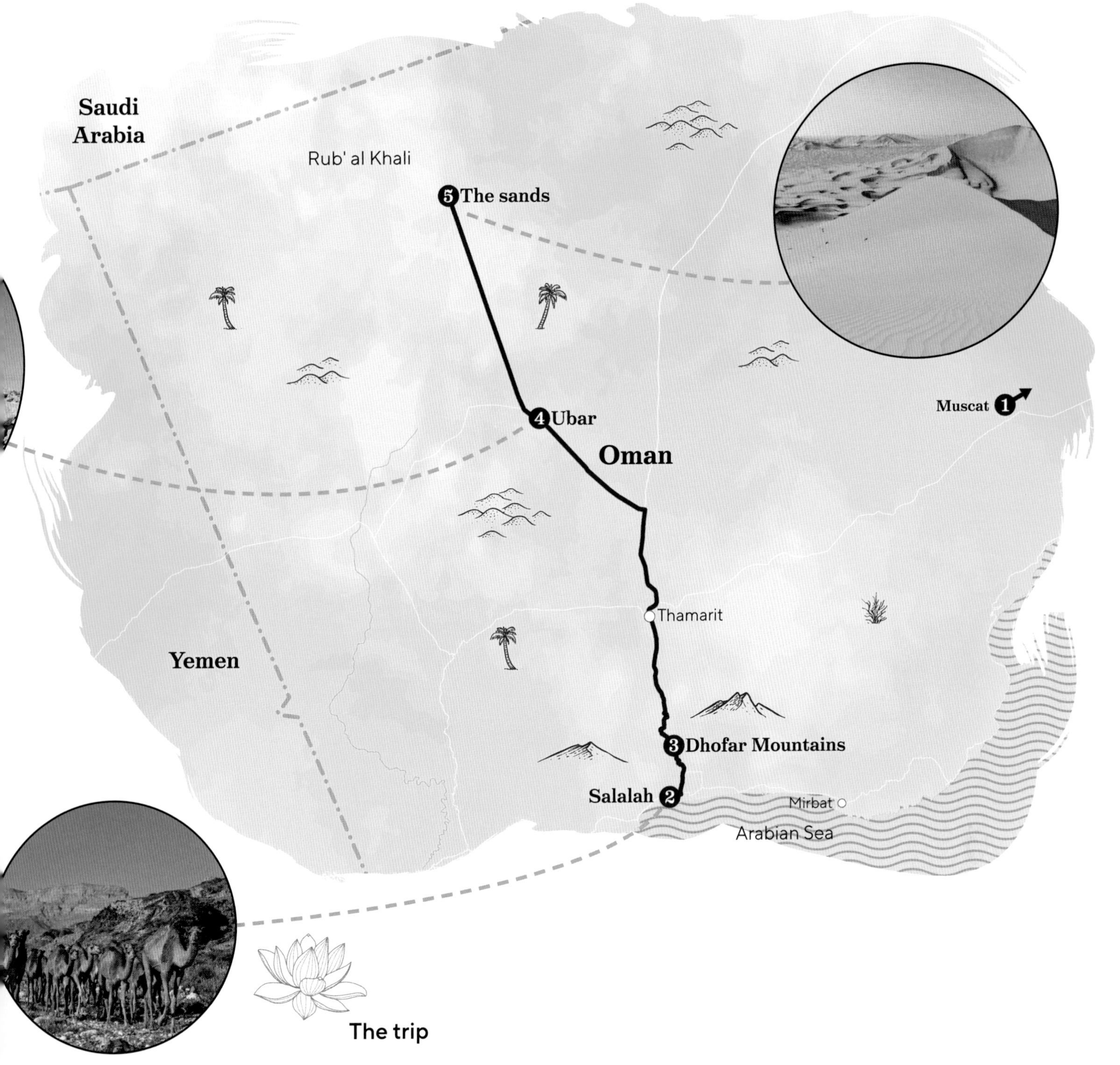

The trip

Distance: Around 300km (190 miles) one-way
Mode of transport: 4WD vehicle
Difficulty: Easy

1 Muscat Wedged between mountains and ocean, the characterful Omani capital will likely be your entry point into the country

2 Salalah The seaside city of Salalah is the capital of the southern Dhofar region

3 Dhofar Mountains These desert-brown mountains turn green during the *khareef* (rainy) season

4 Ubar This ancient 'Lost City' is one of the last signs of civilisation before you enter the desert

5 The sands The undulating Empty Quarter is a sandy desert roughly the size of France, spread across four countries

Clockwise, from left: sunset over the shifting sands; ancient Ubar, the 'Lost City' at the edge of the desert; Empty Quarter dunes; camels at Salalah

Mussallem Hassan, desert guide

The desert gives you peace, gives you silence, gives you happiness. Your mind feels free. When we see a village and see houses, the doors are normally closed – so we don't go there. But when we see a camp in the desert and we know there are nomads – we can go there and be welcome, and that is our tradition. With a tent, the doors are always open.

but even on a short trip, you can return with a sense of its oceanic vastness.

Into the dunes

A good place to start is the port of Salalah in Oman – a town characterised by its seafaring associations with East Africa. Stock up on supplies in its souk before joining up on a guided tour of the Empty Quarter and venturing inland. Very soon after leaving the city, you'll find yourself amongst the frankincense groves of the Dhofar Mountains. Once a year, around June, this corner of Arabia miraculously flowers with the advent of the *khareef* – when the southeastern monsoon shrouds the hills, and new life springs from parched soil. This landscape offers a last glimpse of greenery before you enter the interminable desert. Right on the cusp of the dunes, tours usually stop at the remains of Ubar – an ancient settlement sometimes known as the 'Atlantis of the Sands'. Eventually the tarmac gives out: dunes rise on the horizon, and drivers must deflate their vehicle's tires to steer through the wastes.

Digging deep

A journey into the Empty Quarter is analogous to a sea crossing in its scope; and like an ocean voyage, good planning is essential. People often travel in fleets of two or more 4WD vehicles in case of mechanical malfunction – spare fuel, spare parts and having the wherewithal to conduct maintenance are all essential. So too must expeditions have the means of sustaining human life in a place far from help: spare food, jerry cans full of water and satellite phones in case of emergency are also stashed on board. There are many dangers, but by far the greatest is disorientation – the shifting waveforms of

© KATIEKK / SHUTTERSTOCK

> "The desert gives you peace, gives you silence, gives you happiness. Your mind feels free"

the dunes are constantly changing. It is a map that is forever redrawing itself with the temperament of the winds. All serious guides will have their own mental map of the desert, updated with every journey. This cognitive cartography is a subtle craft that visitors can begin to hone for themselves.

Guides can also read the more granular tendencies of sands in the way sea captains can read the eddies of the oceans. The Empty Quarter is a place more or less bereft of landmarks – so tours normally see 4WDs rev up the biggest dunes, make rollercoaster-like descents down sheer slopes, crest ridges and range freely across the landscape. It looks effortless but it takes immense skill – on some tours participants themselves can drive and learn to judge the drifts. Whoever is behind the wheel, though, it is almost inevitable the vehicle will at some point get stuck – and passengers will have to grab a shovel and help carve a path out of the rut. Sometimes it can take minutes to free the wheels; occasionally it's hours. There are stories of vehicles that were stuck for days, and of some which never returned to tarmac roads. It is at this point that the desert teaches its most valuable lessons: self-reliance, staying cool under pressure, keeping faith that there is a way out.

Desert sunsets

At the end of the day the temperature cools, and the contours of the dunes are thrown into sharp relief – your shadow grows longest with the last flare of the sun. Around this time camp is set up: cardamom coffee boils on a fire, the wind ruffles goatskin tents and the night-time desert stirs with hopping jerboas and slithering snakes. Soon the sky is bejewelled by the constellations, and stories of djinns (genies) are shared before sleep. Throughout history deserts have been places of revelation: saints, prophets and hermits came to these empty landscapes to hear the word of God most clearly. Even for those of no faith, the desert is a fitting place for a time of inner reflection – a rare and precious pocket of space and silence in an ever more crowded world.

From left: a 4WD is the only way to get deep into the dunes; the capital city Muscat is your first port of call

Practicalities

Region: Oman
Start: Salalah
Finish: The sands

Getting there and back: The capital, Muscat, is home to Oman's main international airport – from here flights to Salalah are extremely regular, taking 1hr 40min. You can also connect to Salalah from airports in the UAE and elsewhere across the Gulf.

When to go: Summer temperatures in the Empty Quarter can be in excess of 50°C (122°F), so it's a good idea to travel in the cooler months from October to April.

What to take: Long sleeves, sunscreen and a sun hat are all essential precautions against the midday sun. At night, the desert can get surprisingly chilly – bring plenty of warm layers, too.

Where to stay: Salalah has a wide range of hotels (often fully booked throughout the *khareef* season). The Al Baleed Resort by Anantara is one of the most coveted, with whitewashed villas perched between a lagoon and the tides of the Indian Ocean.

Tours: Arabian Sand Tours (arabiansandtoursservices.com) are specialists in Empty Quarter travel, offering bespoke one- to ten-night itineraries out of Salalah, including all meals/drinks, and accommodation in traditional tents.

Essential things to know: Everyone venturing into the Empty Quarter should unequivocally do so as part of a guided tour – every year a few unfortunate souls make a trip into the sands in private vehicles and never return. Empty Quarter tours are also available from the UAE (typically out of Liwa in the Emirate of Abu Dhabi) and, in recent times, from Saudi Arabia.

Pilot a canoe down the Zambezi River

Paddle your way through some of Africa's wildest territory on a Zambezi canoe safari, steering among hippos by day and camping on remote river islands by night.

Canoeing in the domain of the Big Five is not for the fainthearted – you'll need fortitude to stay calm in the face of lurking crocodiles and pods of hippos.

Being low to the water offers a wholly different perspective on this African landscape – you can get closer to wildlife, and cultivate a far deeper understanding of their habitat.

After a few days on the Zambezi, you can become a confident paddler – you'll likely end your odyssey with slightly bigger biceps, and a squint honed by the river's watery horizons.

Safari is a tightrope-walk between danger and comfort. There are the primeval thrills of sharp teeth, swiping claws and trampling feet. But there is always the knowledge that, if this dose of danger gets too much, your guide can hit the pedal in their 4WD and whisk you to safety (and probably an ice-cold G&T back at camp).

No such option exists on a canoe safari. This is a visceral and intimate business – sometimes more hair-raising than being in a car, but more rewarding too. You can glide soundlessly in the near-midst of prowling predators; down low in the water you are dwarfed by elephants, whose crinkly flesh towers above. You function as the driver (and the engine) of your vehicle – the hushed voice of a guide is your prompt to paddle faster or to jam an oar in the current and make a retreat. You are master of your own destiny in a watery wilderness.

Destiny could present nowhere better for a canoe expedition than the Zambezi, which rises in the border country of the Democratic Republic of Congo and Angola, thunders over Victoria Falls and – 2574km (1599 miles) after it began as a trickle – disgorges into the Indian Ocean. For canoeists, the best stretch is near the midway point, west of Lake Kariba, where the river braids among myriad islands.

The trip

Distance: 70km (43 miles)
Mode of transport: Canoe
Difficulty: Difficult

1 Lusaka With its dusty, tree-lined streets and thriving markets, the Zambian capital is your gateway to adventures on the Zambezi

2 Chirundu Begin your Zambezi journey by casting off with your canoe in this easygoing riverside town

3 Mana Pools National Park Zimbabwe's Unesco-listed reserve is named after pools formed by the Zambezi's meandering course

4 Chongwe River The confluence of the Chongwe and the Zambezi is a good place to make your exit onto dry land

5 Lower Zambezi National Park Post-paddle, explore one of Zambia's most prized wilderness areas, which hugs the Zambezi's north bank

Clockwise, from left: Zambezi sunset; a river guide scouting the waters; spot an array of wildlife at the Mana Pools and Lower Zambezi national parks

Tavengwa Kangwara, canoe guide

Some people in Zambia think a canoe safari is dangerous. But I have been paddling here for 21 years – I still have all my arms and legs. I have had a few hippo attacks – but not the dangerous kind. So I feel very much connected to the Zambezi. I love the tranquillity, the peacefulness. The sound of the river sounds to me like a lullaby, or a song.

Casting off

Your start-point is the border town of Chirundu – a hub for trans-African traffic, where two bridges straddle the Zambezi. Canoeists rise early, a thin film of mist clinging to the river's waters as tents, food and supplies in waterproof bags are loaded onto boats. Then the armada launches into the gentle current, a chorus of birdsong ringing from the acacia trees. You'll soon get your bearings: to starboard (the right) is Zimbabwean territory; to port (left) the hills of Zambia rise over the eddies and swirls. The border lies midstream, with river islands divided between the two nations.

In truth, however, the river is its own country. You soon find yourself swept downstream – further from people and places – and enter into a place where animals are sovereign, and political boundaries can sometimes seem of lesser consequence. With no glass, perspex, metal or canvas to separate you from the big beasts of the African bush, you are conscious that you might (theoretically at least) be part of the food chain here. Respect and responsibility are all qualities shaped by the flow of the Zambezi.

Hungry hippos

You'll soon find yourself skirting Zimbabwe's Mana Pools National Park, said to be home to as many as 12,000 elephants. Look out for herds gathering at the water's edge for a drink, or swimming through backchannels using their trunks like a snorkel. Guides rightly insist you keep a safe distance away – though more often than not it is another lumbering mammal that is of greater concern. Hippos dominate this stretch of the river, gathering in pods of between a dozen to 50 or more. They can panic if canoes get too close – it's essential to let them know you're passing by, generally through banging the side of the boat as if you were knocking on someone's door.

> "This is a visceral and intimate business – sometimes more hair-raising than being in a car, but more rewarding too"

Canoeists quickly come to understand there is a hierarchy for river traffic – some species have priority or command right of way, and humans must wait patiently for the way to clear before they progress. It is at once daunting – hearing the splash of a Nile crocodile slipping into the river as you paddle by – but oddly reassuring to be ensconced in your fibreglass canoe, a roving pocket of safety (as long as it remains buoyant, at least).

© PHILIP LEE HARVEY / LONELY PLANET

Making camp

Along the way you'll occasionally notice luxury safari lodges on the riverbank – some with swimming pools, or restaurants serving gourmet three-course meals. Canoeists, by contrast, are a self-sustaining species. The day's paddling ends with team members hauling their boats onto remote river islands, and erecting a huddle of tents around a campfire before a blood-red sun slips over the horizon. You'll need to muck in with camp chores and cooking, keeping an eye out for night-time interlopers with a vigilant sweep of a torch. But the rewards are magnificent: eating dinner beside a crackling campfire, falling asleep to the faraway roar of lions on the hunt.

Days follow a similar pattern, rising early to pack away camp, making headway in the cool of the morning as fish eagles glide above. Gradually the river changes character – the tight bends and maze of channels beyond Chirundu turn to a longer, straighter incarnation of the Zambezi as it ribbons eastward. The adventure ends three days later, with paddlers loading their canoes onto trailers where the little Chongwe River right-angles into the stately procession of the Zambezi. You complete your journey having entered into a kind of communion with the river: you have been a passenger on its current, a peer among water-going hippos and crocodiles. It might not be a total exaggeration to say you have entered a state of oneness with its flow. But just remember – on no account whatsoever should you ever get in the water.

© PHILIP LEE HARVEY / LONELY PLANET

From top: overnight camp on a Zambezi River island; look for lions in Zambia's Lower Zambezi National Park

Practicalities

Region: Zambia/Zimbabwe
Start: Chirundu
Finish: Chongwe River

Getting there and back: Lusaka's International Airport is well served by flights from within Africa and beyond – it's about a 3hr drive from Chirundu. Alternatively, expect a 6hr drive to Chirundu from Harare International Airport, over the border in Zimbabwe.

When to go: Canoeing season on the Zambezi generally runs from March to December, with a break to coincide with the rains. Temperatures are coolest from May to August.

What to take: Although most equipment will be provided by your canoeing operator, certain personal items will be helpful – notably lightweight clothing, a headtorch for camp, and a sun hat (shade is minimal to nonexistent on the river).

Where to stay: Zambezi Breezers campsite is a time-honoured launching point for canoe safaris, set on a country lane a short distance out of Chirundu. Accommodation takes the form of tented chalets, and may be covered as part of canoe safari packages.

Tours: With decades of experience, River Horse Safaris (riverhorsesafaris.com) are the preeminent outfit running canoe safaris out of Chirundu, with itineraries ranging from two to seven nights. Tours see participants camping on both Zambian and Zimbabwean sides of the river (note that if you're entering from Zambia and staying only on Zimbabwean river islands, you may not need to worry about a visa for Zimbabwe).

Essential things to know: While paddling on the Zambezi isn't technically difficult, some canoeing experience is recommended. Most of all, it's absolutely essential to travel with an experienced guide, who knows the river well and can manage the hazards posed by hippos and crocodiles.

Index

M

YOUR TRIP STARTS HERE
October 2023
Published by Lonely Planet Global Limited
CRN 554153
www.lonelyplanet.com
1 2 3 4 5 6 7 8 9 10
Printed in Malaysia
ISBN 9781837580064

Written by Alexis Averbuck, Andrew Bain, Amy C Balfour, Sarah Barrell, Robin Barton, Joe Bindloss, Garth Cartwright, Harmony Difo, Jennifer Hattam, Carolyn B Heller, Virginia Jealous, Patrick Kinsella, Brian Kluepfel, Alex Leviton, Bradley Mayhew, Craig McLachlan, Etain O'Carroll, Lorna Parkes, Ashley Parsons, Sarah Reid, Simon Richmond, Regis St Louis, Oliver Smith, Kerry Walker, Steve Waters, Adam Weymouth, Nicola Williams

Publishing Director: Piers Pickard

Senior Editor: Robin Barton

Editors: Polly Thomas, Nick Mee, Rory Goulding

Image Researcher: Claire Guest

Print Production: Nigel Longuet

Lonely Planet Global Limited Office

Digital Depot, Roe Lane (off Thomas St),
Digital Hub, Dublin 8, D08 TCV4
Ireland

STAY IN TOUCH
lonelyplanet.com/contact

Paper in this book is certified against the Forest Stewardship Council™ standards. FSC™ promotes environmentally responsible, socially beneficial and economically viable management of the world's forests.